Mechanisms of Neurotoxicity

Mechanisms of Neurotoxicity

Guest Editors

Paola Alberti
Eleonora Pozzi

MDPI

Basel • Beijing • Wuhan • Barcelona • Belgrade • Novi Sad • Cluj • Manchester

Guest Editors

Paola Alberti	Eleonora Pozzi
School of Medicine and Surgery	Dipartimento di Medicina e Chirurgia
University of Milano-Bicocca	University of Milano-Bicocca
Monza	Monza
Italy	Italy

Editorial Office
MDPI AG
Grosspeteranlage 5
4052 Basel, Switzerland

This is a reprint of the Special Issue, published open access by the journal *International Journal of Molecular Sciences* (ISSN 1422-0067), freely accessible at: https://www.mdpi.com/journal/ijms/special_issues/TC2EN88RQK.

For citation purposes, cite each article independently as indicated on the article page online and as indicated below:

Lastname, A.A.; Lastname, B.B. Article Title. *Journal Name* **Year**, *Volume Number*, Page Range.

ISBN 978-3-7258-5595-7 (Hbk)
ISBN 978-3-7258-5596-4 (PDF)
https://doi.org/10.3390/books978-3-7258-5596-4

Contents

About the Editors

Paola Alberti

Paola Alberti, Experimental Neurology Unit, School of Medicine and Surgery, University of Milano-Bicocca, 20900, Monza, Italy, and NeuroMI (Milan Center for Neuroscience), 20126, Milan, Italy. She is currently Assistant Professor in Human Anatomy at the University of Milano-Bicocca (Monza, Italy) and Consultant Neurologist at Fondazione IRCCS San Gerardo dei Tintori (Monza, Italy). Her expertise is related to peripheral neuropathies, especially chemotherapy-induced peripheral neurotoxicity (CIPN). She oversees both clinical and preclinical CIPN research projects. At a preclinical level she has developed a line of research that applies morpho-functional techniques to study axonal damage mechanisms, with particular attention to ion channel dysfunctions. To this end she learned nerve excitability testing from Prof Hugh Bostock, PhD, in London (UCL). At a clinical level she is Principal Investigator of national and international studies aimed at investigating various clinical and epidemiological aspects of CIPN: ICAVS (NCT04633655), CIPN COST (NCT04986891) and NEUPER study (NCT05088681). She was a visiting research fellow at Johns Hopkins University (2015) and the University of Maryland (2018). She is currently a member of the board of the Italian Association of Peripheral Neuropathies (ASNP), chair of the junior committee of the Peripheral Nerve Society (PNS) and representative of this committee within the Toxic Neuropathy Consortium of PNS, and chair of the neurological complication study group of the Multinational Association for Supportive Care in Cancer (MASCC).

Eleonora Pozzi

Eleonora Pozzi, Experimental Neurology Unit, School of Medicine and Surgery, University of Milano-Bicocca, 20900, Monza, Italy, and NeuroMI (Milan Center for Neuroscience), 20126, Milan, Italy. She graduated in Biotechnology (Bachelor's degree) in 2013 and Biology (Master's degree, cum laude) in 2016 at the University of Milano-Bicocca. In January 2020, she received her PhD in Neuroscience. Since 2014 she has been collaborating in in vivo research projects on Chemotherapy-Induced Peripheral Neurotoxicity at Experimental Neurology Unit (University of Milano-Bicocca), where she now works as a postdoctoral researcher. Her main expertise is in the development and use of in vivo experimental models to study the peripheral neuropathies. She is author of several scientific papers in international peer-reviewed journals. She is an active member of Italian and International scientific associations.

International Journal of *Molecular Sciences*

MDPI

Editorial

Special Issue "Mechanisms of Neurotoxicity"

Paola Alberti [1,2,*] and Eleonora Pozzi [1,2]

1 Experimental Neurology Unit, School of Medicine and Surgery, University of Milani-Bicocca, 20900 Monza, Italy; eleonora.pozzi@unimib.it
2 NeuroMI (Milan Center for Neuroscience), 20126 Milan, Italy
* Correspondence: paola.alberti@unimib.it; Tel.: +39-02-6448-81-54

An exciting and relevant topic is addressed in this paper collection encompassing both peripheral and central nervous system mechanisms of damage. This is of particular interest since neurons are a perennial cell population and, therefore, neurotoxicity understanding and management is a relevant challenge to treat/prevent neurological disorders [1–3]. There are many different potentially neurotoxic agents which include chemotherapy drugs, environmental pollution, et cetera. Mechanisms of damage that involve the peripheral and/or central nervous system are explored to pave the way to potential treatment strategies relying on a robust biological rational.

Since neurons are excitable cells, ion channels/transporters can be a pivotal element leading to axonal damage and neuronal death [4,5]. A clear-cut review of their involvement, exploiting as a playground chemotherapy-induced peripheral neurotoxicity (CIPN) is provided as well as an in depth reasoning on how ion channels/transporters are quire susceptible, in their functioning, to changes in the environment the cell is exposed to which can be also triggering neurotoxicity.

The central nervous system is not overlooked too in this paper collection, as mentioned above, and the role of excitotoxicity [6,7] is also dissected in the in depth in the neurodegenerative disorder and in pain modulation.

Another topic that is presented via research data is the role of oxidative stress in determining alterations of the nervous system [8–11], exploiting zebrafish models, as well as its role in determining neurotoxicity acting against glial cells.

Last, but not least, our Special Issue is enriched by a detailed review of mechanisms leading to an entity that is becoming more and more relevant: HIV-associated neurocognitive disorder (HAND) [12,13].

Author Contributions: Writing—original draft preparation, P.A.; writing—review and editing, E.P. All authors have read and agreed to the published version of the manuscript.

Funding: This research received no external funding.

Conflicts of Interest: The authors declare no conflicts of interest.

https://doi.org/10.3390/ijms26209992

References

1. Libertini, G.; Ferrara, N. Aging of Perennial Cells and Organ Parts According to the Programmed Aging Paradigm. *Age* **2016**, *38*, 35. [CrossRef] [PubMed]
2. Cserép, C.; Pósfai, B.; Dénes, Á. Shaping Neuronal Fate: Functional Heterogeneity of Direct Microglia-Neuron Interactions. *Neuron* **2021**, *109*, 222–240. [CrossRef] [PubMed]
3. Magrassi, L.; Leto, K.; Rossi, F. Lifespan of Neurons Is Uncoupled from Organismal Lifespan. *Proc. Natl. Acad. Sci. USA* **2013**, *110*, 4374–4379. [CrossRef] [PubMed]
4. Harris, K.; Won, S.J.; Uruk, G.; Mai, N.; Ogut, D.; Zhao, Y.; Xie, L.; Baxter, P.; Everaerts, K.; Sah, R.; et al. Excitotoxic Neuronal Death Requires Superoxide Entry into Neurons through Volume-Regulated Anion Channels. *Sci. Adv.* **2025**, *11*, eadw0424. [CrossRef] [PubMed]
5. Liu, Y.; Jin, W.; Zhou, H.; Wang, X.; Ren, H.; Yang, X.; Luo, K.; Dou, X. Role of Mitochondrial Ca^{2+} in Stroke: From Molecular Mechanism to Treatment Strategy (Review). *Mol. Med. Rep.* **2025**, *32*, 271. [CrossRef] [PubMed]
6. Wu, W.L.; Gong, X.X.; Qin, Z.H.; Wang, Y. Molecular Mechanisms of Excitotoxicity and Their Relevance to the Pathogenesis of Neurodegenerative Diseases-an Update. *Acta Pharmacol. Sin.* **2025**. [CrossRef] [PubMed]
7. Dong, X.X.; Wang, Y.; Qin, Z.H. Molecular Mechanisms of Excitotoxicity and Their Relevance to Pathogenesis of Neurodegenerative Diseases. *Acta Pharmacol. Sin.* **2009**, *30*, 379–387. [CrossRef] [PubMed]
8. Lee, J.W.; Choi, H.; Hwang, U.K.; Kang, J.C.; Kang, Y.J.; Kim, K.I.; Kim, J.H. Toxic Effects of Lead Exposure on Bioaccumulation, Oxidative Stress, Neurotoxicity, and Immune Responses in Fish: A Review. *Environ. Toxicol. Pharmacol.* **2019**, *68*, 101–108. [CrossRef] [PubMed]
9. Kumar, V.; Gill, K.D. Oxidative Stress and Mitochondrial Dysfunction in Aluminium Neurotoxicity and Its Amelioration: A Review. *Neurotoxicology* **2014**, *41*, 154–166. [CrossRef] [PubMed]
10. Prakash, C.; Soni, M.; Kumar, V. Mitochondrial Oxidative Stress and Dysfunction in Arsenic Neurotoxicity: A Review. *J. Appl. Toxicol.* **2016**, *36*, 179–188. [CrossRef] [PubMed]
11. Ma, Y.; Su, Q.; Yue, C.; Zou, H.; Zhu, J.; Zhao, H.; Song, R.; Liu, Z. The Effect of Oxidative Stress-Induced Autophagy by Cadmium Exposure in Kidney, Liver, and Bone Damage, and Neurotoxicity. *Int. J. Mol. Sci.* **2022**, *23*, 13491. [CrossRef] [PubMed]
12. Eggers, C.; Arendt, G.; Hahn, K.; Husstedt, I.W.; Maschke, M.; Neuen-Jacob, E.; Obermann, M.; Rosenkranz, T.; Schielke, E.; Straube, E. HIV-1-Associated Neurocognitive Disorder: Epidemiology, Pathogenesis, Diagnosis, and Treatment. *J. Neurol.* **2017**, *264*, 1715–1727. [CrossRef] [PubMed]
13. Clifford, D.B. HIV-Associated Neurocognitive Disorder. *Curr. Opin. Infect. Dis.* **2017**, *30*, 117–122. [CrossRef] [PubMed]

International Journal of
Molecular Sciences

MDPI

Article

Interactions of VMAT2 with CDCrel-1 and Parkin in Methamphetamine Neurotoxicity

Heli Chauhan [1], Nicholas J. Carruthers [2,3], Paul M. Stemmer [1,2], Bernard L. Schneider [4] and Anna Moszczynska [1,*]

1 Department of Pharmaceutical Sciences, Eugene Applebaum College of Pharmacy and Health Sciences, Wayne State University, 259 Mack Ave., Detroit, MI 48201, USA; pmstemmer@wayne.edu (P.M.S.)
2 Institute of Environmental Health Sciences and Proteomics Core Facility, 540 East Canfield Ave., Detroit, MI 48202, USA
3 Bioinformatics Core, Michigan Medicine, University of Michigan, NCRC Building 14, 2800 Plymouth Rd., Ann Arbor, MI 48109, USA
4 Bertarelli Platform for Gene Therapy, École Polytechnique Fédérale de Lausanne, School of Life Sciences, Ch. Des Mines 9, CH-1202 Geneva, Switzerland
* Correspondence: amosz@wayne.edu

Abstract: In recent years, methamphetamine (METH) misuse in the US has been rapidly increasing, and there is no FDA-approved pharmacotherapy for METH use disorder (MUD). In addition to being dependent on the drug, people with MUD develop a variety of neurological problems related to the toxicity of this drug. A variety of molecular mechanisms underlying METH neurotoxicity has been identified, including the dysfunction of the neuroprotective protein parkin. However, it is not known whether parkin loss of function within striatal dopaminergic (DAergic) terminals translates into decreased DA storage capacity. This study examined the relationship between parkin, its substrate cell division cycle related-1 (CDCrel-1) associated with synaptic vesicles, and vesicular monoamine transporter-2 (VMAT2) responsible for packaging DA in an in vivo model of METH neurotoxicity. To assess the individual differences in response to METH's neurotoxic effects, a large group of male Sprague Dawley rats were treated with binge METH or saline and sacrificed 1 h or 24 h later. This study is the first to show that CDCrel-1 interacts with VMAT2 in the rat striatum and that binge METH can alter this interaction as well as the levels and subcellular localization of CDCrel-1. The proteomic analysis of VMAT-2-associated proteins revealed the upregulation of several proteins involved in the exocytosis/endocytosis cycle and responses to stress. The results suggest that DAergic neurons are engaged in counteracting METH-induced toxic effects, including attempts to increase endocytosis and autophagy at 1 h after the METH binge, with the responses varying widely between individual rats. Studying CDCrel-1, VMAT2, and other proteins in large groups of outbred rats can help define individual genetic and molecular differences in responses to METH neurotoxicity, which, in turn, may aid treating humans suffering from MUD and its neurological consequences.

Keywords: methamphetamine neurotoxicity; vesicular monoamine transporter-2; parkin; CDCrel-1; exocytosis/endocytosis cycle

1. Introduction

Methamphetamine (METH) is a highly addictive and powerful central nervous system psychostimulant that induces a feeling of intense euphoria. Therefore, METH misuse is a worldwide health problem. In the United States, there are close to 2,000,000 people who misuse METH, and deaths from METH overdose are rapidly rising [1,2]. METH overdose is not the only danger to health. METH is highly neurotoxic, and its misuse causes a variety of serious neurological problems and is associated with a risk of developing Parkinson's disease [3–5]. A deficiency in dopamine (DA) within the nigrostriatal DA pathway is believed to underlie this risk [6]. Despite numerous clinical trials conducted to date, there is no FDA-approved medication for METH use disorder (MUD) or the neurological problems associated with the disorder. Such medication is particularly needed for people with severe MUD (people who use METH heavily) as they have the most difficulty quitting METH use, are at high risk of overdosing on the drug, and suffer the most from the neurological consequences of METH neurotoxicity.

In experimental animals, high METH doses (4×5–20 mg/kg or single > 30 mg/kg) are neurotoxic to DAergic and serotonergic nerve terminals in selective brain areas, with the dorsal striatum being the most affected [7,8]. At the molecular level, METH neurotoxicity to DAergic and serotonergic terminals in experimental animals is manifested by persistent (long-lasting after METH cessation) reductions in DAergic and serotonergic markers (DA, serotonin and their metabolites, DA and serotonin transporters, vesicular monoamine transporter-2 (VMAT2), tyrosine hydroxylase, and tryptophan hydroxylase) [9–12], microglial activation and gliosis [13,14], as well as morphological and structural abnormalities such as brain edema, increased blood–brain barrier permeability, swollen neuronal axons, and damaged myelin [15–17]. The METH-mediated neurotoxic effects and their manifestation in individuals addicted to METH are similar to those observed in experimental animals; however, the actual degeneration of DAergic and five serotonergic nerve terminals, or cell bodies, has been questioned [6]. Nevertheless, many human METH users suffer from persistent DAergic and serotonergic deficits as well as from brain structural and metabolic abnormalities in the same brain areas as those affected by METH in experimental animals, suggesting the functional dysregulation of these areas (reviewed in [7,8,18]).

Our previous studies showed that parkin, a neuroprotective ubiquitin-protein ligase, protected DAergic terminals in rat dorsal striatum from METH neurotoxicity [19] and might be a potential new drug target in severe MUD [20]. Others have shown an essential role of VMAT2 in METH neurotoxicity and MUD [21–24]. It is not known whether there is an interaction between parkin and VMAT2 in METH neurotoxicity.

The VMAT2 is a proton pump localized on synaptic monoaminergic storage vesicles and controls intracellular DA levels, transporting DA from the cytoplasm into the vesicles in exchange for protons [25,26]. METH reverses VMAT2 function and disrupts the proton gradient of the vesicles, mechanisms through which it stimulates the efflux of DA from the storage vesicles to the cytoplasm [27,28]. The release of DA by METH is a first step in METH neurotoxicity as DA rapidly autoxidizes in the cytoplasm, leading to oxidative stress and damage to proteins, lipids, and organelles within DAergic terminals [29]. In addition, neurotoxic METH doses rapidly redistribute rat striatal VMAT2 immunoreactivity from synaptic vesicle-enriched, nonmembrane subcellular fractions to a location not retained in the preparation of the synaptosomes while leaving VMAT2 membrane expression unchanged [30] or increased [31]. This impaired recycling of synaptic VMAT2 vesicles from the plasma membrane to the cytosol suggested a possibility that high-dose METH might render membrane VMAT2 vesicles trapped at the plasma membrane and nonfunctional while removing other VMAT2 vesicles from DAergic terminals via retrograde axonal transport. These changes render DAergic terminals with a decreased ability to sequester DA, thus increasing the pro-oxidant effects of the neurotransmitter.

Parkin is an E3 ligase in the ubiquitin-proteasome system that protects DA neurons from diverse cellular insults, including METH- and DA-induced oxidative stress [19,32]. This protection may be partially exerted via the regulation of DAergic neurotransmission. Thus, parkin has been found to regulate DA transporter trafficking between the plasma membrane and cytosol in vitro [33]. It is not known whether parkin has a similar effect on VMAT2 under physiological or neurotoxic conditions in vivo. The available literature data suggest that parkin may regulate VMAT2 vesicle trafficking via parkin substrates involved in the exocytosis/endocytosis cycle, namely, α-synuclein [34], α- and β-tubulin [35], synaptotagmins IV and 11 [36,37], synphilin-1 [38], endophilin, dynamin, synaptojanin-1 [39], and CDCrel-1 (septin 5) [40,41]. Without parkin, these substrates would abnormally accumulate. Song and colleagues reported that parkin modulated the endo-lysosomal pathway in cultured cells [42]. Feng and colleagues reported that parkin stabilized microtubules (MTs) and facilitated the transport of misfolded proteins along the MTs for degradation [35].

It is not known whether the METH neurotoxicity-induced deficit in striatal parkin levels that our group demonstrated in [12] mediates the redistribution of VMAT2 vesicles away from striatal terminals observed by the aforementioned previous studies [30,31].

CDCrel-1 is a filament-forming protein involved in the inhibition of the exocytotic process [43,44]. The overexpression of CDCrel-1 produces DA neuron-specific cell death [45], which can be prevented by increasing parkin levels [46]. Both parkin and VMAT2 were shown to be oxidatively damaged and decreased after the administration of high METH doses [12,47,48], while the effects of METH on CDCrel-1 have not been studied. Moreover, little is known about the interactions between parkin, VMAT2, and CDCrel-1 under METH-induced neurotoxic conditions. Knowledge of the effects of neurotoxic METH doses on endocytic and exocytic proteins is also limited. We tested the hypothesis that METH-induced parkin deficit increases CDCrel-1 levels in rat striatal neuronal terminals, which, in turn, alters the localization of VMAT2 vesicles and decreases DA storage within the terminals.

2. Results

2.1. The 4 × 8 mg/kg METH Binge Elicits Variable Thermal Responses in Wild-Type Rats

In experimental animals, high doses of METH are known to cause hyperthermia, which is one of the mediators and indicators of METH neurotoxicity [29,49]. To confirm that METH binge-induced hyperthermia in the rats, their core body temperature was measured at 1 h after each METH or saline injection. Significant increases in core body temperatures were observed in all METH-treated wild-type (WT) rats compared to saline controls (two-way RT ANOVA followed by the Sidak post hoc test) (Figure 1b,c). All the WT rats reached 39 °C at least once. There was a significant main effect of the treatment (METH vs. saline) in both experimental cohorts, i.e., rats sacrificed at 1 h and those sacrificed at 24 h after the last METH injection (1 h cohort: treatment $F(1,28) = 52.6$, $p < 0.0001$, time ($F(2.42,67.8) = 10.7$, $p < 0.0001$, and treatment × time interaction ($F(4,112) = 16.2$, $p < 0.0001$; 24 h cohort: treatment $F(1,16) = 60.6$, $p < 0.0001$), time ($F(2.42,37.6) = 5.99$, $p < 0.01$, and treatment × time interaction ($F(4,62) = 11.6$, $p < 0.0001$)). It was observed that despite the same METH regimen, the thermal responses to METH were highly variable. The METH-treated rats could be divided into two groups depending on the severity of hyperthermia: those with high hyperthermia (HH) and those with low hyperthermia (LH) (Figure 1d,e). Significant differences in core body temperatures were observed between the saline and METH treatments as well as between the subgroups (1 h cohort: treatment $F(2,27) = 106$, $p < 0.0001$, time ($F(3.15,83.5) = 53.1$, $p < 0.0001$, and treatment × time interaction ($F(8,106) = 29.3$, $p < 0.0001$; 24 h cohort: treatment $F(2,15) = 61.1$, $p < 0.0001$), time ($F(3.31,49.7) = 23.5$, $p < 0.0001$, and treatment × time interaction ($F(8,60) = 18.3$, $p < 0.0001$)). Areas under the curve (AUC) were calculated for all METH-treated rats.

2.2. The Effects of the 4 × 8 mg/kg METH Binge on α-Tubulin and β-Actin Immunoreactivity in Striatal Synaptosomal Fractions

There is surprisingly little literature on the effect of METH on the cytoskeleton. The available data suggest that METH can disorganize both MTs and actin filaments [50]. Therefore, we compared β-actin and α-tubulin immunoreactivities in all striatal synaptosomal fractions at 1 h and 24 h after the last dose of METH vs. saline. β-Actin immunoreactivity was significantly decreased in all three synaptosomal fractions at 1 h after METH (total fraction: −17%, membrane/endosomal fraction: −19%, and vesicular/cytosolic fraction: −26%, $p < 0.05$, multiple unpaired t-tests with Holm–Sidak correction for multiple comparisons, $n = 7$–11) (Figure 2a). No significant changes were detected in β-actin immunoreactivity at 24 h after METH (Figure 2b). No statistically significant changes were detected at 1 h or 24 h after METH in the levels of α-tubulin; however, individual variability in its immunoreactivity was observed in the total and membrane/endosomal synaptosomal fractions from binge METH-treated rats at the 1 h time point (Figure 2c,d), thus suggesting that METH influenced α-tubulin levels differently in different rats. Consequently, to control for METH effects on the cytoskeleton, we used total protein levels (measured by Ponceau S dye) as a loading control in the subsequent quantification of immunoreactivity on Western blots.

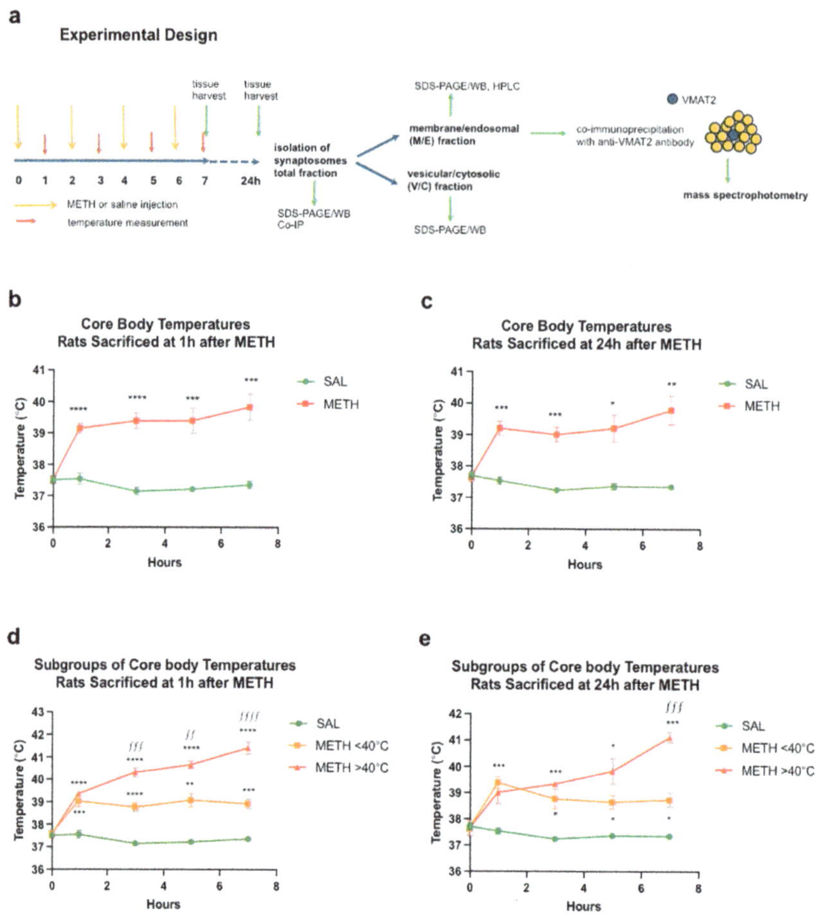

Figure 1. Experimental design and assessment of METH neurotoxicity. (**a**) A total of 8.0 mg/kg free base METH or saline (1 mL/kg) was administered to rats every 2 h in four successive intraperitoneal (i.p.) injections. Core body temperatures (°C) were measured before the first METH or saline injection and 1 h after each METH or saline injection. The rats were sacrificed 1 h or 24 h after the last injection of METH or saline. Stratal synaptosomes were isolated, separated into total, membrane/endosomal, and vesicular/cytosolic fractions, and analyzed. Proteomic analysis was performed on VMAT2-associated proteins coimmunoprecipitated from membrane/endosomal fractions. (**b**,**c**) Core body temperatures of rats euthanized at 1 h (**b**) or 24 h (**c**) after METH or saline. (**d**,**e**) The METH-treated rats were divided into two groups depending on the severity of hyperthermia: those with high hyperthermia and those with low hyperthermia (average of 4 temperature readings >40 °C and <40 °C, respectively) and euthanized at 1 h (**d**) or 24 h (**e**) after METH or saline. Significant differences between saline and METH rats: * $p < 0.05$, ** $p < 0.01$, *** $p < 0.001$, and **** $p < 0.0001$. Significant differences between HH and LH rats: ^{ff} $p < 0.01$, ^{fff} $p < 0.001$, and ^{ffff} $p < 0.0001$. Values expressed as mean ± SEM. Abbreviations: METH, methamphetamine; SAL, saline.

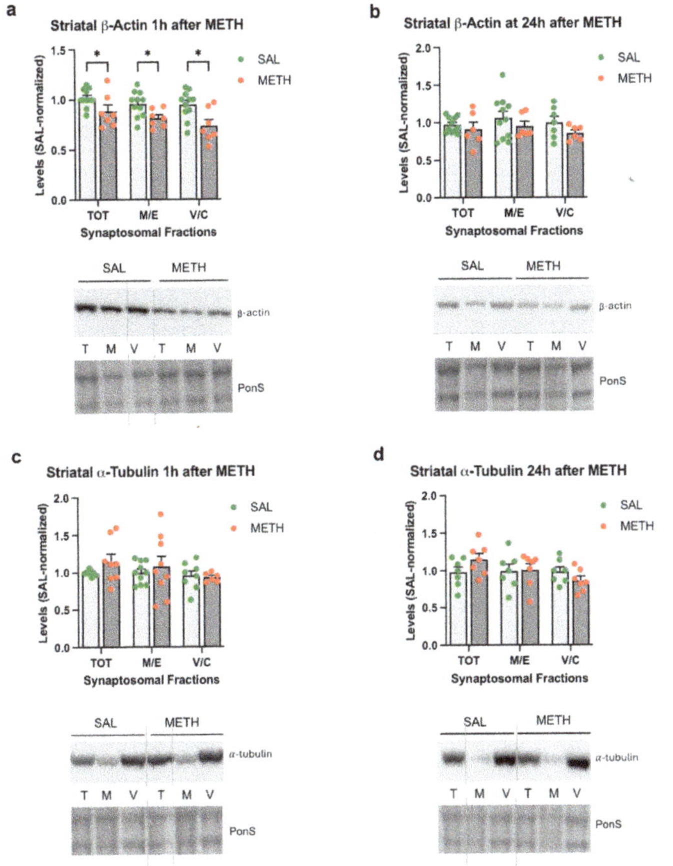

Figure 2. The effects of the 4 × 8 mg/kg METH binge on α-tubulin and β-actin immunoreactivity in striatal synaptosomal fractions. Immunoreactivity of β-actin (**a**,**b**) or α-tubulin (**c**,**d**) in the total (TOT or T), membrane/endosomal (M/E or M), and vesicular/cytosolic (V/C or C) synaptosomal fractions of the striatum in rats that were euthanized at 1 h (**a**,**c**) or 24 h (**b**,**d**) after the last dose of saline or METH. * $p < 0.05$, n = 7–11. Values are expressed as mean ± SEM. Vertical grey lines show where the blot was cut for rearrangement. Abbreviations: METH, methamphetamine; SAL, saline; PonS, Ponceau S.

2.3. The Effects of the 4 × 8 mg/kg METH Binge on Parkin Immunoreactivity in Striatal Synaptosomal Fractions

We previously determined that 4 × 10 mg/kg METH decreased parkin immunoreactivity in the total striatal synaptosomal fraction at 1 h and 24 h after the last dose of the drug, with parkin immunoreactivity returning to the baseline by 48 h [12]. In this study, we used 4 × 8 mg/kg METH to decrease the mortality rate. The 4 × 8 mg/kg dose did not significantly decrease parkin immunoreactivity in synaptosomal fractions compared to the saline controls; however, a trend toward statistical significance was observed in the total and vesicular/cytosolic synaptosomal fractions (−18% and −20%, respectively, p = 0.09, unpaired multiple t-tests with Holm–Sidak correction for multiple comparisons, n = 10–17) at 1 h after the last METH dose (Figure 3a). No significant changes in parkin levels (or trends toward statistical significance) were detected at 24 h after METH compared to the saline controls (Figure 3b). As with core body temperature, high variability in parkin immunoreactivity in METH-treated rats was observed at both time points (Figure 3a,b).

When the HH and LH rats from the 1 h group were assessed separately, a significant deficit in parkin immunoreactivity was detected in the LH rats in all synaptosomal fractions (two-way ANOVA followed by the Holm–Sidak post hoc test; main effect of treatment: $F(1,52) = 15.74$, $p < 0.001$; total fraction: −23%, $p < 0.05$; membrane/endosomal fraction: −28%, $p < 0.01$; vesicular/cytosolic fraction: −23%, $p < 0.05$), with HH rats displaying unchanged parkin immunoreactivity (Figure 3d). However, there was no significant correlation between core body temperature and parkin immunoreactivity in synaptosomal fractions, indicating an influence of additional factor(s) on parkin immunoreactivity in the METH-treated rats. In METH neurotoxicity-resistant cerebellum, a significant increase in synaptosomal parkin was detected compared to the saline controls (Supplementary Figure S1).

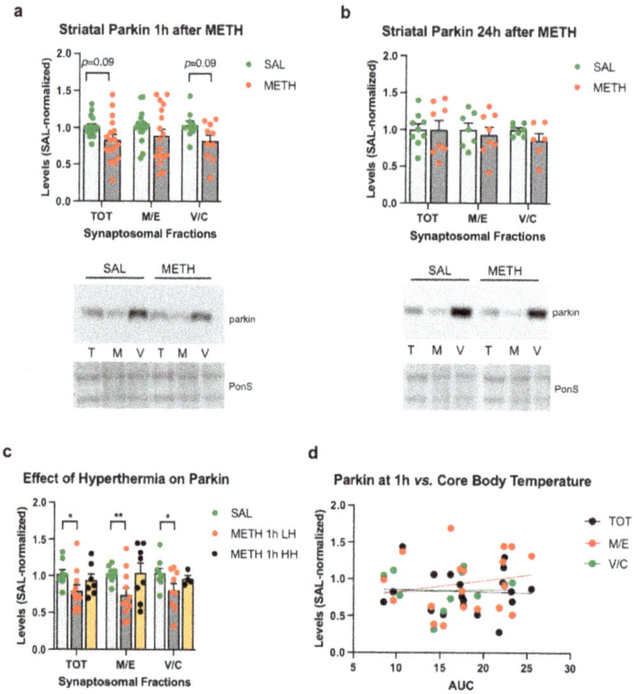

Figure 3. The effects of the 4 × 8 mg/kg METH binge on parkin immunoreactivity in striatal synaptosomal fractions. Immunoreactivity of parkin in the total (TOT or T), membrane/endosomal (M/E or M), and vesicular/cytosolic (V/C or C) synaptosomal fractions of the striatum in rats euthanized at (**a**) 1 h or (**b**) 24 h. (**c**) Immunoreactivity of parkin in striatal synaptosomal fractions in high-hyperthermia (HH) and low-hyperthermia (LH) rats. (**d**) Correlations of parkin immunoreactivity in synaptosomal fractions with core body temperature (area under the curve AUC) of the rats sacrificed 1 h after the last METH dose. * $p < 0.05$, ** $p < 0.01$, n = 10–17. Values are expressed as mean ± SEM. Abbreviations: METH, methamphetamine; SAL, saline; PonS, Ponceau S.

2.4. The Effects of the 4 × 8 mg/kg METH Binge on VMAT2 Immunoreactivity in Striatal Synaptosomal Fractions

Of the three isoforms of VMAT2, our anti-VMAT2 antibody easily identified the partially glycosylated form (~55 kDa) and the glycosylated form (68–75 kDa) of the protein; the non-glycosylated VMAT2 (45 kDa) was detected at low levels (Figure 4). At 1 h after METH, we did not detect significant changes in VMAT2 immunoreactivity in synaptosomal fractions as compared to the saline controls (Figure 4a). At 24 h after METH, a significant decrease in VMAT2 immunoreactivity was detected in the vesicular/cytosolic fraction

(−34%, $p < 0.05$, unpaired multiple t-tests with Holm–Sidak correction for multiple comparisons, $n = 9–11$) in METH-treated rats relative to the saline-treated rats (Figure 4b). As with individual parkin levels, high variability in individual synaptosomal VMAT2 immunoreactivity was observed after the METH binge at both time points, 1 h and 24 h. Subgrouping VMAT2 values from the 1 h METH group according to the levels of hyperthermia did not produce significant changes in VMAT2 immunoreactivity as compared to the corresponding saline group (Figure 4c). There was no significant correlation between core body temperature and VMAT2 immunoreactivity in the vesicular/membrane fraction in the METH-treated rats (Figure 4d); however, the correlation between core body temperature and membrane/endosomal VMAT2 displayed a strong trend toward statistical significance ($p = 0.055$, $R^2 = 0.21$, Pearson's correlation test).

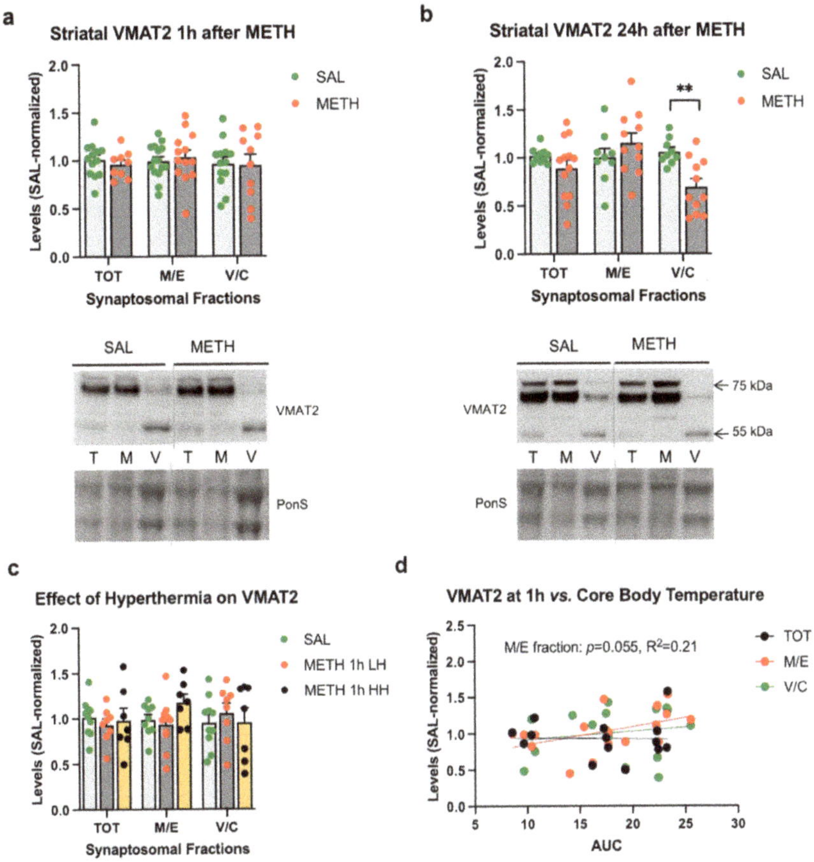

Figure 4. The effects of the 4 × 8 mg/kg METH binge on VMAT2 immunoreactivity in striatal synaptosomal fractions. Immunoreactivity of VMAT2 in the total (TOT or T), membrane/endosomal (M/E or M), and vesicular/cytosolic (V/C or C) synaptosomal fractions of the striatum in rats euthanized at (**a**) 1 h or (**b**) 24 h. ** $p < 0.01$, $n = 9–11$. (**c**) Immunoreactivity of VMAT2 in striatal synaptosomal fractions in high-hyperthermia (HH) and low-hyperthermia (LH) rats. (**d**) Correlations of VMAT2 immunoreactivity in synaptosomal fractions with core body temperature (area under the curve, AUC) of rats sacrificed 1 h after the last METH dose. A strong trend toward statistical significance was detected in the membrane/endosomal fraction ($p = 0.055$). Values are expressed as mean ± SEM. Vertical grey lines show where the blot was cut for rearrangement. Abbreviations: METH, methamphetamine; SAL, saline; PonS, Ponceau S.

2.5. Parkin Does Not Affect VMAT2 Immunoreactivity in Striatal Synaptosomal Fractions

Since parkin binds with MTs and actin filaments [51,52] and the available evidence suggests its involvement in the exocytosis/endocytosis cycle, parkin may play a role in the intracellular trafficking of VMAT2-containing vesicles. To test this hypothesis, we assessed VMAT2 immunoreactivity in striatal synaptosomes from rats overexpressing parkin in the nigrostriatal DA pathway. Parkin overexpressed about 5-fold in all synaptosomal fractions (Figures S2b and S3). The binge METH-treated parkin-overexpressing rats displayed a similar profile of thermal response to METH as the WT rats did (treatment $F(1,10) = 102$, $p < 0.0001$, time ($F(1.77,17.7) = 18.1$, $p < 0.01$, and treatment × time interaction ($F(4,40) = 4.92$, $p < 0.01$, one-way RT ANOVA followed by the Sidak post hoc test) (Figure 5a). As in the first METH experiment, binge METH treatment caused a small decrease in parkin immunoreactivity in WT rat striatum that displayed a trend toward statistical significance when analyzed by Student's t-test (−26%, $p = 0.09$, $n = 5$) (Figure 4b; a representative blot is presented in Supplementary Figure S3). The trend was lost when WT and parkin overexpression data were analyzed using two-way ANOVA. METH treatment did not significantly change parkin immunoreactivity in parkin-overexpressing rats (treatment and treatment × parkin levels $p > 0.05$; parkin levels $F(1,16) = 31.4$, $p < 0.0001$) (Figure 5b). The overexpression of parkin did not affect VMAT2 immunoreactivity in any synaptosomal fraction in the saline-treated rats ($p > 0.05$, $n = 5$, multiple unpaired t-tests with Holm–Sidak correction for multiple comparisons) (Figure 5c). Binge METH did not decrease VMAT2 immunoreactivity in striatal synaptosomal fractions from the WT or parkin-overexpressing rats at 1 h after the last METH injection ($p > 0.02$, $n = 5$, two-way ANOVA with the Holm–Sidak post hoc test) (Figure 5d). There was no significant correlation between total parkin immunoreactivity and total, membrane/endosomal, or vesicular/cytosolic immunoreactivity of VMAT2 in METH-treated rats at 1 h after the treatment ($p > 0.05$, Pearson's correlation test) (Figure 5e).

2.6. The Effects of the 4 × 8 mg/kg METH Binge on CDCrel-1 Immunoreactivity in Striatal Synaptosomal Fractions

To determine the effect of the METH binge on CDCrel-1 levels in rat striatal synaptosomes, we assessed CDCrel-1 immunoreactivity in the WT rats treated with saline or 4 × 8 mg/kg binge METH. As compared to the saline controls, no significant changes in CDCrel-1 immunoreactivity in striatal synaptosomal fractions were detected in the WT rats at 1 h after the last METH injection (Figure 6a), whereas a significant decrease in CDCrel-1 immunoreactivity was detected in the total and vesicular/cytosolic fractions at 24 h after METH as compared to the saline controls (−13% and −22%, respectively, $p < 0.05$, $n = 9$–10, multiple unpaired t-tests with Holm–Sidak correction for multiple comparisons) (Figure 6b). As with other measured indices, a wide range of individual responses to the drug was observed at 1 h, with some rats having very high CDCrel-1 immunoreactivity (Figure 6a). After dividing the METH group into two subgroups (group 1 with total synaptosomal CDCrel-1 immunoreactivity at a >120% increase and group 2 with total synaptosomal CDCrel-1 immunoreactivity at a <120% increase, with no change, or at a decrease), significant METH-induced increases in the total and membrane/endosomal fractions in group 1 were observed (+58%, $p < 0.0001$ and +55%, $p < 0.001$, $n = 6$–13, multiple unpaired t-tests with Holm–Sidak correction for multiple comparisons) and significant METH-induced decreases in the total and vesicular/cytosolic fractions in group 2 emerged (−10% and −21%, respectively, $p < 0.05$, $n = 11$–13, multiple unpaired t-tests with Holm–Sidak correction for multiple comparisons) (Figure 6c,d). CDCrel-1 immunoreactivity was not significantly changed or markedly varied in the control cerebellum (Supplementary Figure S1). After dividing the METH rats according to their thermal responses to METH (low or high hyperthermia: LH or HH), a small statistically significant decrease in CDCrel-1 immunoreactivity was observed in the vesicular/cytosolic fraction in the HH rats as compared to the saline controls (−22%, $p < 0.05$, multiple unpaired t-tests with Holm–Sidak correction for multiple comparisons) (Figure 6e). There was no significant correlation between the core body temperature and CDCrel-1 immunoreactivity in synaptosomal fractions, suggesting that core body temperature does not have marked influence on CDCrel-1 levels in striatal synaptosomes ($p > 0.05$, Pearson's correlation test) (Figure 6f).

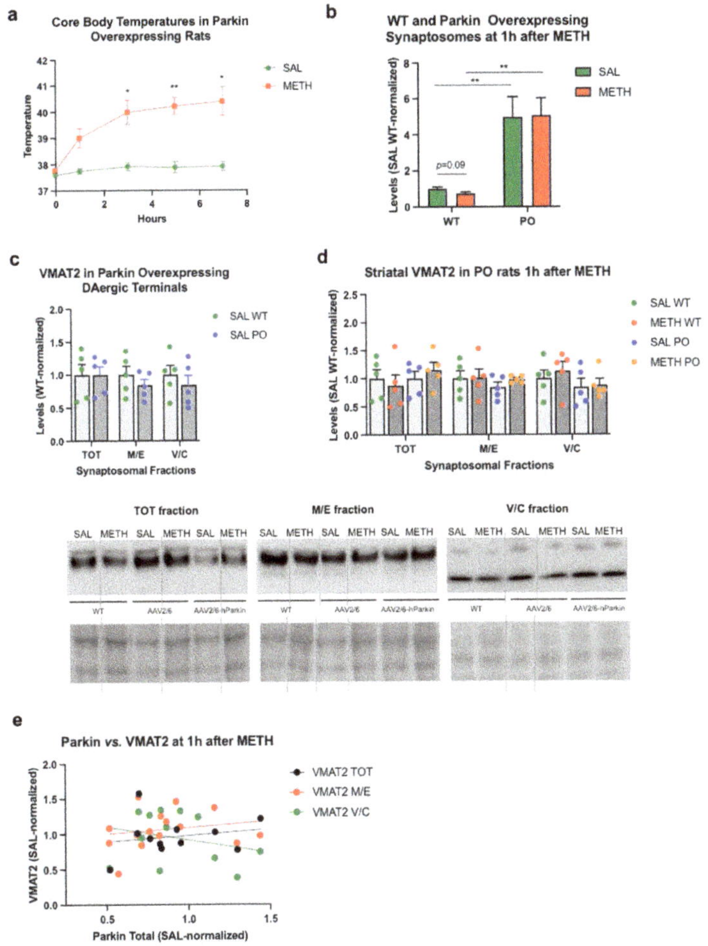

Figure 5. Parkin does not affect VMAT2 immunoreactivity in striatal synaptosomal fractions. (**a**) Core body temperatures (°C) of the wild-type rats and rats overexpressing parkin in the nigrostriatal dopamine pathway. * $p < 0.05$, ** $p < 0.01$. (**b**) Immunoreactivity of parkin in the total synaptosomal fraction of the striatum in rats euthanized at 1 h after METH or saline. ** $p < 0.01$ $n = 5$. Parkin overexpression was about 5-fold in both treatment groups. METH did not significantly alter parkin immunoreactivity in the wild-type ($p = 0.09$) or parkin-overexpressing ($p > 0.1$) striatal synaptosomes at 1 h after the treatment. (**c**) Immunoreactivity of VMAT2 in the total (TOT), membrane/endosomal (M/E), and vesicular/cytosolic (V/C) synaptosomal fractions in the striatum from wild-type and parkin-overexpressing saline-treated rats. (**d**) Immunoreactivity of VMAT2 in striatal synaptosomal fractions in the wild-type and parkin-overexpressing rats treated with METH or saline at 1 h after the treatment. (**e**) Correlations of VMAT2 immunoreactivity in synaptosomal fractions with total synaptosomal parkin immunoreactivity. Values are expressed as mean ± SEM. Vertical grey lines show where the blot was cut for rearrangement. Abbreviations: AAV2/6, adeno-associated viral vector 2/6; METH, methamphetamine; SAL, saline; PO, parkin-overexpressing; WT, wild-type; PonS, Ponceau S; AAV2/6-parkin, parkin-encoding AAV2/6; AAV6, non-coding AAV2/6.

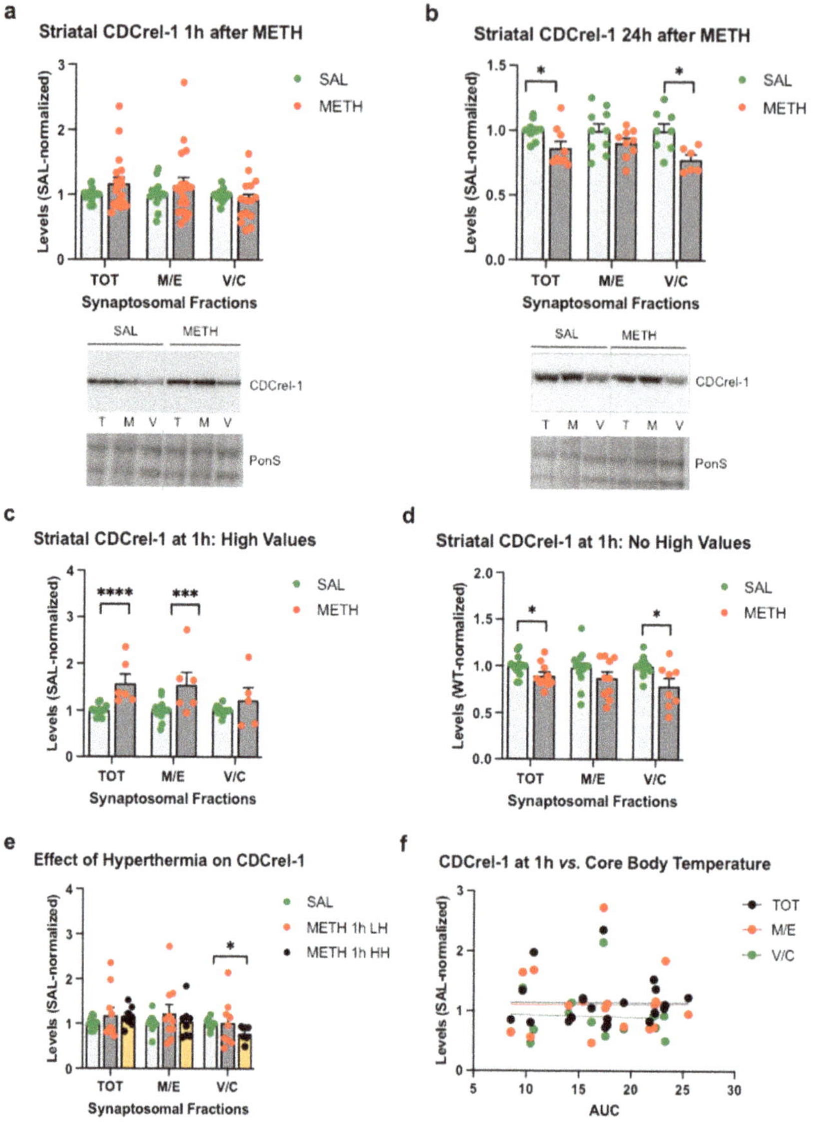

Figure 6. The effects of the 4 × 8 mg/kg METH binge on CDCrel-1 immunoreactivity in striatal synaptosomal fractions. Immunoreactivity of CDCrel-1 in the total (TOT or T), membrane/endosomal (M/E or M), and vesicular/cytosolic (V/C or C) synaptosomal fractions of the striatum in rats euthanized at (**a**) 1 h or (**b**) 24 h. * $p < 0.05$, $n = 9$–10. (**c**,**d**) Immunoreactivity of CDCrel-1 in striatal synaptosomal fractions separated into two subgroups based on individual variability to METH. * $p < 0.05$, *** $p < 0.001$, **** $p < 0.0001$, $n = 6$–13. (**e**) Immunoreactivity of CDCrel-1 in striatal synaptosomal fractions in high-hyperthermia (HH) and low-hyperthermia (LH) rats. * $p < 0.05$, $n = 6$–10. (**f**) Correlations of CDCrel-1 immunoreactivity in synaptosomal fractions with core body temperature (area under the curve, AUC) of the wild-type rats sacrificed 1 h after the last METH dose. Values are expressed as mean ± SEM. Vertical grey lines show where the blot was cut for rearrangement. Abbreviations: METH, methamphetamine; SAL, saline; PonS, Ponceau S.

2.7. CDCrel-1 Interactions with Parkin in Striatal Synaptosomal Fractions

In vitro, parkin regulates the levels of CDCrel-1, a protein that regulates cytoskeleton organization and inhibits exocytosis [41,43]. CDCrel-1 also induces the neurodegeneration of DA neurons when overexpressed in the nigrostriatal DA pathway [45]. Consequently, the METH-induced deficit in parkin could have increased CDCrel-1 levels and contributed to METH-induced neurodegeneration of DAergic terminals in the rat striatum, at least in some of the METH-treated rats. To determine the effect of parkin on CDCrel-1 levels in rat striatal synaptosomes, we first performed a coimmunoprecipitation experiment. As shown in Figure 7a, the anti-parkin antibody coimmunoprecipitated a small amount of CDCrel-1 with parkin in striatal synaptosomes from the WT rats. Next, we assessed CDCrel-1 immunoreactivity in parkin-overexpressing rats. The overexpression of parkin in the nigrostriatal DA pathway resulted in trends toward statistical significance in the total and vesicular/cytosolic fractions in METH-naïve rats (−18%, $p = 0.09$ and −43%, $p = 0.09$, $n = 5$, multiple unpaired t-tests with Holm–Sidak correction for multiple comparisons) (Figure 7b). CDCrel-1 immunoreactivity showed a trend toward a statistically significant decrease in the vesicular/cytosolic fraction after binge METH administration in the WT rats (−33%, $p = 0.079$, $n = 5$, multiple unpaired t-tests with Holm–Sidak correction for multiple comparisons) (Figure 7b). The trend towards statistical significance was lost when all treatment data were analyzed using two-way ANOVA (Figure 7c). No METH-induced high CDCrel-1 immunoreactivity was observed in this experiment, likely because of the smaller sample size as compared to the first experiment with the WT rats. These results agree with those shown in Figure 7d (a subgroup with no high CDCrel-1 values). As aforementioned, the METH-naïve parkin-overexpressing rats displayed a 33% decrease in vesicular/cytosolic CDCrel-1 as compared to the WT rats. The administration of the METH binge did not further decrease CDCrel-1 immunoreactivity in this fraction in the rats overexpressing parkin in the nigrostriatal DA pathway (PO rats) (Figure 7c). The correlation of parkin immunoreactivity with CDCrel-1 immunoreactivity was positive and statistically significant in the total synaptosomal fraction at 1 h after METH ($p < 0.05$, $R^2 = 0.259$, Pearson's correlation test) (Figure 7d).

2.8. CDCrel-1 Interactions with VMAT2 in Striatal Synaptosomes

We next examined whether there is CDCrel-1: VMAT2 interaction in striatal synaptosomes in the METH-naïve and METH-exposed rats. Anti-VMAT2 antibody immunoprecipitated CDCrel-1 from untreated synaptosomes and METH-treated synaptosomes, with more CDCrel-1 being pulled down at 1 h after METH treatment (Figure 8a,b). Furthermore, there was a significant correlation between CDCrel-1 and VMAT2 immunoreactivity in the membrane/endosomal fraction (Figure 8c).

2.9. The Effects of the 4 × 8 mg/kg METH Binge on Dopamine Levels in the Synaptosomal Membrane/Endosomal Fraction

Neurotoxic binge METH largely depletes DA from all VMAT2 vesicle pools in striatal synaptosomes by 1 h. To assess how much DA remained in membrane-bound VMAT2 vesicles, we assessed DA content in synaptosomal membrane/endosomal fractions at 1 h after the last dose of METH or saline. There was a significant main effect of the treatment (METH vs. saline) but no significant main effect of time (1 h vs. 24 h) on DA content in the membrane/endosomal fraction ($F(1,18) = 22.3$, $p < 0.001$, $n = 5$–7, two-way ANOVA with the Holm–Sidak post hoc test). DA content was decreased by 29% and 66% at 1 h and 24 h post-METH, respectively (Figure 8d,e).

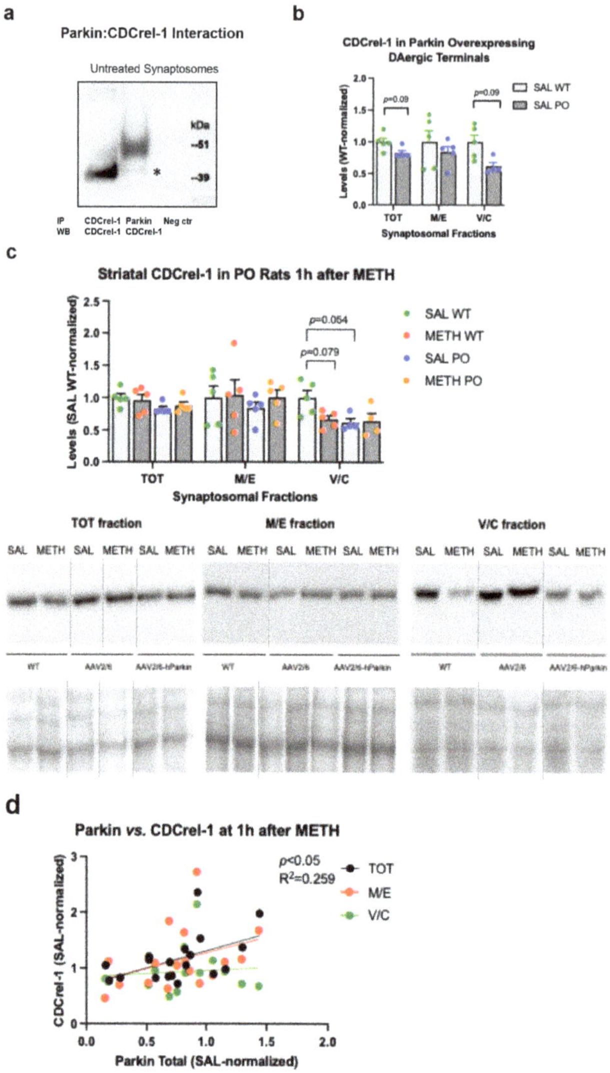

Figure 7. CDCrel-1 interactions with parkin in striatal synaptosomal fractions. (**a**) Anti-parkin antibody immunoprecipitated CDCrel-1 from untreated striatal synaptosomes (asterisk). (**b**) There was a weak trend toward statistical significance for the overexpression of parkin in the nigrostriatal dopamine pathway decreasing CDCrel-1 levels in the total and vesicular/cytosolic synaptosomal fractions ($p = 0.09$, $n = 5$). (**c**) Trends toward statistical significance for parkin and METH decreasing CDCrel-1 immunoreactivity in the wild-type rats at 1 h after the last dose of the drug were detected ($p = 0.054$ and $p = 0.079$, respectively, $n = 5$). METH treatment did not decrease CDcrel-1 levels in parkin-overexpressing rats. (**d**) Correlations of CDCrel-1 immunoreactivity in synaptosomal fractions with parkin immunoreactivity of the wild-type rats sacrificed 1 h after the last METH dose. A statistically significant positive correlation was found in the total synaptosomal fraction. Values are expressed as mean ± SEM. Abbreviations: METH, methamphetamine; SAL, saline; PonS, Ponceau S; AAV2/6-parkin, parkin-encoding AAV2/6; AAV6, non-coding AAV2/6; IP, immunoprecipitation; WB, Western blotting.

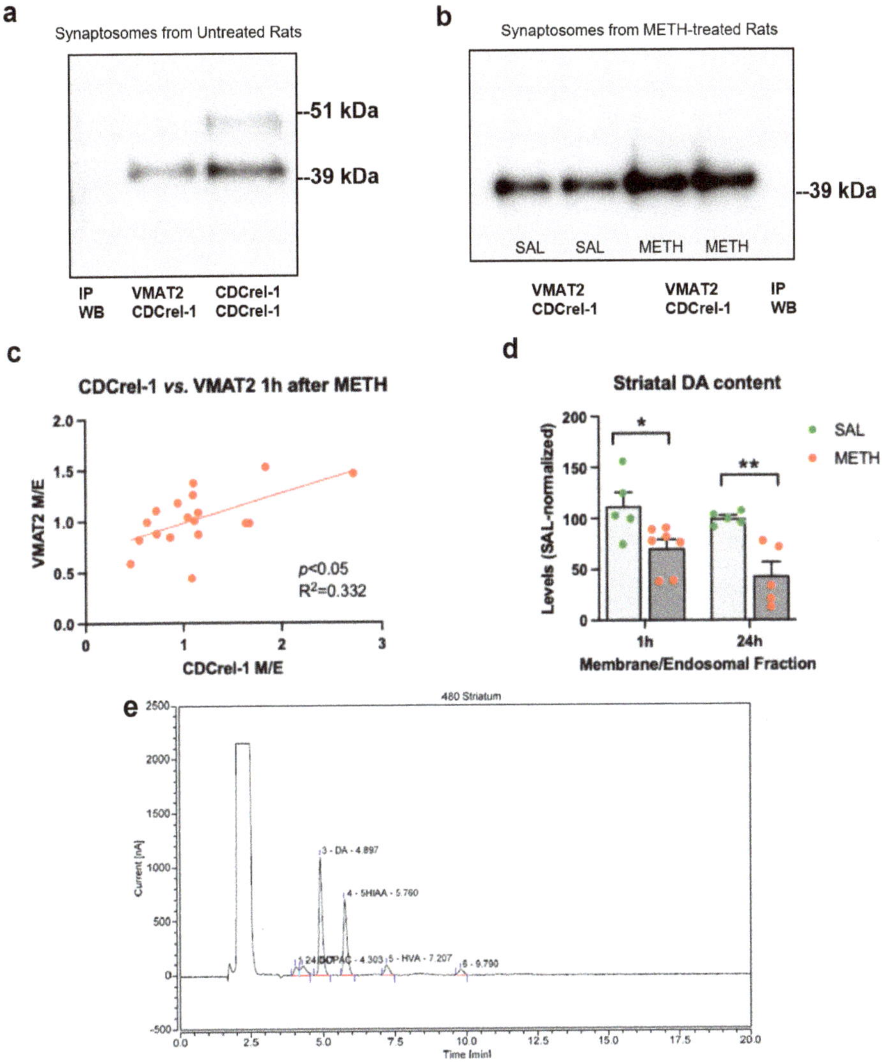

Figure 8. CDCrel-1 interactions with VMAT2 in striatal synaptosomes. (**a**) Anti-VMAT2 antibody immunoprecipitated CDCrel-1 from untreated striatal synaptosomes. (**b**) METH treatment increased the amount of immunoprecipitated CDCrel-1 at 1 h after the last dose of the drug. (**c**) There was a significant positive correlation between CDCrel-1 and VMAT2 immunoreactivity in the membrane/endosomal fraction of striatal synaptosomes ($p < 0.05$). (**d**) Striatal VMAT2 vesicles associated with the membrane/endosomal fraction had significantly lower dopamine content in METH-treated rats than the saline-treated rats at 1 h after the treatment (−29%). Over the next 24 h, dopamine content decreased to 66%. (**e**) Representative chromatogram for (**d**). * $p < 0.05$, ** $p < 0.01$, $n = 5$–7. Values are expressed as mean ± SEM. Abbreviations: DA, dopamine; METH, methamphetamine; SAL, saline; IP, immunoprecipitation; WB, Western blotting.

2.10. The Effects of the 4 × 8 mg/kg METH Binge on VMAT2-Associated Proteins in the Synaptosomal Membrane/Endosomal Fraction

The intracellular transport of vesicles within terminals is mediated not only by MTs but also by actin microfilaments [53]. Both MTs and actin filaments, as well as filamentous CDCrel-1, play essential roles in the neuronal exocytosis/endocytosis cycle. To examine whether a METH binge alters exocytic and endocytic proteins associated with VMAT2 vesicles, we immunoprecipitated VMAT2 and its interacting partners from membrane/endosomal synaptosomal fractions of saline- and METH-treated rats sacrificed at 1 h after their treatment and analyzed the immunoprecipitated proteins by mass spectrometry (n = 4). Among the 1281 proteins detected, an abundance of 10 proteins was significantly altered at FDR < 0.15: 9 increased and 1 decreased (Table 1). Of the 10 proteins, 6 were involved in axonal/intracellular transport and the exocytosis/endocytosis cycle: α-tubulin N-acetyltransferase 1 (ATAT1), protein kinase C, and casein kinase substrate in neurons protein 1 (PACN1, syndapin-1), MT-associated protein RP/EB family member 2 (MARE2), receptor-type tyrosine-protein phosphatase-like N (PTPRN), glia maturation factor beta (GMFB), and sorting nexin-17 (SNX17). Ten differentially expressed proteins are presented in Table 1).

Table 1. List of differentially changed (FDR < 0.15, fold change > 1.5) proteins after binge METH treatment.

Accession #	Protein Symbol	Protein Name	Function	Fold Change
Q6MG11	ATAT_RAT	Alpha-tubulin N-acetyltransferase 1	Acetylates α-tubulin on MTs, promotes MT destabilization, and accelerates MT dynamics	9.2
Q9Z0W5	PACN1_RAT	Protein kinase C and casein kinase substrate in neurons protein 1	Reorganizes MT and actin cytoskeleton, decreases MT polymerization and stability, and is required for bulk endocytosis	2.7
Q3B8Q0	MARE2_RAT	Microtubule-associated protein RP/EB family member 2	Unknown; may be involved in MT polymerization and dynamics	1.9
Q63259	PTPRN_RAT	Receptor-type tyrosine-protein phosphatase-like N	Neurotransmitter loading into dense-core synaptic vesicles	2.4
Q63228	GMFB_RAT	Glia maturation factor beta	Role in the growth, differentiation, and stress responses in neurons and glia, and actin-mediated endocytosis	11
Q6AYS6	SNX17_RAT	Sorting nexin-17	Regulates endocytic trafficking of several proteins, plays a role in protein sorting and autophagy	0.45
Q63065	PDK1_RAT	Pyruvate dehydrogenase (acetyl-transferring) kinase isozyme 1, mitochondrial	Regulation of glucose and fatty acid metabolism, and responses to cellular stresses	3.3
P0C546	S2542_RAT	Mitochondrial coenzyme A transporter SLC25A42	Transports coenzyme A into mitochondria in exchange for ADP	2.0
P06399	FIBA_RAT	Fibrinogen alpha chain	Role in responses to injury and inflammation, and axonal repair	2.5
P23593	APOD_RAT	Apolipoprotein D	Role in lipid trafficking, food intake, inflammation, antioxidative responses, and axon regeneration	5.8

3. Discussion

The major findings of this study are as follows: (1) there are wide individual differences in measured indices among outbred Sprague Dawley rats in responses to a 4 × 8 mg/kg METH binge, (2) CDCrel-1 interacts with VMAT2 in vivo and may have a role in VMAT2 trafficking under METH-induced neurotoxicity, (3) parkin does not have a major role in the regulation of VMAT2 or CDCrel-1 levels or trafficking in striatal neuronal terminals, and (4) changes in exocytosis/endocytosis cycle proteins are early responses to the METH binge.

Previous studies have determined that binge administration of high METH doses causes the degeneration of DAergic terminals in the dorsal striatum via oxidative stress, excitotoxicity, and hyperthermia [7]. Hyperthermia is a critical determinant of METH neurotoxicity, with higher temperatures causing more damage [54]. The rats employed in our study were outbred but were of the same strain, sex, and age, and they were housed and assessed under the same conditions. Therefore, the wide individual differences in the thermal and molecular responses to the 4 × 8 mg/kg METH binge were likely due to genetic differences such as variations in genes encoding thermal receptors, antioxidant enzymes, VMAT2, or DA transporters (higher function or levels of DA transporters at the membrane; more METH gets into DAergic terminals). Most studies on METH neurotoxicity used a 4 × 10 mg/kg METH binge. The results from these studies are, for the most part, consistent. The results from this study indicate that individual differences emerge when the sample sizes are sufficiently increased (our sample sizes of the 1 h cohorts were larger than the sample sizes employed in previous studies on METH neurotoxicity). In support of the latter scenario, we found less variability in parkin and CDCrel-1 immunoreactivity in the second experiment, which employed five rats per group (WT vs. PO rat experiment; saline vs. METH). Neuronal responses to oxidative stress and other stressors depend on the severity of the stress. Low-to-moderate oxidative stress increases the levels of proteins involved in stress responses, while severe oxidative stress damages these proteins [55]. Whether stress response proteins are increased or decreased following stress also depends on the length of the post-stress period. The observed U-shape-like responses to METH suggest that, after exposure to the 4 × 8 mg/kg METH binge, some rats could still upregulate their stress response pathways, while some could not.

We previously determined that a 4 × 10 mg/kg METH decreased parkin levels in striatal synaptosomes at 1 h and 24 h after the last dose of the drug, with parkin levels returning to the baseline by 48 h [12]. The decrease was caused by oxidative modification and the degradation of parkin. In this study, 4 × 8 mg/kg METH decreased parkin levels in the total and vesicular/cytosolic synaptosomal fractions at 1 h but not at 24 h after METH. It can be speculated that oxidative stress was less severe in the present study, allowing parkin levels to recover within 24 h. If that was the case, the recovery was likely due to the de novo synthesis of parkin in neuronal perikaryal and axonal transport to the terminals [56]. The discrepancy was more likely due to the different populations of rats used in this study, which expressed more individual differences than those used in the previous study. As with body temperature, high variability in striatal parkin levels was observed in METH-treated rats: a decrease in some rats and no change, or even an increase, in others. Interestingly, significant parkin deficit was detected in rats with low hyperthermia, while highly hyperthermic rats displayed mostly unchanged parkin levels. Hyperthermia itself can cause oxidative stress and protein aggregation [57]. Oxidative stress can decrease the number of viable MTs and inhibit axonal transport [58]. It can also damage degradative systems [59]. Parkin is sensitive to DA-mediated oxidative modifications and prone to misfolding and aggregating under oxidative stress [12,60–63]. Under conditions of mild oxidative stress, oxidized proteins are rapidly degraded by the 20S proteasome [64–66]. When oxidative stress becomes severe, proteasomal function decreases either because of direct oxidative damage, because the accumulation of oxidized proteins exceeds the capacity of proteasomes to clear them, or because proteins are so severely altered that they are no longer recognized as substrates [65,67]. The 26S proteasome plays a predominant role in the normal turnover of parkin, whereas the 20S proteasome and lysosome degrade oxidized and aggregated parkin, respectively [12,60,68]. In view of this knowledge, it might be concluded that oxidatively modified parkin was successfully removed from striatal DAergic terminals in rats with low hyperthermia but not in highly hyperthermic rats, in which severe oxidative stress impaired the proteasome, lysosome, and retrograde axonal transport involved in the removal of damaged proteins from terminals [69]. Another possibility is that oxidative stress changed parkin conformation, obscuring the antibody-binding epitope. Neither explanation agrees with our previous observation of parkin deficit after 4 × 10 mg/kg METH, which would cause more DA-mediated oxidative stress than 4 × 8 mg/kg METH. The deficit suggests viable 20S and lysosomal function and/or MT-mediated axonal transport after 4 × 10 mg/kg METH. A potential explanation of our results is an oxidative stress-independent hyperthermia effect on parkin immunoreactivity.

It has been established that the levels of antioxidant defenses increase in response to mild oxidative stress and decrease in response to severe oxidative stress [70–72]. Our finding of increased parkin levels in synaptosomes from METH neurotoxicity-resistant cerebellum agrees with this evidence and suggests that a similar compensatory increase in parkin levels could have taken place in non-DAergic synaptosomes, which are in the majority in striatal synaptosomes, masking parkin deficits in DAergic synaptosomes, and adding to the confounding effect of hyperthermia.

There are three pools of VMAT2 vesicles in the DAergic terminals: (1) the reserve pool, which is away from the active zone at the plasma membrane; (2) the recycling pool that is closer to the active zone; and (3) the readily releasable pool docked and ready for exocytosis [73]. The vesicles are transported into terminals via anterograde axonal transport and can be removed from them via retrograde axonal transport. It has been reported that METH neurotoxicity is associated with impaired VMAT2 trafficking in striatal synaptosomes at 1 h and VMAT2 deficit at 24 h [30,31,47,48]. Specifically, decreases in partially glycosylated VMAT2 were reported in the vesicular/cytosolic fraction at 1 h after 4 × 10 mg/kg METH binge, with membrane/endosomal VMAT2 levels reported to be increased by a study in mice and not changed by a study in rats [30,31]. These results suggest the degradation of VMAT2 or its mobilization to a non-synaptosomal compartment, potentially by the retrograde transport of damaged VMAT2 to the cell bodies in the substantia nigra pars compacta. This study detected a significant decrease in vesicular/cytosolic VMAT2 only at 24 h. Yamamoto's group reported a decrease in glycosylated VMAT2 in the vesicular/cytosolic fraction at 1 h and it decreased across all synaptosomal fractions at 24 h after 4 × 10 mg/kg METH [47,48]. Given that we used a lower dose of METH binge in this study than that used in the other studies (4 × 8 mg/kg vs. 4 × 10 mg/kg), METH-induced DA-mediated oxidative stress likely needed more time to rise to the threshold necessary to damage VMAT2 and decreased VMAT2 levels in the cytoplasm more than 1 h after the METH binge. Since there was high variability in VMAT2 levels in the METH rats, an alternative explanation for the lack of expected VMAT2 deficit at 1 h is an adaptive increase in VMAT2 levels in some rats and a decrease in others, resulting in a lack of overall change in VMAT2 levels. Despite unchanged levels, VMAT2 function could have been impaired by nitrosylation at 1 h time point in this study. Yamamoto's group demonstrated the modification of VMAT2 by nitrosylation at 1 h after a 4 × 10 mg/kg METH binge [48]. No significant changes in VMAT2 levels were detected in high-hyperthermia or low-hyperthermia rats. Still, a positive correlation between membrane/endosomal VMAT2 and core body temperature was observed in rats with high hyperthermia, suggesting retention at or mobilization of VMAT2 vesicles to the plasma membrane in high-hyperthermia rats.

Parkin could influence axonal VMAT2 vesicle trafficking by its known interaction with MTs and their stabilization [51]. Moreover, parkin could influence VMAT2 vesicle trafficking between their pools via interactions with α-synuclein [34,74], actin [52], and CDCrel-1, the last of which was reported to be a substrate for parkin in cultured cells [41]. CDCrel-1 is a filamentous protein found attached to membrane and synaptic vesicles [75]. CDCrel-1 binds to syntaxin-1, a component of the soluble NSF attachment protein (SNAP) receptor complex (SNARE), a complex crucial in exocytosis. Through this interaction, CDCrel-1 prevents the docking of vesicles to the membrane and decreases exocytosis [43]. We expected that METH neurotoxicity would decrease parkin levels with a consequent increase in CDCrel-1 levels and an altered distribution of VMAT2 vesicles between the vesicular/cytosolic and membrane/endosomal fractions in striatal synaptosomes. We also expected that parkin overexpression would change VMAT2 distribution between synaptosomal fractions. We found evidence to the contrary, suggesting that parkin does not significantly affect VMAT2 levels or its intracellular trafficking in rat striatal terminals in vivo. Our previous finding of unchanged VMAT2 levels in the METH-naïve striatum in parkin knockout rats supports this conclusion [76].

CDCrel-1 was reported to be a substrate for parkin in vitro [41]. We detected only a small effect of parkin on CDCrel-1 levels in striatal synaptosomes in the METH-naïve rats overexpressing parkin. This finding suggests that parkin is not a major E3 ligase degrading CDCrel-1 in the rat nigrostriatal pathway and agrees with the finding of unchanged CDCrel-1 levels in the striatum of parkin knockout mice [77]. Alternatively, as our synaptosomal preparations contained non-DAergic synaptosomes, which also express CDCrel-1, the effect of parkin overexpression on CDCrel-1 could have been "diluted" in DAergic synaptosomes. The METH binge had a variable effect on striatal synaptosomal CDCrel-1 levels at 1 h. Presently, it is unclear why some rats displayed significant increases in CDCrel-1 levels in striatal synaptosomes while others displayed significant decreases (U-shaped response to METH). This variability did not depend on hyperthermia; however, our correlation analysis showed that high CDCrel-1 levels were accompanied by high parkin levels in METH rats. The METH binge caused a deficit in CDCrel-1 in the WT rats at 1 h, and parkin overexpression appeared to protect vesicular/cytosolic CDCrel-1 from being decreased by METH. These results suggest that parkin is not engaged in the degradation of CDCrel-1 in vivo, but the proteins might interact and that METH damages both proteins via oxidative stress and alters their interaction. At 24 h, CDCrel-1 levels were decreased in the total and vesicular/cytosolic fractions, suggesting the removal of damaged CDCrel-1 from cytosol following the METH binge.

The most apparent interaction found in this study was between CDCrel-1 and VMAT2. This interaction was higher in METH-treated rats than in the saline-treated rats at the 1 h time point and was significant in the membrane/endosomal fraction. CDCrel-1 has been suggested to negatively regulate synaptic vesicle release at presynaptic terminals by forming filamentous barricades at the presynaptic membrane upon interaction with syntaxin-1 [43]. It is possible that, in some rats, METH binge-induced oxidative stress leads to CDCrel-1 accumulation at the plasma membrane, which disables the release of syntaxin-1 and VMAT2 vesicle docking and exocytosis [41]. Such "entrapment" could also prevent the recycling of VMAT2 vesicles and proper sequestration of cytosolic DA.

In summary, multiple toxic effects of METH (hyperthermia, oxidative stress, and excitotoxicity) differently influence the levels and interactions of parkin, VMAT2, and CDCrel-1 in striatal terminals in different rats. This individual variability in responses to binge METH indicates variability in sensitivity to METH neurotoxicity and stress responses in Sprague Dawley rats.

Tubulin and actin are cytoskeletal proteins found in abundance in the cytosol and synaptic active zones, [78] where they are involved in intra-neuronal transport and the exocytosis/endocytosis cycle. The deficits in β-actin immunoreactivity observed in all synaptosomal fractions at 1 h after METH suggest that METH damages actin filaments in striatal terminals. This conclusion is supported by reports of oxidative stress and heat shock inducing the collapse of actin filaments and MTs [79,80]. As parkin binds to both actin filaments and MTs, a deficit in its levels might have contributed to the disorganization of the cytoskeleton in METH binge-treated rats.

β-Actin mediates exocytosis and all kinetically distinguishable forms of endocytosis via membrane pit formation [81,82]. Consequently, METH-induced deficit in β-actin would negatively impact intra-synaptosomal transport as well as the exocytosis/endocytosis cycle of synaptic vesicles, including VMAT2 vesicles. There is evidence for METH altering the exocytosis/endocytosis cycle. Thus, in addition to non-exocytic DA release via the DA transporter, METH upregulates vesicular DA release in the striatum [83,84]. Proteomic/genetic studies have reported altered brain levels of proteins with functions related to the cytoskeleton, transport, endocytosis, and exocytosis after exposure to repeated low and medium METH doses [85,86]. A functional study showed that amphetamines increase endocytosis of the DA transporter [87] and glutamate transporter EAAT3 [88] in DA neurons. Oxidative stress was reported to decrease or increase endocytosis in neurons depending on the stressor and experimental conditions (reviewed in [89]). Finally, our group demonstrated that binge METH administration decreases the function of 26S proteasome and the levels of parkin ([12] and Figure 3a) at 1 h after the last METH injection. Parkin is a component of the ubiquitin-proteasome system. The function of the 26S proteasome is important for exocytosis/endocytosis and DAergic neurotransmission [90], and several parkin substrates are involved in the mediation of exocytosis and/or endocytosis, namely, α-synuclein [34], α- and β-tubulin [35], synaptotagmins IV and 11 [36,37], synphilin-1 [38], endophilin, dynamin, synaptojanin-1 [39], and CDCrel-1 [40,41]. These reports, however, do not provide a complete understanding of METH neurotoxicity-induced alterations in the VMAT2 vesicle exocytosis/endocytosis cycle in DAergic terminals in vivo. Our proteomic data provide more insight into these alterations. The endocytosis-/exocytosis-related parkin substrates mentioned above were found among proteins associated with membrane/endosomal VMAT2; however, their levels were not significantly changed by the METH binge, likely due to the small sample sizes (n = 4). Nevertheless, ten proteins were found differentially expressed by the binge METH treatment at 1 h: nine increased and one decreased in abundance. Six of these proteins are involved in axonal/intracellular transport and exocytosis/endocytosis—ATAT1, PACN1, (syndapin-1), MARE2, PTPRN, GMFB (increased), and SNX17 (decreased)—suggesting that the 4 × 8 mg/kg METH binge altered these processes.

ATAT1 localizes to clathrin-coated pits and acetylates α-tubulin. It promotes MT destabilization and accelerates MT dynamics [91]. Tubulin hyperacetylation is associated with cellular responses to stresses, including oxidative stress [92], and increased cell survival through the induction of autophagy [93]. Protein syndapin-1 also decreases MT stability, assists in actin polymerization, and regulates actin cytoskeleton dynamics [94]. Syndapin-1 is required for activity-dependent bulk endocytosis (ADBE) of synaptic vesicles [94]. ADBE generates many synaptic vesicles after intense neuronal activity or increased temperature, replenishing the reserve pool of the vesicles [95,96]. The observed increase in syndapin-1 levels likely occurred in response to METH binge-induced increase in synaptic exocytosis and, therefore, a need for ADBE and new vesicles for the storage of DA released into the cytosol. PTPRN is an important transmembrane protein in dense-core synaptic vesicles (DCVs), involved in cargo loading and the exocytosis of DA in PC12 cells [97]. The function of MARE2 is unknown. The available evidence suggests it may play a role in MT dynamics [98] and axonal delivery of DCVs [99]. The upregulation of PTPRN and MARE2 suggests the mobilization of DCVs to the plasma membrane of DAergic terminals. VMAT2 is associated mainly with small synaptic vesicles but is also found in DCVs [100]. Furthermore, DA was reported to be co-stored with cholecystokinin in DCVs [101]. It is plausible that DCV vesicles were mobilized from substantia nigra pars compacta cell bodies to striatal terminals to increase DA storage by replacing vesicles with dysfunctional nitrosylation-damaged VMAT2. Nexin-17 interacts with many receptors, including integrins, in a sequence-specific manner to regulate their recycling [102–104]. Furthermore, nexin-17 participates in the endocytic trafficking and processing of potentially harmful proteins and is linked to autophagy [104]. Autophagy is required to maintain pre-synaptic machinery during neuronal activity under physiological and stressful conditions [105,106]. In presynaptic terminals, the process involves the formation of the autophagosome around the organelle marked for destruction and MT-mediated retrograde transport to cell bodies for fusion with lysosomes and degradation. Nexin-17 is essential for autophagosome component recycling [107]; therefore, the observed downregulation of nexin-17 at 1 h after the METH binge may reflect an impairment of autophagy despite ATAT1 upregulation. In support of this, it has been shown that METH impairs autophagy at several levels, including autophagosome formation and maturation (reviewed in [105]).

FIBA (fibrinogen α-chain) is a blood plasma protein reported to be expressed by astrocytes and neurons under neuroinflammatory or neurodegenerative conditions [108]. FIBA also interacts with its receptors on neurons as well as with integrins and may be involved in axonal repair [108]. For example, FIBA binds to integrin α5β1, which is present on endothelial cells [103,109]. This integrin is involved in inflammatory responses and neuronal regeneration [110]. Another protein involved in responses to stress that was found to increase at 1 h after the METH binge was APOD. APOD is a member of the lipocalin superfamily involved in lipid trafficking, inflammation, and antioxidative responses; it is involved in the axon regeneration process as a lipid transporter [111].

PDK1 and SLC25A42 are mitochondrial proteins. As both mitochondria and synaptic vesicles are attached to actin and plasma membrane, they could have been pulled down by anti-VMAT2 antibody via this connection. PDK1 is a kinase that inhibits the formation of acetyl-coenzyme A and decreases mitochondrial respiration. By this mechanism, it protects cells against oxidative stress and apoptosis [112]. SLC25A42 transports coenzyme A into mitochondria in exchange for ADP. Coenzyme A is necessary for the Krebs cycle function and, consequently, for mitochondria function. Several steps in the exocytosis/endocytosis cycle require ATP, including ADBE, which relies on ATP-dependent actin polymerization [112]. Increases in both PDK1 and SLC25A42 would have opposing effects on mitochondrial function, with a net result being decreased levels of ATP as it was found in deficit at 1 h after binge METH [113]. PDK1 levels likely increased in response to METH-induced oxidative stress and inflammation, while SLC25A42 levels increased to provide ATP for ADBE. Mitochondrial function and ATP levels could also decrease in response to increased nigral neuron firing caused by METH.

GMFB is a protein expressed in glia and some neurons, including substantia nigra neurons [114]. GMFB is upregulated in several neurodegenerative conditions where it predominantly plays detrimental roles [115]. For example, DAergic neurons and astrocytes in GMFβ$^{-/-}$ mice have reduced sensitivity to oxidative stress [116]. On the other hand, overexpression of GFMB increases the expression of the antioxidant enzyme CuZn-SOD [115]. GMFB is also a modulator of the actin cytoskeleton [116]. GMFB overexpression reduces the levels of actin-related protein 2/3 complex (Arp2/3), the facilitator of actin polymerization [117]. The increase in GMFB suggests reduced actin polymerization at 1 h after the METH binge, which agrees with previous findings of METH's ability to modulate actin polymerization and increased actin cycling [50].

Our proteomic data suggest an increased rate of VMAT2 endocytosis/exocytosis cycle in striatal synaptosomes at 1 h after the 4 × 8 mg/kg METH binge, which does not result in significant changes in VMAT2 levels between the studied synaptosomal fractions. VMAT2 cycling could have changed only between the plasma membrane and endosomes. It is also possible that VMAT2 traveled between the terminals and cell body with the rate of its removal equal to its replacement rate.

Overall, the results suggest that at 1 h after a 4 × 8 mg/kg METH binge, DAergic terminals are engaged in counteracting METH-induced toxic effects, including attempts to increase VMAT2 vesicle endocytosis and autophagy. The levels and trafficking of parkin, VMAT2, and CDCrel-1 are altered differently between the rats, depending on the level of hyperthermia and oxidative stress each animal could combat based on genetic makeup.

4. Materials and Methods

4.1. Animals

This study employed young adult male Spraque Dawley rats (~55 days old at the beginning of this study) from Harlan Laboratories (now Envigo, Indianapolis, IN, USA). Upon arrival, the animals were pair-housed and maintained under standard environmental conditions in an AAALAC-accredited vivarium. Thus, the animals were maintained on a 12 h light/dark cycle with continuous ad libitum access to food and water.

Rats overexpressing parkin in the nigrostriatal dopamine pathway (PO rats) were generated in our laboratory according to a previously published protocol [19]. Briefly, rat parkin-encoding AAV2/6 gene transfer vector (AAV2/6-parkin) was microinjected into the left substantia nigra pars compacta (1×10^7 transduction units); whole non-coding AAV2/6 vector was microinjected into the right one at the following coordinates: −5.6 mm (AP) from Bregma, −2 mm (ML) from Bregma, and −7.6 mm (V) from the dura according to the Paxinos and Watson's rat brain atlas. After 3 weeks, the rats were treated with binge METH or saline. The non-coding AAV2/6 with a DNA segment cloned upstream of the pgk promoter to adapt the size of vector genome to the AAV packaging capacity (AAV2/6) and rat parkin-encoding AAV2/6 (AAV2/6-parkin) were a kind gift from Dr. Bernard Schneider at the Swiss Federal Institute of Technology Lausanne (EPFL), Switzerland.

A total of 95 rats were used in this study. The first experiment started with 40 rats that were divided into the following four groups based on treatment (METH or saline) and time of sacrifice (1 h or 24 h after the last injection of METH or saline): SAL/1 h, METH/1 h, SAL/24 h, and METH/24 h. More rats were added to the groups because high variability was observed in measured indices after METH administration. A few rats died from METH overdose and were excluded from neurochemical analyses. The final number of rats included in the analyses were as follows: SAL/1 h—n = 18, METH/1 h—n = 18, SAL/24 h—n = 10, and METH/24 h—n = 10. The second experiment started with 20 rats. This cohort was divided into METH and saline group. Within each group, some rats were microinjected with adeno-associated viral vector 2/6 coding for parkin (AAV2/6-parkin), some were microinjected with non-coding AAV2/6, and some were microinjected with saline, in the substantia nigra pars compacta. Rats that died following the METH binge were excluded from further analyses. The final number of samples in each subgroup was n = 5. To generate tissue for the HPLC and mass spectrometry analyses, 8 more rats were treated with saline and 8 with binge METH.

4.2. Administration of Methamphetamine

(+)-Methamphetamine hydrochloride (METH, 8.0 mg/kg free base) (Sigma-Aldrich, St. Louis, MO, USA) or saline (1 mL/kg) was administered to the rats every 2 h in four successive intraperitoneal (i.p.) injections, ~0.25 mL each. Many previous studies, including ours, have established this treatment paradigm to induce oxidative stress at 1 h and the degeneration of DAergic terminals in rat striatum several days later [7,12]. METH neurotoxicity is associated with hyperthermia, which peaks at approximately 1 h after each injection. Therefore, core body temperatures of the rats were measured via a rectal probe digital thermometer (Thermalert TH-8; Physitemp Instruments, Clifton, NJ, USA) before the beginning of the treatment (baseline temperatures) and at 1 h after each METH or saline injection. Non-anesthetized rats were killed by decapitation at 1 h or 24 h after the last injection of the drug or saline (these time points are commonly used in METH research to assess the short-term effects of binge METH). The brains were removed, and the METH neurotoxicity-sensitive dorsal striatum (referred to hereafter as the striatum) and the control METH neurotoxicity-resistant cerebellum were dissected out, flash-frozen on dry ice, and stored at −80 °C until assayed. The experimental design is presented in Figure 1a.

4.3. SDS-PAGE and Western Blotting

Synaptosomal fractions (total synaptosomal, membrane/endosomal, and vesicular/cytosolic) were prepared from striatal and cerebellar tissue by differential centrifugation as previously described [19]. Specifically, tissue pieces were homogenized in 0.5 mL 0.32 M sucrose with protease inhibitor cocktail and centrifuged at 800× g for 24 min at 4 °C. The supernatant was then centrifuged at 22,000× g for 17 min at 4 °C. The pellet was resuspended in 150 μL ice-cold distilled water and retained as the total synaptosomal fraction. Part of the total synaptosomal fraction was further centrifuged at 22,000× g for 17 min at 4 °C, and the supernatant was retained as the vesicular/cytosolic fraction, while the pellet was resuspended in 150 μL ice-cold distilled water and retained as the membrane/endosomal fraction. Sample protein concentrations were determined using a Bradford protein assay, using bovine serum albumin as the standard.

Synaptosomal fractions were subjected to reducing sodium dodecyl sulfate-polyacrylamide gel electrophoresis (SDS–PAGE). The 10–20 μg proteins per lane were loaded on 4–12% Bis-Tris gels (Life Technologies, Grand Island, NY, USA). Electrophoresis and Western blotting were performed as previously described [76], utilizing the following primary antibodies: CDCrel-1 (1:1000, overnight at 4 °C) (MAB5358; Chemicon International, Temecula, CA, USA); parkin (1:1000; overnight at 4 °C) (Prk8; Cell Signaling Technology, Danvers, MA, USA); β-actin (1:1000; 1 h at room temperature) (8H10D10; Signaling Technology, Danvers, MA, USA); α-tubulin (1:1000, overnight at 4 °C) (sc58668; Santa Cruz Biotech, Santa Cruz, CA, USA), and VMAT2 (1:3000, overnight at 4 °C) (NBP1-69750H, Novus Biologicals, Toronto, ON, Canada), as well as appropriate secondary antibodies. Blots were developed using ECL detection and an LAS4000 bioimager (GE Healthcare, Piscataway, NJ, USA). Immunoreactivity was quantified using ImageJ software v.1.50 (National Institutes of Health, Bethesda, MD, USA). For standardization across the blots, each blot contained all experimental groups. The Western blot data were expressed as relative optical density units and normalized to the controls on each blot. This approach normalized differences in the development of the blot and across blots.

4.4. Immunohistochemistry

Midbrains were post-fixed with 4% paraformaldehyde, frozen in isopentane, and kept at −80 °C for further immunohistochemistry procedures. Coronal brain sections (30 μm thick) containing substantia nigra pars compacta were sliced on a cryostat (Thermo Fisher Waltham, MA, USA) and processed as previously described [19]. Citrate buffer antigen retrieval (ThermoFisher, Waltham, MA, USA) was applied to all tissue sections. The sections were incubated overnight at 4 °C with the anti-parkin antibody (Prk8; Cell Signaling Technology) and antibody against DAergic marker tyrosine hydroxylase) (AB152, EMD Millipore Corp., Billerica, MA, USA) both diluted 1:100 in the blocking buffer. The sections were then incubated for 2.5 h at room temperature with Alexa Fluor-488 conjugated secondary antibody (Invitrogen, Carlsbad, CA, USA). DRAQ5 (Invitrogen) was used to stain nuclei. The sections were then mounted using Fluoromount mounting medium (Southern Biotech, Birmingham, AL, USA). The immunostaining on each slice (3 sections per slice) was imaged using the Leica TCS SPE-II laser scanning confocal microscope (Leica, Wetzlar, Germany) and averaged per rat.

4.5. Co-Immunoprecipitation

Striatal synaptosomal fractions were prepared from rats euthanized at 1 h after the METH binge. Dynabeads (Life Technologies, Grand Island, NY, USA) were incubated with 2 μL of either the rabbit polyclonal VMAT2 primary antibody (NBP1-69750H, Novus Biologicals), mouse monoclonal CDCrel-1 primary antibody (MAB5358; Chemicon International, Temecula, CA, USA), parkin mouse monoclonal primary antibody (Prk8, 1:1000; Cell Signaling Technology), or radioimmunoprecipitation assay (RIPA) buffer (negative control) for 6–12 h at 4 °C. This was followed by the addition of synaptosomal fractions (100–200 μg) and a second round of incubation (12 h at 4 °C). For all quantitative coimmunoprecipitation studies, equal protein content of the synaptosomal fraction was incubated with beads for each representing group (saline or METH). Following each incubation, the beads were washed three times using phosphate-buffered saline containing 0.02% Tween-20 (PBST). Following immunoprecipitation, parkin-, CDCrel-1-, and VMAT2-associated proteins were separated from the beads using SDS Tris–Glycine sample buffer (Bio-Rad, Hercules, CA, USA) and heating (10 min at 70 °C). The supernatants were run on 4–12% Bis-Tris gels under reducing conditions and subjected to Western blot analysis using the mouse CDCrel-1 primary antibody and anti-mouse secondary antibody, as described above.

4.6. Mass Spectrometry

Membrane/endosomal fractions of synaptosomes were incubated with Dynabeads conjugated with anti-VMAT2 antibody and processed as described for coimmunoprecipitation. The resulting eluted samples were washed and subjected to SDS-PAGE. After staining with Sypro Ruby stain to visualize proteins, multiple gel pieces were excised, washed with 25 mM NH_4HCO_3/50% acetonitrile (for 15 min with each solution), dehydrated in 100% acetonitrile (ACN), rehydrated in 50 mM NH_4HCO_3, and dehydrated again in 100% ACN. Subsequently, the gel pieces were vacuum-dried for 5 min. Next, the following reactions were performed: reduction with 5 mM dithiothreitol in 50 mM NH_4HCO_3 and alkylation with 15 mM iodoacetic acid in 50 mM NH_4HCO_3. The gel pieces were again dehydrated with ACN, the solvent was removed, and sequencing-grade trypsin (Promega, Madison, WI, USA) in 25 mM NH_4HCO_3/10% ACN was added to rehydrate the gel slices and initiate overnight digestion. Following digestion, peptides were extracted from the gel plugs using 50% ACN/0.05% formic acid. The free peptides were then speed-vacuumed to dryness and solubilized in 2% ACN/0.1% formic acid.

The peptides were separated by reverse phase chromatography (Acclaim PepMap RSLC C18 column, Thermo Scientific, Waltham, MA, USA), followed by ionization with the Nanospray Flex Ion Source (Thermo Scientific), and introduced into a Q Exactive mass spectrometer (Thermo Scientific). Abundant species were fragmented with high-energy collision-induced dissociation (HCD). Data analysis was performed using Proteome Discoverer 1.4 (Thermo Fisher, Waltham, MA, USA), which was incorporated with the Mascot algorithm (Matrix Science, Boston, MA, USA). The Swiss Prot rat protein database downloaded on March 2013 was used, and a reverse decoy protein database was run simultaneously for false discovery rate (FDR) determination. Secondary analysis was performed using Scaffold 4.2.1 (Proteome Software, Portland, OR, USA). The minimum protein identification probability was set at ≥95% with 2 unique peptides at ≥99% minimum peptide identification probability. Only those proteins identified with at least 2 unique peptides, a $p < 0.05$, a Benjamin–Hochberg FDR < 0.15, and a fold change >1.5 were considered differentially expressed.

4.7. High-Performance Liquid Chromatography

The synaptosomal membrane/endosomal pellet was obtained as described in Chu et al. [118]. Briefly, striatal tissue was homogenized in ice-cold 0.32 M sucrose solution. The homogenate was centrifuged at 800× g for 24 min at 4 °C. The supernatant was centrifuged at 22,000× g for 17 min at 4 °C to obtain the synaptosomal pellet. The pellet was resuspended in 75 µl of ice-cold water, and osmolarity was immediately restored by adding an equal volume of pH 7.5 buffer containing 25 mM HEPES and 100 mM potassium tartrate. The resulting solution was fractionated at 22,000× g for 17 min at 4 °C into the membrane/endosomal and vesicular/cytosolic fractions. The membrane/endosomal pellet was resuspended in ice-cold perchloric acid (final concentration 0.3 N). The resulting solution was then centrifuged at 22,000× g for 30 min 4 °C to obtain the precipitated protein pellet. The supernatant was analyzed for DA content using high-performance liquid chromatography (HPLC), using a Shimadzu Prominence HPLC system with electrochemical detection (Shimadzu Scientific Instruments, Columbia, MD, USA) as published previously [12]. Samples (20 µL) were injected onto a 3 µm C-18 reverse phase column (150 × 3.2 mm, 3 µM particle size, Thermo Scientific). DA was eluted with a mobile phase consisting of 90 mM sodium dihydrogen phosphate monohydrate, 50 mM citric acid, 1.7 mM 1-octane sulfonic acid, 50 µM EDTA, and 10% acetonitrile (pH 3.8). The analytes were detected using an electrochemical detector (Dionex Coulochem III, E1 = −150 mV, E2 = +220 mV). The protein pellet was resuspended in 1 N NaOH overnight at 4 °C, and its concentration was determined using the Bradford assay. To determine DA content, 20 µL of the supernatant was injected into a C-18 reverse phase column (ThermoScientific). The mobile phase consisted of 50 mM sodium citrate, 50 mM sodium phosphate, 200 µM EDTA, 1.5 mM heptane sulphonic acid, and 14% methanol adjusted to pH 3.8. The concentration of

DA was quantified by interpolating peak areas relative to those generated by a range of appropriate standards (Sigma Aldrich, St. Louis, MO, USA). The DA values were normalized to protein content in each sample and expressed as pg of analyte per µg of protein.

4.8. Statistical Analyses

Statistical analyses were performed using the program GraphPad Prism (GraphPad Software, version 10.3.1, San Diego, CA, USA). The comparisons made in this study were pre-planned. We established a priori that the striatum was the METH-affected brain region while the cerebellum was unaffected by METH based on existing knowledge of the effects of binge METH in the rat brain. Based on previously published results, samples were assumed to come from populations with the same standard deviations. Two-way or mixed model (if values were missing) repeated measures (RT) ANOVA followed by the Sidak post hoc test was performed on body core temperature data. The Greenhouse–Geisser correction was used if there was a lack of sphericity. Differences between the control and METH groups were analyzed separately in each synaptosomal fraction by multiple unpaired *t*-tests and the Holm–Sidak method to correct for multiple comparisons. Two-way ANOVA followed by the Holm–Sidak post hoc test was employed when the data had two variables. Correlations were determined using simple regression analysis and Pearson's correlation test. The Western blotting data were expressed as relative optical density units on each gel normalized to the controls. This approach normalized differences across blots, allowing standardization across the treatment groups. The data are presented as mean ± standard error (SEM). Statistical significance was set at $p < 0.05$.

5. Limitations of This Study

There are several limitations of this study. The sample sizes were small in the second (PO vs. WT rats) and third (HPLC, DA content) experiments. Given the individual differences that emerged in the first experiment, substantially increasing the sizes of all experimental groups would provide more information on individual responses to METH neurotoxicity. Oxidative stress indices were not measured. An assessment of the severity of oxidative stress in each experimental animal would provide information on whether it was a factor driving individual differences in responses to the METH binge. This study was largely observational and did not establish causal relationships between the measured variables. Another limitation of our research is that a mix of DAergic and non-DAergic synaptosomes was assessed. However, this was because the population of DAergic synaptosomes in the striatum constitutes <1% of the overall synaptosomal population and is difficult to separate. The presence of non-DAergic terminals (mostly glutamatergic) introduced a confounder and complicated data interpretation.

6. Conclusions and Future Directions

The major conclusion of this study is that studying large groups of outbred rats such as Sprague Dawley rats can help define individual molecular and genetic differences in responses to a neurotoxic METH binge. Our research shows, for the first time, that individual differences include the trafficking of VMAT2 and CDCrel-1 in striatal terminals as well as core body temperature. Our results also indicate that a neurotoxic METH binge induces changes to multiple proteins involved in the exocytosis/endocytosis cycle in striatal, most likely DAergic, terminals. This has not been shown before. Finally, we demonstrate that outbred rats, such as Sprague Dawley rats, can serve as a model for studying individual differences in responses to METH and other neurotoxins.

Future experiments should employ large sample sizes in longitudinal studies to delineate the subgroups differing in responses to METH neurotoxicity. Adding more time points will reveal changes over time, providing a more comprehensive understanding of striatal terminal responses to METH neurotoxicity. Future genetic and proteomic analyses of these responses will shed light on molecular mechanisms driving these responses. Variations in genes encoding thermal receptors, antioxidant enzymes, VMAT2, or the DA transporter should be of particular focus. The incorporation of experimental designs for establishing causal relationships between variables should follow. Establishing a causal relationship between genetic makeup and individual ability to handle METH-induced stresses in rodents will aid the development of individual treatment strategies for humans suffering from METH use disorder and its neurological consequences.

Supplementary Materials: The following supporting information can be downloaded at https://www.mdpi.com/article/10.3390/ijms252313070/s1.

Author Contributions: A.M. designed the experiments, analyzed the data, prepared the figures, and wrote the manuscript. H.C. conducted the experiments, quantified the data, and helped with manuscript preparation. P.M.S. and N.J.C. performed the proteomic analysis. B.L.S. provided the AAV2/6 vectors. All authors have read and agreed to the published version of the manuscript.

Funding: This work was supported by NIH grant DA023085 and NIH grants P30ES036084, P30CA022453, and S10OD030484.

Institutional Review Board Statement: All animal experiments were approved by the Wayne State University Institutional Animal Care and Use Committee (16-03-067, approved 23 May 2016) and conducted in compliance with the ARRIVE guidelines.

Informed Consent Statement: Not applicable.

Data Availability Statement: Raw data and western blots are available from the corresponding author upon request. The original or representative blots have been provided to the publisher.

Acknowledgments: We thank Akhil Sharma and Bryan Killinger for their help with the animal work. We acknowledge the assistance of the Wayne State University Proteomics Core.

Conflicts of Interest: The authors declare no conflicts of interest.

References

1. Jalal, H.; Buchanich, J.M.; Roberts, M.S.; Balmert, L.C.; Zhang, K.; Burke, D.S. Changing dynamics of the drug overdose epidemic in the United States from 1979 through 2016. *Science* **2018**, *361*, 1218. [CrossRef]
2. NIDA. *Overdose Death Rates*; National Institute on Drug Abuse: Bethesda, MD, USA, 2021.
3. Rusyniak, D.E. Neurologic manifestations of chronic methamphetamine abuse. *Neurol. Clin.* **2011**, *29*, 641–655. [CrossRef]
4. Callaghan, R.C.; Cunningham, J.K.; Sajeev, G.; Kish, S.J. Incidence of Parkinson's disease among hospital patients with methamphetamine-use disorders. *Mov. Disord.* **2010**, *25*, 2333–2339. [CrossRef]
5. Callaghan, R.C.; Cunningham, J.K.; Sykes, J.; Kish, S.J. Increased risk of Parkinson's disease in individuals hospitalized with conditions related to the use of methamphetamine or other amphetamine-type drugs. *Drug Alcohol. Depend.* **2012**, *120*, 35–40. [CrossRef]
6. Kish, S.J.; Boileau, I.; Callaghan, R.C.; Tong, J. Brain dopamine neurone 'damage': Methamphetamine users vs. Parkinson's disease—A critical assessment of the evidence. *Eur. J. Neurosci.* **2017**, *45*, 58–66. [CrossRef]
7. Moszczynska, A.; Callan, S.P. Molecular, Behavioral, and Physiological Consequences of Methamphetamine Neurotoxicity: Implications for Treatment. *J. Pharmacol. Exp. Ther.* **2017**, *362*, 474–488. [CrossRef] [PubMed]
8. Jayanthi, S.; Daiwile, A.P.; Cadet, J.L. Neurotoxicity of methamphetamine: Main effects and mechanisms. *Exp. Neurol.* **2021**, *344*, 113795. [CrossRef] [PubMed]
9. Wagner, G.C.; Ricaurte, G.A.; Seiden, L.S.; Schuster, C.R.; Miller, R.J.; Westley, J. Long-lasting depletions of striatal dopamine and loss of dopamine uptake sites following repeated administration of methamphetamine. *Brain Res.* **1980**, *181*, 151–160. [CrossRef] [PubMed]
10. Broening, H.W.; Pu, C.; Vorhees, C.V. Methamphetamine selectively damages dopaminergic innervation to the nucleus accumbens core while sparing the shell. *Synapse* **1997**, *27*, 153–160. [CrossRef]
11. Harvey, D.C.; Lacan, G.; Melegan, W.P. Regional heterogeneity of dopaminergic deficits in vervet monkey striatum and substantia nigra after methamphetamine exposure. *Exp. Brain Res.* **2000**, *133*, 349–358. [CrossRef] [PubMed]
12. Moszczynska, A.; Yamamoto, B.K. Methamphetamine oxidatively damages parkin and decreases the activity of 26S proteasome in vivo. *J. Neurochem.* **2011**, *116*, 1005–1017. [CrossRef] [PubMed]
13. Pu, C.; Fisher, J.E.; Cappon, G.D.; Vorhees, C.V. The effects of amfonelic acid, a dopamine uptake inhibitor, on methamphetamine-induced dopaminergic terminal degeneration and astrocytic response in rat striatum. *Brain Res.* **1994**, *649*, 217–224. [CrossRef]
14. Thomas, D.M.; Walker, P.D.; Benjamins, J.A.; Geddes, T.J.; Kuhn, D.M. Methamphetamine neurotoxicity in dopamine nerve endings of the striatum is associated with microglial activation. *J. Pharmacol. Exp. Ther.* **2004**, *311*, 1–7. [CrossRef]
15. Lorez, H. Fluorescence histochemistry indicates damage of striatal dopamine nerve terminals in rats after multiple doses of methamphetamine. *Life Sci.* **1981**, *28*, 911–916. [CrossRef] [PubMed]

16. Ricaurte, G.A.; Guillery, R.W.; Seiden, L.S.; Schuster, C.R.; Moore, R.Y. Dopamine nerve terminal degeneration produced by high doses of methylamphetamine in the rat brain. *Brain Res.* **1982**, *235*, 93–103. [CrossRef]
17. Sharma, H.S.; Kiyatkin, E.A. Rapid morphological brain abnormalities during acute methamphetamine intoxication in the rat: An experimental study using light and electron microscopy. *J. Chem. Neuroanat.* **2009**, *37*, 18–32. [CrossRef] [PubMed]
18. Davidson, C.; Gow, A.J.; Lee, T.H.; Ellinwood, E.H. Methamphetamine neurotoxicity: Necrotic and apoptotic mechanisms and relevance to human abuse and treatment. *Brain Res. Brain Res. Rev.* **2001**, *36*, 1–22. [CrossRef]
19. Liu, B.; Traini, R.; Killinger, B.; Schneider, B.; Moszczynska, A. Overexpression of parkin in the rat nigrostriatal dopamine system protects against methamphetamine neurotoxicity. *Exp. Neurol.* **2013**, *247*, 359–372. [CrossRef]
20. Sharma, A.; Harutyunyan, A.; Schneider, B.L.; Moszczynska, A. Parkin regulates drug-taking behavior in rat model of methamphetamine use disorder. *Transl. Psychiatry* **2021**, *11*, 293. [CrossRef]
21. Truong, J.; Rau, K.; Hanson, G.; Fleckenstein, A. Pramipexole increases vesicular dopamine uptake: Implications for treatment of Parkinson's neurodegeneration. *Eur. J. Pharmacol.* **2003**, *474*, 223–226. [CrossRef]
22. Truong, J.; Hanson, G.; Fleckenstein, A. Apomorphine increases vesicular monoamine transporter-2 function: Implications for neurodegeneration. *Eur. J. Pharmacol.* **2004**, *492*, 143–147. [CrossRef]
23. Dwoskin, L.; Crooks, P. A novel mechanism of action and potential use for lobeline as a treatment for psychostimulant abuse. *Biochem. Pharmacol.* **2002**, *63*, 89–98. [CrossRef]
24. Harrod, S.; Dwoskin, L.; Crooks, P.; Klebaur, J.; Bardo, M. Lobeline attenuates d-methamphetamine self-administration in rats. *J. Pharmacol. Exp. Ther.* **2001**, *298*, 172–179.
25. Martin, K.; Schafer, H.; Weihe, E.; Eiden, L. Localization and Expression of VMAT2 Aross Mammalian Species: A Translational Guide for Its Visualization and Targeting in Health and Disease. *Adv. Pharmacol.* **2013**, *68*, 319–334.
26. Fleckenstein, A.E.; Hanson, G.R. Impact of psychostimulants on vesicular monoamine transporter function. *Eur. J. Pharmacol.* **2003**, *479*, 283–289. [CrossRef]
27. Sulzer, D.; Rayport, S. Amphetamine and other psychostimulants reduce pH gradients in midbrain dopaminergic neurons and chromaffin granules: A mechanism of action. *Neuron* **1990**, *5*, 797–808. [CrossRef]
28. Pifl, C.; Drobny, H.; Reither, H.; Hornykiewicz, O.; Singer, E. Mechanism of the dopamine-releasing actions of amphetamine and cocaine: Plasmalemmal dopamine transporter versus vesicular monoamine transporter. *Mol. Pharmacol.* **1995**, *47*, 368–373.
29. Yamamoto, B.; Moszczynska, A.; Gudelsky, G. Amphetamine toxicities: Classical and emerging mechanisms. *Ann. N. Y. Acad. Sci.* **2010**, *1187*, 101–121. [CrossRef]
30. Riddle, E.L.; Topham, M.K.; Haycock, J.W.; Hanson, G.R.; Fleckenstein, A.E. Differential trafficking of the vesicular monoamine transporter-2 by methamphetamine and cocaine. *Eur. J. Pharmacol.* **2002**, *449*, 71–74. [CrossRef]
31. Ugarte, Y.V.; Rau, K.S.; Riddle, E.L.; Hanson, G.R.; Fleckenstein, A.E. Methamphetamine rapidly decreases mouse vesicular dopamine uptake: Role of hyperthermia and dopamine D2 receptors. *Eur. J. Pharmacol.* **2003**, *472*, 165–171. [CrossRef] [PubMed]
32. Jiang, H.; Ren, Y.; Zhao, J.; Feng, J. Parkin protects human dopaminergic neuroblastoma cells against dopamine-induced apoptosis. *Hum. Mol. Genet.* **2004**, *13*, 1745–1754. [CrossRef]
33. Moszczynska, A.; Saleh, J.; Zhang, H.; Vukusic, B.; Lee, F.J.; Liu, F. Parkin disrupts the alpha-synuclein/dopamine transporter interaction: Consequences toward dopamine-induced toxicity. *J. Mol. Neurosci.* **2007**, *32*, 217–227. [CrossRef]
34. Shimura, H.; Schlossmacher, M.G.; Hattori, N.; Frosch, M.P.; Trockenbacher, A.; Schneider, R.; Mizuno, Y.; Kosik, K.S.; Selkoe, D.J. Ubiquitination of a new form of alpha-synuclein by parkin from human brain: Implications for Parkinson's disease. *Science* **2001**, *293*, 263–269. [CrossRef] [PubMed]
35. Feng, J. Microtubule: A common target for parkin and Parkinson's disease toxins. *Neuroscientist* **2006**, *12*, 469–476. [CrossRef] [PubMed]
36. Kabayama, H.; Tokushige, N.; Takeuchi, M.; Kabayama, M.; Fukuda, M.; Mikoshiba, K. Parkin promotes proteasomal degradation of synaptotagmin IV by accelerating polyubiquitination. *Mol. Cell Neurosci.* **2017**, *80*, 89–99. [CrossRef]
37. Wang, C.; Kang, X.; Zhou, L.; Chai, Z.; Wu, Q.; Huang, R.; Xu, H.; Hu, M.; Sun, X.; Sun, S.; et al. Synaptotagmin-11 is a critical mediator of parkin-linked neurotoxicity and Parkinson's disease-like pathology. *Nat. Commun.* **2018**, *9*, 81. [CrossRef]
38. Chung, K.K.; Thomas, B.; Li, X.; Pletnikova, O.; Troncoso, J.C.; Marsh, L.; Dawson, V.L.; Dawson, T.M. S-nitrosylation of parkin regulates ubiquitination and compromises parkin's protective function. *Science* **2004**, *304*, 1328–1331. [CrossRef]
39. Cao, M.; Milosevic, I.; Giovedi, S.; De Camilli, P. Upregulation of Parkin in endophilin mutant mice. *J. Neurosci.* **2014**, *34*, 16544–16549. [CrossRef] [PubMed]
40. Cookson, M.R. Parkin's Substrates and the Pathways Leading to Neuronal Damage. *Neuromol. Med.* **2003**, *3*, 1–13. [CrossRef]
41. Zhang, Y.; Gao, J.; Chung, K.K.; Huang, H.; Dawson, V.L.; Dawson, T.M. Parkin functions as an E2-dependent ubiquitin- protein ligase and promotes the degradation of the synaptic vesicle-associated protein, CDCrel-1. *Proc. Natl. Acad. Sci. USA* **2000**, *97*, 13354–13359. [CrossRef]
42. Song, P.; Trajkovic, K.; Tsunemi, T.; Krainc, D. Parkin Modulates Endosomal Organization and Function of the Endo-Lysosomal Pathway. *J. Neurosci.* **2016**, *36*, 2425–2437. [CrossRef] [PubMed]
43. Beites, C.L.; Xie, H.; Bowser, R.; Trimble, W.S. The septin CDCrel-1 binds syntaxin and inhibits exocytosis. *Nat. Neurosci.* **1999**, *2*, 434–439. [CrossRef] [PubMed]
44. Kartmann, B.; Roth, D. Novel roles for mammalian septins: From vesicle trafficking to oncogenesis. *J. Cell Sci.* **2001**, *114 Pt 5*, 839–844. [CrossRef]

45. Dong, Z.; Ferger, B.; Paterna, J.C.; Vogel, D.; Furler, S.; Osinde, M.; Feldon, J.; Bueler, H. Dopamine-dependent neurodegeneration in rats induced by viral vector-mediated overexpression of the parkin target protein, CDCrel-1. *Proc. Natl. Acad. Sci. USA* **2003**, *100*, 12438–12443. [CrossRef]
46. Son, J.H.; Kawamata, H.; Yoo, M.S.; Kim, D.J.; Lee, Y.K.; Kim, S.; Dawson, T.M.; Zhang, H.; Sulzer, D.; Yang, L.; et al. Neurotoxicity and behavioral deficits associated with Septin 5 accumulation in dopaminergic neurons. *J. Neurochem.* **2005**, *94*, 1040–1053. [CrossRef] [PubMed]
47. Eyerman, D.J.; Yamamoto, B.K. A rapid oxidation and persistent decrease in the vesicular monoamine transporter 2 after methamphetamine. *J. Neurochem.* **2007**, *103*, 1219–1227. [CrossRef] [PubMed]
48. Eyerman, D.J.; Yamamoto, B.K. Lobeline attenuates methamphetamine-induced changes in vesicular monoamine transporter 2 immunoreactivity and monoamine depletions in the striatum. *J. Pharmacol. Exp. Ther.* **2005**, *312*, 160–169. [CrossRef]
49. Bowyer, J.F.; Davies, D.L.; Schmued, L.; Broening, H.W.; Newport, G.D.; Slikker, J.W.; Holson, R.R. Further studies of the role of hyperthermia in methamphetamine neurotoxicity. *J. Pharmacol. Exp. Ther.* **1994**, *268*, 1571–1580.
50. Pandey, S.; Miller, C.A. Targeting the cytoskeleton as a therapeutic approach to substance use disorders. *Pharmacol. Res.* **2024**, *202*, 107143. [CrossRef]
51. Yang, F.; Jiang, Q.; Zhao, J.; Ren, Y.; Sutton, M.D.; Feng, J. Parkin stabilizes microtubules through strong binding mediated by three independent domains. *J. Biol. Chem.* **2005**, *280*, 17154–17162. [CrossRef]
52. Huynh, D.P.; Scoles, D.R.; Ho, T.H.; Del Bigio, M.R.; Pulst, S.M. Parkin is associated with actin filaments in neuronal and nonneural cells. *Ann. Neurol.* **2000**, *48*, 737–744. [CrossRef] [PubMed]
53. Chenouard, N.; Xuan, F.; Tsien, R.W. Synaptic vesicle traffic is supported by transient actin filaments and regulated by PKA and NO. *Nat. Commun.* **2020**, *11*, 5318. [CrossRef] [PubMed]
54. Bowyer, J.F.; Tank, A.W.; Newport, G.D.; Slikker, W., Jr.; Ali, S.F.; Holson, R.R. The influence of environmental temperature on the transient effects of methamphetamine on dopamine levels and dopamine release in rat striatum. *J. Pharmacol. Exp. Ther.* **1992**, *260*, 817–824.
55. Szechynska-Hebda, M.; Ghalami, R.Z.; Kamran, M.; Van Breusegem, F.; Karpinski, S. To Be or Not to Be? Are Reactive Oxygen Species, Antioxidants, and Stress Signalling Universal Determinants of Life or Death? *Cells* **2022**, *11*, 4105. [CrossRef]
56. Nakahara, T.; Kuroki, T.; Ohta, E.; Kajihata, T.; Yamada, H.; Yamanaka, M.; Hashimoto, K.; Tsutsumi, T.; Hirano, M.; Uchimura, H. Effect of the neurotoxic dose of methamphetamine on gene expression of parkin and Pael-receptors in rat striatum. *Park. Relat. Disord.* **2003**, *9*, 213–219. [CrossRef]
57. Ahmed, K.; Zaidi, S.F.; Mati Ur, R.; Rehman, R.; Kondo, T. Hyperthermia and protein homeostasis: Cytoprotection and cell death. *J. Therm. Biol.* **2020**, *91*, 102615. [CrossRef]
58. Fang, C.; Bourdette, D.; Banker, G. Oxidative stress inhibits axonal transport: Implications for neurodegenerative diseases. *Mol. Neurodegener.* **2012**, *7*, 29. [CrossRef] [PubMed]
59. Samardzic, K.; Rodgers, K.J. Oxidised protein metabolism: Recent insights. *Biol. Chem.* **2017**, *398*, 1165–1175. [CrossRef]
60. LaVoie, M.J.; Cortese, G.P.; Ostaszewski, B.L.; Schlossmacher, M.G. The effects of oxidative stress on parkin and other E3 ligases. *J. Neurochem.* **2007**, *103*, 2354–2368. [CrossRef]
61. LaVoie, M.J.; Ostaszewski, B.L.; Weihofen, A.; Schlossmacher, M.G.; Selkoe, D.J. Dopamine covalently modifies and functionally inactivates parkin. *Nat. Med.* **2005**, *11*, 1214–1221. [CrossRef]
62. Bazylianska, V.; Sharma, A.; Chauhan, H.; Schneider, B.; Moszczynska, A. Dopamine and Methamphetamine Differentially Affect Electron Transport Chain Complexes and Parkin in Rat Striatum: New Insight into Methamphetamine Neurotoxicity. *Int. J. Mol. Sci.* **2021**, *23*, 363. [CrossRef] [PubMed]
63. Winklhofer, K.F.; Henn, I.H.; Kay-Jackson, P.C.; Heller, U.; Tatzelt, J. Inactivation of parkin by oxidative stress and C-terminal truncations: A protective role of molecular chaperones. *J. Biol. Chem.* **2003**, *278*, 47199–47208. [CrossRef] [PubMed]
64. Grune, T.; Reinheckel, T.; Joshi, M.; Davies, K.J. Proteolysis in cultured liver epithelial cells during oxidative stress. Role of the multicatalytic proteinase complex, proteasome. *J. Biol. Chem.* **1995**, *270*, 2344–2351. [CrossRef]
65. Grune, T.; Reinheckel, T.; Davies, K.J. Degradation of oxidized proteins in mammalian cells. *FASEB J.* **1997**, *11*, 526–534. [CrossRef]
66. Shringarpure, R.; Grune, T.; Davies, K.J. Protein oxidation and 20S proteasome-dependent proteolysis in mammalian cells. *Cell Mol. Life Sci.* **2001**, *58*, 1442–1450. [CrossRef] [PubMed]
67. Reinheckel, T.; Sitte, N.; Ullrich, O.; Kuckelkorn, U.; Davies, K.J.; Grune, T. Comparative resistance of the 20S and 26S proteasome to oxidative stress. *Biochem. J.* **1998**, *335 Pt 3*, 637–642. [CrossRef]
68. Davies, K.J. Degradation of oxidized proteins by the 20S proteasome. *Biochimie* **2001**, *83*, 301–310. [CrossRef]
69. Perlson, E.; Maday, S.; Fu, M.M.; Moughamian, A.J.; Holzbaur, E.L. Retrograde axonal transport: Pathways to cell death? *Trends Neurosci.* **2010**, *33*, 335–344. [CrossRef]
70. Harold, C.; Wallace, T.; Friedman, R.; Gudelsky, G.; Yamamoto, B. Methamphetamine selectively alters brain glutathione. *Eur. J. Pharmacol.* **2000**, *400*, 99–102. [CrossRef]
71. Dias da Silva, D.; Silva, E.; Carmo, H. Cytotoxic effects of amphetamine mixtures in primary hepatocytes are severely aggravated under hyperthermic conditions. *Toxicol. Vitr.* **2013**, *27*, 1670–1678. [CrossRef]
72. Moszczynska, A.; Turenne, S.; Kish, S.J. Rat striatal levels of the antioxidant glutathione are decreased following binge administration of methamphetamine. *Neurosci. Lett.* **1998**, *255*, 49–52. [CrossRef] [PubMed]
73. Rizzoli, S.O.; Betz, W.J. Synaptic vesicle pools. *Nat. Rev. Neurosci.* **2005**, *6*, 57–69. [CrossRef] [PubMed]

74. Choi, P.; Golts, N.; Snyder, H.; Chong, M.; Petrucelli, L.; Hardy, J.; Sparkman, D.; Cochran, E.; Lee, J.M.; Wolozin, B. Co-association of parkin and alpha-synuclein. *Neuroreport* **2001**, *12*, 2839–2843. [CrossRef]
75. Kinoshita, A.; Noda, M.; Kinoshita, M. Differential localization of septins in the mouse brain. *J. Comp. Neurol.* **2000**, *428*, 223–239. [CrossRef] [PubMed]
76. Gemechu, J.M.; Sharma, A.; Yu, D.; Xie, Y.; Merkel, O.M.; Moszczynska, A. Characterization of Dopaminergic System in the Striatum of Young Adult Park2(-/-) Knockout Rats. *Sci. Rep.* **2018**, *8*, 1517. [CrossRef]
77. Goldberg, M.S.; Fleming, S.M.; Palacino, J.J.; Cepeda, C.; Lam, H.A.; Bhatnagar, A.; Meloni, E.G.; Wu, N.; Ackerson, L.C.; Klapstein, G.J.; et al. Parkin-deficient mice exhibit nigrostriatal deficits but not loss of dopaminergic neurons. *J. Biol. Chem.* **2003**, *278*, 43628–43635. [CrossRef]
78. Volknandt, W.; Karas, M. Proteomic analysis of the presynaptic active zone. *Exp. Brain Res.* **2012**, *217*, 449–461. [CrossRef]
79. Velichko, A.K.; Markova, E.N.; Petrova, N.V.; Razin, S.V.; Kantidze, O.L. Mechanisms of heat shock response in mammals. *Cell Mol. Life Sci.* **2013**, *70*, 4229–4241. [CrossRef]
80. Gardiner, J.; Overall, R.; Marc, J. The Nervous System Cytoskeleton under Oxidative Stress. *Diseases* **2013**, *1*, 36–50. [CrossRef]
81. Muller, M.T.; Schempp, R.; Lutz, A.; Felder, T.; Felder, E.; Miklavc, P. Interaction of microtubules and actin during the post-fusion phase of exocytosis. *Sci. Rep.* **2019**, *9*, 11973. [CrossRef]
82. Wu, L.G.; Chan, C.Y. Multiple Roles of Actin in Exo- and Endocytosis. *Front. Synaptic Neurosci.* **2022**, *14*, 841704. [CrossRef] [PubMed]
83. Paladini, C.A.; Fiorillo, C.D.; Morikawa, H.; Williams, J.T. Amphetamine selectively blocks inhibitory glutamate transmission in dopamine neurons. *Nat. Neurosci.* **2001**, *4*, 275–281. [CrossRef] [PubMed]
84. Covey, D.P.; Juliano, S.A.; Garris, P.A. Amphetamine elicits opposing actions on readily releasable and reserve pools for dopamine. *PLoS ONE* **2013**, *8*, e60763. [CrossRef] [PubMed]
85. Yang, M.H.; Kim, S.; Jung, M.S.; Shim, J.H.; Ryu, N.K.; Yook, Y.J.; Jang, C.G.; Bahk, Y.Y.; Kim, K.W.; Park, J.H. Proteomic analysis of methamphetamine-induced reinforcement processes within the mesolimbic dopamine system. *Addict. Biol.* **2008**, *13*, 287–294. [CrossRef]
86. Hamamura, M.; Okouchi, J.; Ozawa, H.; Kimuro, Y.; Iwaki, A.; Fukumaki, Y. Amphiphysin I but not dynamin I nor synaptojanin mRNA expression increased after repeated methamphetamine administration in the rat cerebrum and cerebellum. *J. Neural Transm.* **2013**, *120*, 1039–1052. [CrossRef]
87. Hong, W.C.; Amara, S.G. Differential targeting of the dopamine transporter to recycling or degradative pathways during amphetamine- or PKC-regulated endocytosis in dopamine neurons. *FASEB J.* **2013**, *27*, 2995–3007. [CrossRef]
88. Underhill, S.M.; Wheeler, D.S.; Li, M.; Watts, S.D.; Ingram, S.L.; Amara, S.G. Amphetamine modulates excitatory neurotransmission through endocytosis of the glutamate transporter EAAT3 in dopamine neurons. *Neuron* **2014**, *83*, 404–416. [CrossRef]
89. Lopez-Hernandez, T.; Haucke, V.; Maritzen, T. Endocytosis in the adaptation to cellular stress. *Cell Stress.* **2020**, *4*, 230–247. [CrossRef]
90. Limanaqi, F.; Biagioni, F.; Busceti, C.L.; Ryskalin, L.; Fornai, F. The effects of proteasome on baseline and methamphetamine-dependent dopamine transmission. *Neurosci. Biobehav. Rev.* **2019**, *102*, 308–317. [CrossRef]
91. Iuzzolino, A.; Pellegrini, F.R.; Rotili, D.; Degrassi, F.; Trisciuoglio, D. The alpha-tubulin acetyltransferase ATAT1: Structure, cellular functions, and its emerging role in human diseases. *Cell Mol. Life Sci.* **2024**, *81*, 193. [CrossRef]
92. Mackeh, R.; Lorin, S.; Ratier, A.; Mejdoubi-Charef, N.; Baillet, A.; Bruneel, A.; Hamai, A.; Codogno, P.; Pous, C.; Perdiz, D. Reactive oxygen species, AMP-activated protein kinase, and the transcription cofactor p300 regulate alpha-tubulin acetyltransferase-1 (alphaTAT-1/MEC-17)-dependent microtubule hyperacetylation during cell stress. *J. Biol. Chem.* **2014**, *289*, 11816–11828. [CrossRef] [PubMed]
93. Carmona, B.; Marinho, H.S.; Matos, C.L.; Nolasco, S.; Soares, H. Tubulin Post-Translational Modifications: The Elusive Roles of Acetylation. *Biology* **2023**, *12*, 561. [CrossRef] [PubMed]
94. Quan, A.; Robinson, P.J. Syndapin--a membrane remodelling and endocytic F-BAR protein. *FEBS J.* **2013**, *280*, 5198–5212. [CrossRef] [PubMed]
95. Cheung, G.; Cousin, M.A. Synaptic vesicle generation from activity-dependent bulk endosomes requires a dephosphorylation-dependent dynamin-syndapin interaction. *J. Neurochem.* **2019**, *151*, 570–583. [CrossRef]
96. Cheung, G.; Jupp, O.J.; Cousin, M.A. Activity-dependent bulk endocytosis and clathrin-dependent endocytosis replenish specific synaptic vesicle pools in central nerve terminals. *J. Neurosci.* **2010**, *30*, 8151–8161. [CrossRef]
97. Nishimura, T.; Harashima, S.; Yafang, H.; Notkins, A.L. IA-2 modulates dopamine secretion in PC12 cells. *Mol. Cell Endocrinol.* **2010**, *315*, 81–86. [CrossRef]
98. Nehlig, A.; Molina, A.; Rodrigues-Ferreira, S.; Honore, S.; Nahmias, C. Regulation of end-binding protein EB1 in the control of microtubule dynamics. *Cell Mol. Life Sci.* **2017**, *74*, 2381–2393. [CrossRef]
99. Park, J.; Xie, Y.; Miller, K.G.; De Camilli, P.; Yogev, S. End-binding protein 1 promotes specific motor-cargo association in the cell body prior to axonal delivery of dense core vesicles. *Curr. Biol.* **2023**, *33*, 3851–3864.e7. [CrossRef]
100. Nirenberg, M.J.; Liu, Y.; Peter, D.; Edwards, R.H.; Pickel, V.M. The vesicular monoamine transporter 2 is present in small synaptic vesicles and preferentially localizes to large dense core vesicles in rat solitary tract nuclei. *Proc. Natl. Acad. Sci. USA* **1995**, *92*, 8773–8777. [CrossRef]

101. Llona, I.; Annaert, W.G.; Jacob, W.; De Potter, W.P. Co-storage in large 'dense-core' vesicles of dopamine and cholecystokinin in rat striatum. *Neurochem. Int.* **1994**, *25*, 573–581. [CrossRef]
102. Dhawan, K.; Naslavsky, N.; Caplan, S. Sorting nexin 17 (SNX17) links endosomal sorting to Eps15 homology domain protein 1 (EHD1)-mediated fission machinery. *J. Biol. Chem.* **2020**, *295*, 3837–3850. [CrossRef] [PubMed]
103. Wang, J.; Fedoseienko, A.; Chen, B.; Burstein, E.; Jia, D.; Billadeau, D.D. Endosomal receptor trafficking: Retromer and beyond. *Traffic* **2018**, *19*, 578–590. [CrossRef] [PubMed]
104. Hanley, S.E.; Cooper, K.F. Sorting Nexins in Protein Homeostasis. *Cells* **2020**, *10*, 17. [CrossRef] [PubMed]
105. Limanaqi, F.; Busceti, C.L.; Celli, R.; Biagioni, F.; Fornai, F. Autophagy as a gateway for the effects of methamphetamine: From neurotransmitter release and synaptic plasticity to psychiatric and neurodegenerative disorders. *Prog. Neurobiol.* **2021**, *204*, 102112. [CrossRef]
106. Overhoff, M.; De Bruyckere, E.; Kononenko, N.L. Mechanisms of neuronal survival safeguarded by endocytosis and autophagy. *J. Neurochem.* **2021**, *157*, 263–296. [CrossRef]
107. Zhou, C.; Wu, Z.; Du, W.; Que, H.; Wang, Y.; Ouyang, Q.; Jian, F.; Yuan, W.; Zhao, Y.; Tian, R.; et al. Recycling of autophagosomal components from autolysosomes by the recycler complex. *Nat. Cell Biol.* **2022**, *24*, 497–512. [CrossRef]
108. Golanov, E.V.; Sharpe, M.A.; Regnier-Golanov, A.S.; Del Zoppo, G.J.; Baskin, D.S.; Britz, G.W. Fibrinogen Chains Intrinsic to the Brain. *Front. Neurosci.* **2019**, *13*, 541. [CrossRef]
109. Suehiro, K.; Gailit, J.; Plow, E.F. Fibrinogen is a ligand for integrin alpha5beta1 on endothelial cells. *J. Biol. Chem.* **1997**, *272*, 5360–5366. [CrossRef]
110. Kloss, C.U.; Werner, A.; Klein, M.A.; Shen, J.; Menuz, K.; Probst, J.C.; Kreutzberg, G.W.; Raivich, G. Integrin family of cell adhesion molecules in the injured brain: Regulation and cellular localization in the normal and regenerating mouse facial motor nucleus. *J. Comp. Neurol.* **1999**, *411*, 162–178. [CrossRef]
111. Rassart, E.; Desmarais, F.; Najyb, O.; Bergeron, K.F.; Mounier, C. Apolipoprotein D. *Gene* **2020**, *756*, 144874. [CrossRef]
112. Newington, J.T.; Rappon, T.; Albers, S.; Wong, D.Y.; Rylett, R.J.; Cumming, R.C. Overexpression of pyruvate dehydrogenase kinase 1 and lactate dehydrogenase A in nerve cells confers resistance to amyloid beta and other toxins by decreasing mitochondrial respiration and reactive oxygen species production. *J. Biol. Chem.* **2012**, *287*, 37245–37258. [CrossRef] [PubMed]
113. Chan, P.; Di Monte, D.A.; Luo, J.J.; DeLanney, L.E.; Irwin, I.; Langston, J.W. Rapid ATP loss caused by methamphetamine in the mouse striatum: Relationship between energy impairment and dopaminergic neurotoxicity. *J. Neurochem.* **1994**, *62*, 2484–2487. [CrossRef] [PubMed]
114. Wang, B.R.; Zaheer, A.; Lim, R. Polyclonal antibody localizes glia maturation factor beta-like immunoreactivity in neurons and glia. *Brain Res.* **1992**, *591*, 1–7. [CrossRef] [PubMed]
115. Fan, J.; Fong, T.; Chen, X.; Chen, C.; Luo, P.; Xie, H. Glia maturation factor-beta: A potential therapeutic target in neurodegeneration and neuroinflammation. *Neuropsychiatr. Dis. Treat.* **2018**, *14*, 495–504. [CrossRef] [PubMed]
116. Goode, B.L.; Sweeney, M.O.; Eskin, J.A. GMF as an Actin Network Remodeling Factor. *Trends Cell Biol.* **2018**, *28*, 749–760. [CrossRef]
117. Haynes, E.M.; Asokan, S.B.; King, S.J.; Johnson, H.E.; Haugh, J.M.; Bear, J.E. GMFbeta controls branched actin content and lamellipodial retraction in fibroblasts. *J. Cell Biol.* **2015**, *209*, 803–812. [CrossRef]
118. Chu, P.W.; Hadlock, G.C.; Vieira-Brock, P.; Stout, K.; Hanson, G.R.; Fleckenstein, A.E. Methamphetamine alters vesicular monoamine transporter-2 function and potassium-stimulated dopamine release. *J. Neurochem.* **2010**, *115*, 325–332. [CrossRef]

International Journal of
Molecular Sciences

MDPI

Article

Honey Enriched with Additives Alleviates Behavioral, Oxidative Stress, and Brain Alterations Induced by Heavy Metals and Imidacloprid in Zebrafish

Emanuela Paduraru [1,†], Roxana Jijie [2,†], Ira-Adeline Simionov [3,4], Cristina-Maria Gavrilescu [5], Tudor Ilie [6], Diana Iacob [1], Andreea Lupitu [7], Cristian Moisa [7], Claudia Muresan [7], Lucian Copolovici [7], Dana M. Copolovici [7], Gabriela Mihalache [8], Florin Daniel Lipsa [9], Gheorghe Solcan [10], Gabriela-Alexandra Danelet [10], Mircea Nicoara [1,11], Alin Ciobica [11,12,13,14,*] and Carmen Solcan [10]

1 Doctoral School of Geosciences, Faculty of Geography and Geology, Alexandru Ioan Cuza University of Iasi, No. 20 A Carol I Avenue, 700505 Iasi, Romania; emanuelapaduraru19@yahoo.com (E.P.); dianaelenaiacob84@gmail.com (D.I.); mirmag@uaic.ro (M.N.)

2 Research Center on Advanced Materials and Technologies (RAMTECH), Department of Exact and Natural Sciences, Institute of Interdisciplinary Research, Alexandru Ioan Cuza University of Iasi, No. 11 Carol I Avenue, 700506 Iasi, Romania; roxana.jijie@uaic.ro

3 Department of Food Science, Food Engineering, Biotechnologies and Aquaculture, Dunarea de Jos University of Galati, No. 47 Domnească Street, 800008 Galati, Romania; ira.simionov@gmail.com

4 REXDAN Research Infrastructure, Dunarea de Jos University of Galati, No. 98 George Coșbuc Street, 800385 Galati, Romania

5 Department of Biomedical Sciences, Grigore T. Popa University of Medicine and Pharmacy, No. 16 University Street, 700115 Iasi, Romania; cristina.gavrilescu@umfiasi.ro

6 Synergy Plant Products, No. 12 Milano Street, Prejmer, 507165 Brasov, Romania; tudoriliebv@gmail.com

7 Faculty of Food Engineering, Tourism and Environmental Protection, Institute for Research, Development and Innovation in Technical and Natural Sciences, Aurel Vlaicu University, No. 2 Elena Dragoi Street, 310330 Arad, Romania; pag.andreea@yahoo.com (A.L.); moisa.cristian@yahoo.com (C.M.); claudia.muresan@uav.ro (C.M.); lucian.copolovici@uav.ro (L.C.); dana.copolovici@uav.ro (D.M.C.)

8 Integrated Center of Environmental Science Studies in the North-Eastern Development Region (CERNESIM), Department of Exact and Natural Sciences, Institute of Interdisciplinary Research, Alexandru Ioan Cuza University of Iasi, No. 11 Carol I Avenue, 700506 Iasi, Romania; gabriela.mihalache@uaic.ro

9 Department of Food Technologies, Ion Ionescu de la Brad University of Life Sciences, No. 3 Mihail Sadoveanu Alley, 700490 Iasi, Romania; florin.lipsa@iuls.ro

10 Faculty of Veterinary Medicine, Ion Ionescu de la Brad University of Life Sciences, No. 8 Mihail Sadoveanu Alley, 700489 Iasi, Romania; gsolcan@uaiasi.ro (G.S.); danelet.alexandra95@yahoo.com (G.-A.D.); csolcan@uaiasi.ro (C.S.)

11 Department of Biology, Faculty of Biology, Alexandru Ioan Cuza University of Iasi, No. 20A Carol I Avenue, 700505 Iasi, Romania

12 Center of Biomedical Research, Romanian Academy, No. 8 Carol I Avenue, 700506 Iasi, Romania

13 Academy of Romanian Scientists, No. 54 Independence Street, Sector 5, 050094 Bucharest, Romania

14 "Ioan Haulica" Institute, Apollonia University, No. 11 Pacurari Street, 700511 Iasi, Romania

* Correspondence: alin.ciobica@uaic.ro

† These authors contributed equally to this work.

Abstract: Environmental concerns have consistently been a focal point for the scientific community. Pollution is a critical ecological issue that poses significant threats to human health and agricultural production. Contamination with heavy metals and pesticides is a considerable concern, a threat to the environment, and warrants special attention. In this study, we investigated the significant issues arising from sub-chronic exposure to imidacloprid (IMI), mercury (Hg), and cadmium (Cd), either alone or in combination, using zebrafish (*Danio rerio*) as an animal model. Additionally, we assessed the potential protective effects of polyfloral honey enriched with natural ingredients, also called honey formulation (HF), against the combined sub-chronic toxic effects of the three contaminants. The effects of IMI (0.5 mg·L^{-1}), Hg (15 μg·L^{-1}), and Cd (5 μg·L^{-1}), both individually and in combination with HF (500 mg·L^{-1}), on zebrafish were evaluated by quantifying acetylcholinesterase (AChE) activity, lipid peroxidation (MDA), various antioxidant enzyme activities like superoxide dismutase and glutathione peroxidase (SOD and GPx), 2D locomotor activity, social behavior, histological and immunohistochemical factors, and changes in body element concentrations. Our findings revealed that all concentrations of pollutants may disrupt social behavior, diminish swimming performances (measured by total distance traveled, inactivity, and swimming speed), and elevate oxidative stress (OS) biomarkers of SOD, GPx, and MDA in zebrafish over the 21-day administration period. Fish exposed to IMI and Hg + Cd + IMI displayed severe lesions and increased GFAP (Glial fibrillary acidic protein) and S100B (S100 calcium-binding protein B) protein expression in the optic tectum and cerebellum, conclusively indicating astrocyte activation and neurotoxic effects. Furthermore, PCNA (Proliferating cell nuclear antigen) staining revealed reduced cell proliferation in the IMI-exposed group, contrasting with intensified proliferation in the Hg + Cd group. The nervous system exhibited significant damage across all studied concentrations, confirming the observed behavioral changes. Moreover, HF supplementation significantly mitigated the toxicity induced by contaminants and reduced OS. Therefore, the exposure to chemical mixtures offers a more complete picture of adverse impacts on aquatic ecosystems and the supplementation with bioactive compounds can help to reduce the toxicity induced by exposure to environmental pollutants.

Keywords: honey formulation; environmental contaminants; protective effects; combined exposure; physicochemical properties; antioxidant properties; antimicrobial activity

1. Introduction

The functional food product market is growing due to increased consumer awareness of healthy food, the connection between nutrition and diseases, and the demand for innovative products. Efforts to enhance honey's nutritional value with functional ingredients are ongoing, as honey can be enriched with additional components to boost its health benefits and overall dietary composition [1–4]. Introducing additives to honey not only enhances the sensory properties of honey, such as color, taste, or smell, but also significantly amplifies its pharmacokinetic, physicochemical, and microbiological effects, as evidenced by previous research [2,5–9]. For example, flavored honey has been proven to possess antioxidant, antiviral, antibacterial, antifungal, and hepatoprotective properties [3,5,7,8]. Sowa et al. (2019) found that polyfloral honey enriched with *Melilotus* flowers contains coumarin and that coumarin may hold therapeutic potential [8]. Additionally, honey supplemented with sea buckthorn leaf and black raspberry fruits effectively inhibited *S. aureus* biofilm formation [5]. Furthermore, a synergistic hepatoprotective effect was observed for a mixture of chestnut honey, artichoke (*Cynara cardunculus* var. *scolymus* L.), milk thistle (*Silybum marianum* L.), and borututu (*Cochlospermum angolensis* Welw.) [7]. Moreover, the usefulness of natural supplements has attracted ample interest due to their potential protective role against the adverse effects of pollutants. It has been demonstrated that sesame oil alleviates cypermethrin-induced brain toxicity [10], *Ginkgo biloba* extract provides protective benefits against mercury chloride ($HgCl_2$)-induced oxidative damage [11], dietary supplementation with *Arthrospira platensis* (Gomont, 1892) may restore normal hepatic function disrupted by cadmium chloride ($CdCl_2$) exposure [12], and *Turnera diffusa* (Willd) may play a protective role against testicular toxicity induced by fenitrothion (FNT) and/or hexavalent chromium [13].

The co-occurrence of metals and pesticides in aquatic environments has been demonstrated by environmental surveys [14–16], and their presence in food and drinking water has also been indicated by several studies [17,18]. However, little knowledge exists about their combined toxic effects and potential interactions. Studies have associated acute and chronic exposure to heavy metals and pesticides with testicular toxicity [13], hepatotoxicity [19,20], and neurotoxicity [21,22]. For instance, co-exposure to chlorpyrifos (CPF) and Cd has been found to cause synergistic injury, reducing the viability of Hep G2cells [20]. It has been observed that the combination of CPF and Cd forms a complex that promotes the bioaccumulation of metal, leading to increased toxicity. Similar synergistic impairment has been reported in zebrafish (*Danio rerio*, Hamilton, 1822) embryos due to the combination of nickel sulfate ($NiSO_4$) and buprofezin [23]. Lajmanovich et al. (2019) demonstrated the synergistic toxicity of a glyphosate-based herbicide and arsenite (As (III)) mixture, which negatively affected tadpole development by increasing OS and thyroid hormone levels [24]. In a previous study, it was shown that acute exposure to deltamethrin (DM) and lead (Pb) induces damage to zebrafish, resulting in significant alterations in their behavior [22]. Conversely, the DM-associated behavioral changes and OS were mitigated in the presence of nickel (Ni) and Cd [21]. Thus, evaluating the effects of single contaminants may not fully reflect the real exposure risk to aquatic ecosystems and human health.

The limited availability of data on the toxicity of joint compound exposure concerning animal behavior and cognitive responses is an intriguing aspect worth considering [25]. Most toxicology studies predominantly focus on alterations in physiological and biochemical parameters [26], rather than on the relationship between biological and behavioral processes [22]. As a result of its complex behavioral responses, the zebrafish (*Danio rerio*) has rapidly emerged as a popular model organism for screening the effects of various toxic and bioactive compounds and the associated mechanisms [27,28]. In addition to sharing many molecular, biochemical, cellular, and physiological similarities with mammals, zebrafish have several advantages including a high reproductive rate, external fertilization, optical transparency, small size, rapid development, a short life cycle, easy husbandry, and cost-effectiveness [29,30]. Therefore, by evaluating the impact of various substances on zebrafish, valuable information for both human and fish safety can be obtained.

In our study, a stress condition was induced by exposing zebrafish to a ternary mixture of Cd, Hg, and IMI. Previous research indicates that exposure to these contaminants can result in organ and system damage in zebrafish. For instance, Cd exposure can elicit neurotoxicity, microbiota dysbiosis, disruption of thyroid endocrine and reproductive systems, and developmental abnormalities [27,31–35]. Xia et al. (2020) demonstrated that exposure to 5 $\mu g \cdot L^{-1}$ Cd for 7 days induced changes in locomotor activities and in microbiota diversity and richness in zebrafish [31]. Combined exposure to 10 $\mu g \cdot L^{-1}$ Cd and polystyrene beads (5 μm in diameter) for 3 weeks induced oxidative damage and inflammation in zebrafish tissues [27]. In addition, co-exposure of zebrafish adults to 0.1 $\mu g \cdot L^{-1}$ Cd and tributyltin (TBT) for 90 days can cause neurotoxicity and disruption of the thyroid system as well as developmental impairments in the offspring larvae [34]. Stronger effects were found for zebrafish exposed to ionic Cd compared to Cd-containing nanoparticles (5 nm in diameter), for the same nominal concentration of Cd (10 $\mu g \cdot L^{-1}$) [35]. Likewise, Hg exposure may lead to cumulative impacts on motor and cognitive functions, characterized by increased anxiety-related responses [36] and decreased aggressive behavior [37]. Moreover, the short-term exposure to 7.7 and 38.5 $\mu g \cdot L^{-1}$ $HgCl_2$ resulted in morpho-functional alterations in zebrafish gills and changes in the expression patterns of Na^+/K^+-ATPase and metallothioneins [38]. The reproductive function of adult zebrafish can be impaired by exposure to environmentally relevant concentrations of Hg (0.6–15 $\mu g \cdot L^{-1}$) during early life [39]. Sun et al. (2018) revealed that exposure of zebrafish embryos/larvae to Hg in the range of environmentally relevant concentrations (1–16 $\mu g \cdot L^{-1}$) for 7 days augmented the whole-body thyroid hormone levels and altered the transcription of related hypothalamic–pituitary–thyroid axis genes [40]. Similarly, IMI treatment has been associated with adverse effects on social behavior [41] and oxidative damage [42]. For instance, low concentrations of IMI (100 and 1000 $\mu g \cdot L^{-1}$) caused intestinal histological injury, OS, an inflammatory response, and gut microbiota dysbiosis [42]. Co-exposure to 100 $\mu g \cdot L^{-1}$ IMI and 20 $\mu g \cdot L^{-1}$ polystyrene (PS) microplastics for 21 days led to growth inhibition and alterations in hepatic parameters and OS-related biochemical parameters [43]. Similarly, Hou et al. (2024) showed that exposure to 10–500 $\mu g \cdot L^{-1}$ IMI inhibited the growth of zebrafish, but also altered liver glucose metabolism and decreased plasma insulin levels [44].

Therefore, the present study aims to assess the impact of two heavy metals and one pesticide on zebrafish and to investigate the potential protective effects of HF against toxicity induced by these contaminants. Our study involved a comprehensive analysis of adult zebrafish, including their behavior, biochemical profiles, and body element concentrations, as well as histological and immunohistochemical changes. Furthermore, there is currently a lack of research on the effectiveness of HF in mitigating the toxic effects of a combination of heavy metals and pesticides.

2. Results and Discussion

2.1. Physicochemical Analysis and Antimicrobial Activity of HF

According to the literature, honey's physicochemical and biological properties are related to the botanical and geographical origins and bee species, as well as harvesting, processing, storage, and transport conditions [45]. The physicochemical parameter values for raw polyfloral honey (control) enriched with 12 ingredients (HF) are given in Table 1. Polyfloral honey showed a moisture content of 19.46%, which is below the maximum limit (20%) set by Codex Alimentarius and European standards [46,47]. While higher moisture content was noticed in the HF (22.66%), the addition of aqueous extracts increased the water content. The amount of water present in honey may affect its quality and stability during storage; e.g., honey with higher moisture content is more prone to fermentation and the formation of acetic acid [48]. In general, honey with low moisture content has high total soluble solids [49], which is in agreement with our result. Moreover, the TSS for raw honey was similar to that reported by Albu et al. (2021), who found a TSS of 80.6 ± 1.41 °Brix [50]. The pH mean values for honey samples were acidic and relatively close, ranging between 3.93 ± 0.01 for raw honey and 4.05 ± 0.05 for HF. According to published results, a pH range of 3.2 to 4.5 is considered acceptable for honey samples and is a good indicator of honey's stability and quality [48]. Also, both honey samples met the maximum level of acidity requirement, which is 50 $meq \cdot kg^{-1}$ [46]. Similar results for polyfloral honey acidity were published by Sakač et al. (2019) [48], who found an acidity level of 19.3 ± 1.88 $meq \cdot kg^{-1}$. As the literature describes, free acidity increases during the storage time and fermentation because the sugars and the alcohols from the honey composition are transformed into organic acid by the action of yeasts [51,52]. Another important parameter in honey quality control is electrical conductivity, which correlates with the content of ions, organic acids, and proteins [53,54]. As depicted in Table 1, the obtained mean values of electrical conductivity were below the maximum limit of 0.8 $mS \cdot cm^{-1}$ set by Codex Alimentarius [46] and close to the data previously published by Kunat-Budzyńska and co-authors [55]. On the other hand, with the introduction of the additives, the final product is characterized by a vibrant color and robust flavor (Figure S1 in Supplementary Materials).

The mineral content is one of the parameters used to assess the nutritional value of honey, but also an indicator of environmental pollution. According to literature data, darker honey tends to have slightly higher mineral amounts compared to lighter honey [56]. Our data support this finding, the total mineral content of honey enriched with additives accounts for 871.53 $\mu g \cdot g^{-1}$, which is 2.4 times higher than the value of polyfloral honey (361.17 $\mu g \cdot g^{-1}$). K was the most abundant macroelement in both tested samples, with a mean concentration of 238.56 $\mu g \cdot g^{-1}$ for the control sample and 534.63 $\mu g \cdot g^{-1}$ for HF, respectively. The results obtained are strongly supported by other authors' findings [56–58]. Further, the Pb and Cd concentrations did not exceed the acceptable levels proposed by the World Health Organization (WHO) and the Food and Agriculture Organization (FAO) of 25 $\mu g \cdot kg^{-1}$ for Pb and 7 $\mu g \cdot kg^{-1}$ for Cd [56].

Moreover, the addition of additives enhanced the total content of phenolic compounds (2.6-fold), flavonoids (4.5-fold), and carotenoids (5.7-fold increase) compared to the control sample. Further, the flavonoid fraction of total phenolic compounds increased from 41.6% to 72.7% with the introduction of 12 additives to polyfloral honey. A similar effect of enriching honey with polyphenolic compounds was previously observed for honey supplemented with powdered sea buckthorn leaves and fruits, respectively [5]. Higher enrichment (over 400%) was reported for honey mixed with 4% fruit or 1% leaves of *Rubus* sp. [3].

Table 1. Physicochemical parameters, bioactive compounds, and total antioxidant capacity of honey samples.

Parameters	Honey Sample	
	Polyfloral	Formulation
Physicochemical parameters		
Moisture (%)	19.467 ± 0.115 [b]	22.667 ± 0.115 [a]
TSS (°Brix)	80.533 ± 0.115 [a]	77.333 ± 0.115 [b]
pH	3.93 ± 0.01 [b]	4.05 ± 0.05 [a]
Free acidity (meq·kg^{-1})	20.100 ± 0.100 [b]	40.067 ± 0.115 [a]
Electrical conductivity (mS·cm^{-1})	0.453 ± 0.002 [b]	0.538 ± 0.001 [a]
Metal ion concentrations (mg·g^{-1})		
Bioelements		
K	238.56 ± 22.24 [b]	534.63 ± 38.11 [a]
Ca	63 ± 9.45 [b]	180.3 ± 19.83 [a]
Mg	35.53 ± 1.88 [b]	100.89 ± 5.06 [a]
Na	5.921 ± 0.51 [b]	38.084 ± 5.71 [a]
Zn	11.69 ± 1.25 [a]	9.31 ± 1.01 [b]
Fe	5.62 ± 0.5 [b]	6.52 ± 0.53 [a]
Mn	0.483 ± 0.053 [b]	1.139 ± 0.159 [a]
Cu	0.233 ± 0.028 [b]	0.513 ± 0.035 [a]
Cr	0.134 ± 0.012 [a]	0.143 ± 0.013 [a]
Toxic metals		
Cd	1.9 ± 0.2 (×10^{-3}) [a]	2 ± 0.3 (×10^{-3}) [a]
Pb	9 ± 2 (×10^{-3}) [a]	10 ± 2.5 (×10^{-3}) [a]
Bioactive compounds		
TPC (mg GAE per 100 g sample)	143.415 ± 1.771	371.318 ± 0.740
TFC (mg RE per 100 g sample)	59.751 ± 0.272	269.932 ± 5.203
TAC (mg CGE per 100 g honey)	1.761 ± 0.127	10.086 ± 0.421
Antioxidant capacity (AC)		
DPPH (mg GAE per 100 g honey)	0.584 ± 0.075	4.851 ± 0.058
ABTS (mg TE per 100 g honey)	1214.72 ± 86.036	2315.716 ± 110.638

Note: Data are expressed as mean ± SD, and values sharing the same letter indicate no significant differences at a confidence level of 95% using the *t*-test.

The superiority of the antioxidant activity (DPPH and ABTS) of HF over raw honey was also observed, which is in agreement with other studies [2,59]. For example, a several-fold increase in the antioxidant capacity was found for creamed honey enriched with lavender (~3 times), lemon balm (~6.8 times), nettle (~3.2 times), peppermint (~5.9 times), and ginger (1.8 times) in contrast to polyfloral honey [2]. The results of Wilczyńska et al. (2017) [59] showed that the addition of cinnamon, ginger, and cardamom to polyfloral honey may cause changes in its antioxidant capacity. The highest antioxidant activity was obtained for honey flavored with cinnamon [59].

Among the 12 compounds explored, syringic acid, quercetin, ferulic acid, riboflavin, and pyrogallol were present in various amounts in both studied samples. As illustrated in Table 2, syringic acid was the dominant phenolic acid in the raw honey sample, and pyrogallol in the HF sample, while quercetin was found among flavonoids. Other authors indicate the presence of caffeoylquinic acids in rape honey enriched with mulberry leaves or fruits [1] and coumarin and *o*-coumaric acid in polyfloral honey fortified with *Melilotus officinalis* and *M. albus* [8]. Also, an increase in rutin and quercetin concentrations was observed in honey supplemented with *Sophora* flower [60], while Tomczyk et al. (2019) [1] highlight the potential of using mulberry leaves as a source of phenolic compounds for enriching honey instead of fruits.

Table 2. Identification and quantification of common phenolic acids, flavonoids, and vitamins in honey samples.

Compound	t_R (min)	Control Sample (mg per 100 g Sample)	HF (mg per 100 g Sample)
Phenolic acids			
Gallic acid		Tr	Tr
Pyrogallol	5.94	0.05 ± 0.01	1.76 ± 0.33
Caffeic acid		Tr	Tr
Vanillic acid		Tr	Tr
Syringic acid	8.79	0.78 ± 0.0168	0.27 ± 0.02
p-Coumaric acid		Tr	Tr
Rosmarinic acid		Tr	Tr
Ferulic acid	11.00	0.36 ± 0.01	0.58 ± 0.01
Flavonoids			
Rutin		Tr	Tr
Catechin		Tr	Tr
Quercetin	11.40	0.50 ± 0.01	1.21 ± 0.01
Kaempferol		Tr	Tr
Vitamin			
Riboflavin (Vitamin B_2)	8.19	0.1 ± 0.01	0.13 ± 0.013
Ascorbic acid (Vitamin C)		Tr	Tr

Note: t_R—retention time; Tr—traces (below limit of detection).

Furthermore, the results reveal that the HF at the tested concentrations (0.5, 1, and 1.5 $g \cdot mL^{-1}$) does not exhibit antibacterial activity against Gram-negative and Gram-positive bacteria (Figure S2 in Supplementary Materials). Similarly, no antibacterial activity was observed against *E. coli* [5] and *P. aeruginosa* [9] for other types of flavored honey irrespective of the added additives (e.g., fruit, herbs, and spices). The lack of antibacterial activity of HF might be attributed to the relatively low contents and diversity of polyphenols and flavonoids (Table 2). Alzahrani et al. (2012) have shown that the honey antimicrobial effects decrease as the polyphenols content decreases [61]. In their study, the highest antimicrobial activity was reported for manuka honey with a total polyphenol content of 899.09 mg gallic acid per kg, followed by acacia honey (627.56 mg gallic acid per kg), wild carrot honey (503.09 mg gallic acid per kg), and lavender honey (111.42 mg gallic acid per kg) [61]. Equally important is the composition and content of polyphenols found in honey. For instance, the high antimicrobial effects of manuka honey were attributed to the presence of caffeic acid, pinocembrin, chrysin, and galangin among the polyphenols [62]. However, according to Cheung et al. (2019), the composition of phenolic compounds depends on the geographical origin, while the concentrations of the compounds are influenced by the floral source [63]. Apart from the phytochemical composition, other factors influencing the antibacterial activity of honey include its chemical composition (water content, acidity, level of H_2O_2, non-peroxide components or sugar concentration), origin (geographical, seasonal, and botanical), and harvesting, processing, and preservation conditions. In addition, the low concentrations used in the experiment could have played an important role in the absence of antimicrobial effect. In general, the concentrations reported as having antibacterial activity can range from less than 3% to more than 50% [64–66]. Thus, the effectiveness of different types of honey against bacteria can vary significantly, with differences between them being as much as 100-fold [64].

2.2. *Effects of HF on Mixture of Heavy Metals and IMI-Induced Toxicity in Zebrafish*

2.2.1. Behavioral Analysis

Since animal behavior integrates effects across multiple biological levels, subtle changes can be observed following short-term exposures to low concentrations of compounds. Thus, behavioral assays have drawn increasing attention in toxicity studies, particularly in the ecotoxicology and pharmacology fields, as they offer numerous advantages in assessing the impact of various compounds on organisms [21,22]. Common behavioral assays include the evaluation of locomotor activity, conspecific interaction, aggressiveness, and various sensory responses [67]. In this context, we used both the locomotor activity and conspecific interaction assays to investigate whether sub-chronic exposure to heavy metals alone and in combination with IMI has any detrimental effects on fish locomotion parameters and on the time spent near the social stimulus. In addition, the efficiency of HF supplementation in alleviating behavioral impairments induced by the three toxicants was assessed.

According to the results shown in Figure 1A–C, there is no significant difference in behavioral endpoints between HF-treated zebrafish and the control group, as well as in the baseline performance of animals. Instead, the sub-chronic exposure to heavy metals and IMI mixture caused significant changes in zebrafish swimming behavior compared to the control and individual contaminant groups, which were characterized by an increased period of inactivity and reduced average swim velocity and total distance traveled. For instance, the simultaneous exposure to Hg, Cd, and IMI decreased the total distance traveled by 33% and 38.6% in relation to the control (p = 0.01) and IMI (p < 0.001) groups, while decreasing the mean velocity by 48.6% and 56.9% compared with the control (p = 0.002) and IMI (p < 0.001) groups, respectively. Similarly, the results of other studies indicated that co-exposure to Pb and TiO_2 nanoparticles [68] or carbendazim and chlorpyrifos [69] can lead to slower exploration behavior in zebrafish.

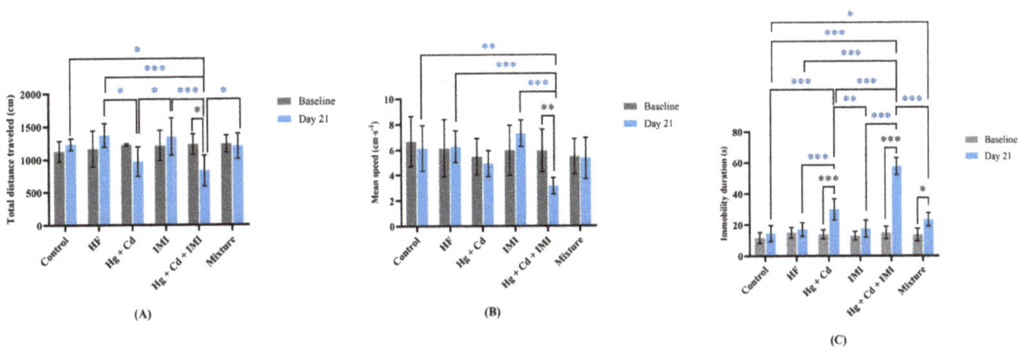

Figure 1. Locomotor activity of control and treated groups. (**A**) Total distance traveled (cm), (**B**) mean speed (cm·s^{-1}), and (**C**) immobility duration (s). Data are expressed as mean ± SD and analyzed by two-way ANOVA followed by Tukey's multiple comparison test. Statistically significant differences are denoted by * $p < 0.05$, ** $p < 0.01$, and *** $p < 0.001$. The experiments were repeated 2 times with 7 animals per group.

In addition, we found a significant increase in the inactivity time of the zebrafish from an average of approximately 8% of the time for both the control and HF-exposed group to an average of 24% and 12.5% for the ternary mixture (p < 0.001) and Hg + Cd (p < 0.001). Similar to our observation, zebrafish exposed to 50 µg·L^{-1} $PbCl_2$ [67], 45 µg·L^{-1} IMI [70], and 0.75 mM acrylamide [71] showed behavioral changes, characterized by a significant increase in immobility duration. A previous study also reported an augmentation in immobile time and a decrease in the total distance traveled by zebrafish after 96 h exposure to various proportions of a hospital effluent [72]. On the other hand, the results indicate that the locomotor impairments induced by the ternary mixture can be partially attenuated by HF supplementation. Interestingly, there were no significant differences when comparing the velocity and distance moved by the animals from the mixture group with those from the HF and control groups, respectively. The period of inactivity for zebrafish exposed to the mixture (Hg + Cd + IMI + HF) was partially restored to the control value (p = 0.043). Consequently, the fish exposed to Hg, Cd, and IMI in combination with HF moved faster

(1.7 times) and a longer distance (1.5 times) and had a lower immobility duration (2.5 times) than those from the Hg + Cd + IMI group. Our results are in agreement with an earlier study by Abdulmajeed et al. (2016) [72] who showed that honey supplementation may mitigate lead-induced memory and locomotor activity impairments.

Furthermore, we found that exposure to singular and combined chemical treatments for 21 days impaired the zebrafish's social behavior, as indicated by the reduced time spent in the zone near conspecific individuals compared to the control group (Figure 2A). The time spent near the social stimulus decreased by 37% when zebrafish were treated with Hg + Cd and IMI ($p = 0.031$) compared to the control group, while the highest magnitude of reduction was observed for the group exposed to heavy metals in combination with IMI ($p = 0.004$). As shown in Figure 2A,B, the HF co-treatment with chemicals restored social preference to the normal level ($p = 0.094$), but with an increase in the time spent in the central segment compared to the control group ($p = 0.002$). Similarly, the results of Robea et al. (2020) showed that the simultaneous administration of vitamin C [73] and vitamin B_{12} [74] with fipronil and pyriproxyfen can partially reverse the social interaction deficits induced by insecticides, but with an increase in the time spent in the right arm compared to the control group on day 14 of exposure.

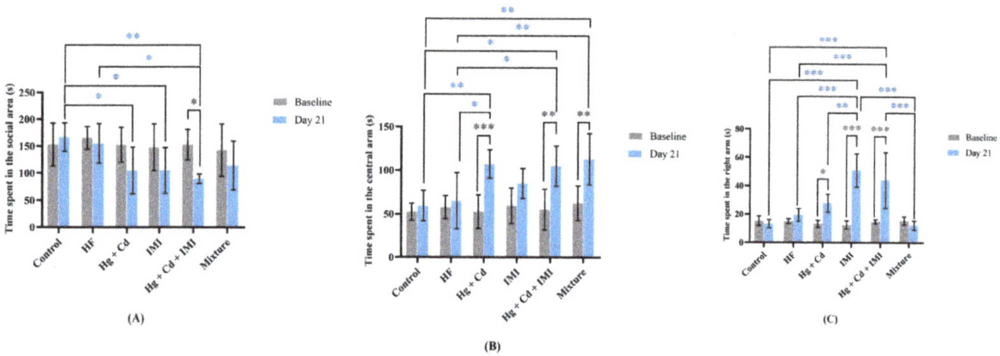

Figure 2. Social interaction behavior observed among zebrafish after chemical compound exposure. Time spent in the left (social area) (**A**), central (**B**), and right (**C**) arms. Data are expressed as mean ± SD ($n = 7$) and analyzed by two-way ANOVA followed by Tukey's multiple comparison test. Statistically significant differences are denoted by * $p < 0.05$, ** $p < 0.01$, and *** $p < 0.001$. The experiments were repeated 2 times with 7 animals per group.

Moreover, the zebrafish exposed to heavy metals alone and in combination with IMI showed a preference for the central segment, spending 107.4 ± 16.3 s and 105.3 ± 22.8 s, respectively. Also, the IMI ($p < 0.001$) and Hg + Cd + IMI ($p < 0.001$) groups showed a marked difference with the control regarding the time spent in the right arm (Figure 2C). Similar to our observation, valproic acid [75,76], tetrabromobisphenol A [77], silver nitrate [78], bisphenol S and estradiol [79], and antidepressant [80] treatments induced social interaction deficits in zebrafish.

2.2.2. Biochemical Analysis

Biochemical responses of organisms are frequently used as early warning signs of environmental contamination [43]. The imbalance between the production and elimination of reactive oxygen species (ROS) in tissues leads to the appearance of OS, which is a key factor for many chronic diseases like neurological, metabolic, and cardiovascular disorders [70,81]. However, organisms have evolved a complex system of defense to neutralize ROS, consisting of a network of enzymatic (e.g., SOD, GPx, CAT, etc.) and non-enzymatic (e.g., vitamin C, uric acid, vitamin E, melatonin, etc.) antioxidants. The antioxidants can act at various levels, inhibiting the formation of free radicals, scavenging them, and repairing or removing the oxidized biomolecules [81]. Antioxidants are substances that counteract the deterioration caused by oxidants such as O_2, OH^-, superoxide, and/or lipid peroxyl radicals [82]. Cells have a defense system against oxidative damage, which includes free radicals and other protective agents such as catalase, SOD, peroxidase, ascorbic acid, tocopherol, and polyphenols [83].

Honey from different floral origins in various countries has been shown to possess high antioxidant properties [82]. The antioxidant activity of honey is primarily attributed to phenolic acids [84] and flavonoids [85–87]. Antioxidant agents stimulate biomolecules like carbohydrates, proteins, lipids, and nucleic acids [83]. This stimulation ultimately triggers an antioxidant response in cells. Honey and honey infused with herbs have been found to exhibit strong antioxidant activity, and this capacity contributes to the prevention of several acute and chronic disorders [83,88]. The precise antioxidant mechanism remains unclear, but proposed pathways include scavenging of free radicals, donation of hydrogen atoms, chelation of metal ions, and the role of flavonoids as substrates for neutralizing hydroxyl and superoxide radicals [83].

The present study aims to evaluate the single and combined sub-chronic effects of IMI, Hg, and Cd treatments on selected OS parameters and AChE activity and to determine whether HF co-administration can alleviate contaminant-induced OS. To assess the oxidative damage induced by pollutants in zebrafish, we measured the activities of SOD and GPx, as well as the MDA contents. As is described in the literature, SOD and GPx are the body's first line of defense against the toxic effects of ROS, while MDA is a marker of oxidative damage caused by these chemical compounds against zebrafish [74]. In addition, the AChE is a key nervous system enzyme often employed as a biomarker for evaluating the effects of various environmental pollutants on aquatic organisms [89]. Figure 3A–D show no significant difference in SO parameters between HF-treated (500 $mg \cdot L^{-1}$) zebrafish and the control group; in contrast, exposure to IMI treatment for 21 days increased MDA content (p = 0.003) and SOD activity (p = 0.047).

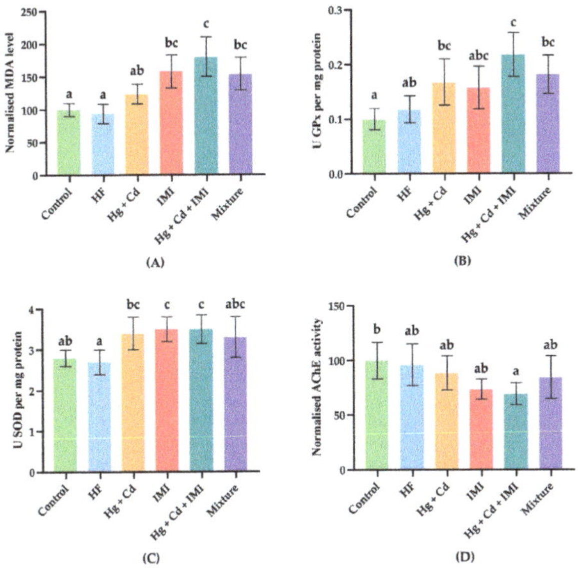

Figure 3. MDA level (**A**), GPx (**B**), SOD (**C**), and AChE (**D**) activities in zebrafish following treatment with Hg (15 $\mu g \cdot L^{-1}$), Cd (5 $\mu g \cdot L^{-1}$), and IMI (0.5 $mg \cdot L^{-1}$) alone and in combination with HF (500 $mg \cdot L^{-1}$) for 21 days. Data are expressed as mean ± SD (n = 5 animals per group with two replicates), and bars sharing the same letter indicate no significant differences at the level of $p < 0.05$ using one-way ANOVA followed by Tukey's multiple comparison test.

In a previous study, neotropical fish (*Prochilodus lineatus* Valenciennes, 1837) exposed to different IMI concentrations showed liver, gill, kidney, and brain lipid peroxidation as well as increased SOD activity in the liver and gills [90]. Also, Ge et al. (2015) showed that 5 mg$\cdot$L^{-1} IMI treatment markedly increased MDA content and SOD activity in zebrafish liver at day 14 [91]. Moreover, sub-chronic exposure to heavy metals caused a significant increase in MDA level (p = 0.003) and GPx activity (p = 0.047), which were about 1.2- and 1.7-fold higher than those in control fish, respectively. Similar to our results, various reports demonstrated that $HgCl_2$ [92], CH_3HgCl [93], $CdCr_2$ [94], and simultaneous $CdCl_2$ and $HgCl_2$ [95] exposure induced lipid peroxidation.

The ternary mixture affected not only the main parameters related to the OS but also the AChE activity. As shown in Figure 3A–D, sub-chronic exposure to heavy metals and IMI promoted a significant inhibition of AChE activity (−30.7% compared to the control group) in the fish brains and a remarkable elevation in antioxidant enzyme activities (p = 0.047 for SOD and p < 0.001 for GPx) and MDA level (p < 0.001) in fish tissues, respectively. According to our findings, an earlier study by Rosales-Pérez et al. (2022) showed that acute exposure to a mixture of contaminants from a hospital effluent markedly inhibited the AChE activity and increased antioxidant enzyme activities (e.g., SOD, CAT, and GPx) in fish brains [96].

The changes in AChE activity concerning behavioral observations provide valuable insights. However, a more comprehensive investigation is needed to understand the underlying mechanisms. The partial reduction in AChE inhibition with HF supplementation suggests reversibility, raising the possibility of adaptive mechanisms mitigating neurotoxic effects in zebrafish. The preservation of swimming behavior despite inhibited AChE activity may point to the activation of compensatory motor control strategies.

In addition, Abdulmajeed et al. (2016) and Azman et al. (2015) found that honey supplementation reduced anxiety, memory impairment, decreased locomotor activity, and OS levels in rats [72,97]. In our study, co-administration of HF partially normalized OS markers in zebrafish, thus indicating a protective role against oxidative damage. Furthermore, the cholinergic system has a critical role in mediating locomotor responses to novel stimuli and facilitating spatial memory, highlighting the impact of disruptions to AChE breakdown due to compromised AChE activity [32,94,95]. While our research found that metal concentrations and pesticides negatively impacted AChE activity, it is essential to note that some studies, such as that of Bui Thi et al. (2020), have reported increased AChE activity in fish following heavy metal exposure [67]. This variability underscores the need for further exploration of the context-dependent compensatory mechanisms that zebrafish may employ in response to environmental neurotoxins, which can enhance our understanding of their resilience to neurotoxic stress at both molecular and behavioral levels.

In addition, the co-administration of HF attenuated ternary mixture-induced changes in OS parameters and AChE activity. For example, in rats, honey supplementation ameliorated the enhanced MDA level and normalized the SOD activity [98]; meanwhile, in the case of AChE, Tualang honey improved cholinergic transmission, protecting against hypoxia-induced neuronal damages [99].

2.2.3. Zebrafish Body Element Concentrations

Cu, Fe, Zn, Mn, Ni, Mg, Na, and Ca constitute essential elements crucial for the optimal functionality of various cellular enzymes and proteins [100,101]. The impact of bioelements on biological processes can turn toxic when their concentrations exceed specific thresholds [100,102]. Notably, certain heavy metals possess the potential to influence metal ion dynamics within aquatic environments, prompting a growing interest in evaluating the efficacy of dietary products in mitigating pollutant toxicity [103–106].

In general, honey or honey-based products are described by elevated concentrations of essential elements [45,101]. In the context of our study, the administration of HF produced significant effects on some of the elemental composition of the zebrafish body (Figure 4A–G). This intervention led to a reduction in Fe^+ levels while elevating the concentrations of Mg^+ and Na^+. This outcome underscores the vital physiological role of Na^+ in nerve transmission [107] and the enhanced metabolism of Mg^+, a factor of particular importance [108]. The observed decrease in Fe^+ levels suggests a potential regulatory effect on Fe^+ metabolism.

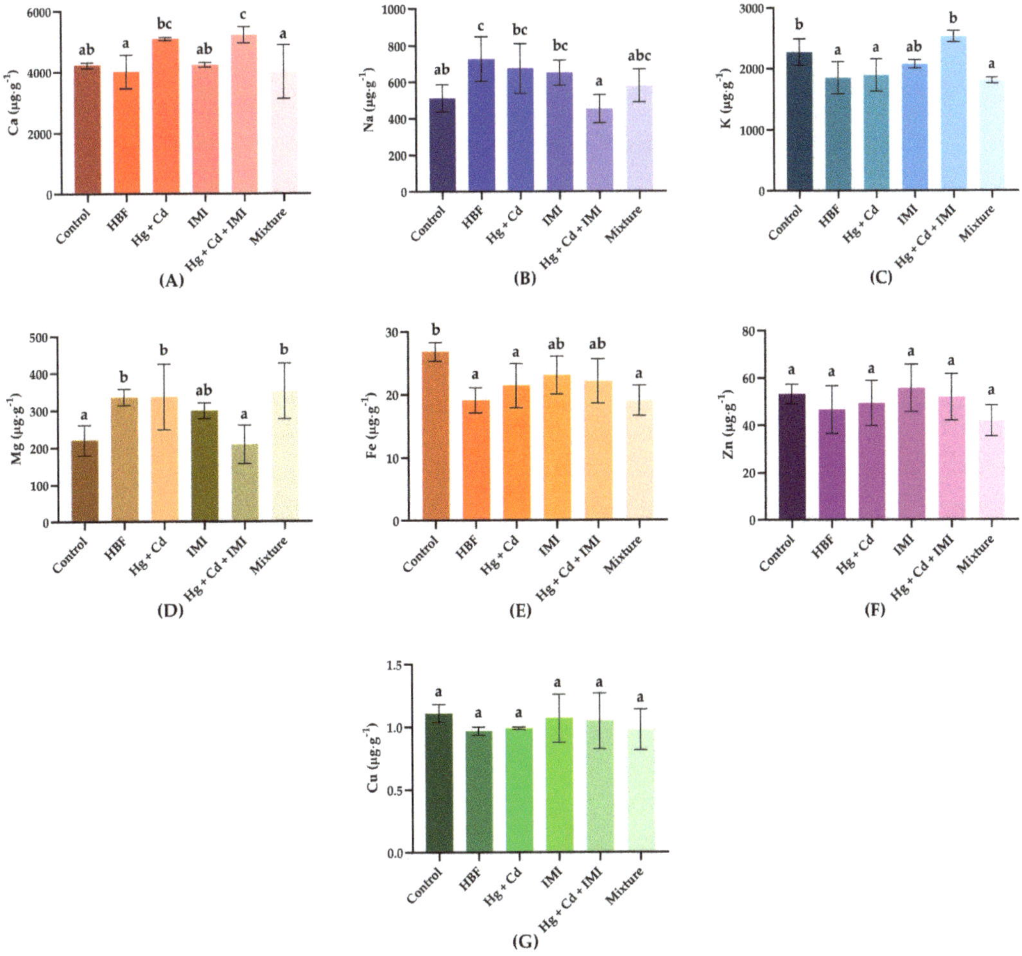

Figure 4. Concentration ($\mu g \cdot g^{-1}$ wet weight) of essential elements ((**A**)—Ca; (**B**)—Na; (**C**)—K; (**D**)—Mg; (**E**)—Fe; (**F**)—Zn; (**G**)—Cu) involved in biological structures and biochemical processes analyzed from entire fish body mass and reported as mean ± SD (n = 5 animals per group), with bars sharing the same letter indicating no significant differences at the level of $p < 0.05$ using one-way ANOVA followed by Tukey's multiple comparison test.

Our study indicates a 10% decrease in Na levels within the zebrafish group exclusively exposed to neonicotinoid insecticides and heavy metals. For example, Araujo et al. (2022) showed a significant reduction in Na levels after 96 h in carbamazepine (CBZ) + acetamiprid (ACT) and ACT + Cd groups [109].

Regarding K levels, no significant differences were observed in the group exposed to the mixture compared to the control; these findings imply that the losses of K^+ and Ca^{2+} were proportional to the ionic disparities between the fish plasma and the surrounding water, suggesting a common mechanism of acute toxicity [110]. The previously mentioned finding is in line with the finding of Araujo (2022), who showed that CBZ + ACT and

CBZ + ACT + Cd treatments did not result in significant differences compared to the control group after 96 h exposure [109]. Similarly, there was no significant difference between K levels in the control groups and zebrafish exposed to 2 mg·L^{-1} of Cr [VI] for 60 days [111].

Our study further demonstrated that exposure to a combination of Hg and Cd led to a 20% increase in Ca levels, while exposure to Cd, Hg, and IMI resulted in a 24% increase in Ca levels. The present study aligns with the findings of Siblerud et al. (2019) [112], linking Hg to elevated intracellular Ca concentrations. Increased Ca levels were measured in zebrafish exposed to CBZ, ACT, Cd, and their combined treatments [109]. Similarly, Shaw et al. reported an increase in Ca amount in the zebrafish treated with 2 mg·L^{-1} of Cr [VI] [111].

The levels of Cu and Zn did not differ significantly between the control group and the fish exposed to Cd, Hg, and IMI; these substances did not cause considerable changes in Cu and Zn metabolism. Devarapogu and Asupatri (2023) suggested a potential role for Zn in alleviating Cd toxicity [113]. Contrary to our results, significant differences in Zn and Cu levels compared to controls were measured in zebrafish exposed to CBZ, ACT, Cd, and their binary combination [109], as well as for Cr [VI]-treated zebrafish [111]. However, further analysis is necessary to understand the possible interactions and effects of mixtures on essential elements comprehensively.

2.2.4. Evaluation of Histologic and IHC Lesions

Zebrafish from the control and HF groups did not present histologic changes in the studied nervous system areas. The lesion severity score was (-). Contrarily, the optic tectum in fish exposed to IMI and Hg + Cd + IMI showed severe spongiosis in the neuronal layer, edema, and vacuolization of the neuropil with multiple areas of extravasated erythrocytes, as shown in Figure 5.

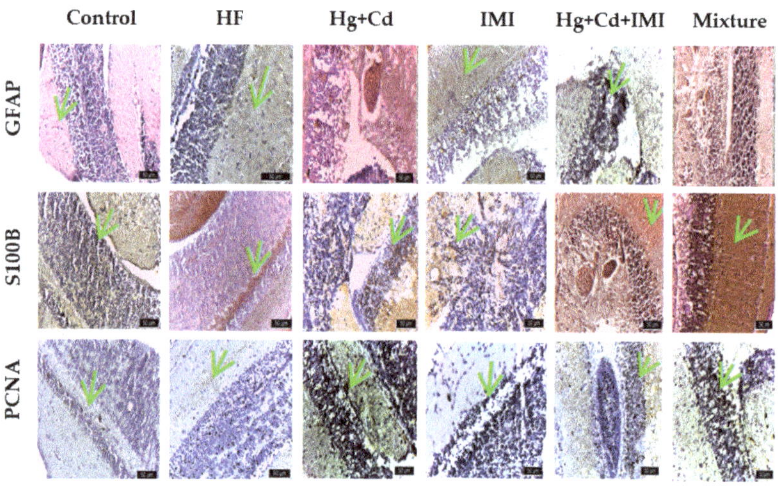

Figure 5. IHC expression of GFAP, S100B, and PCNA markers in the optic tectum for the control and exposed groups green arrow marks IHC positive cells (n = 5 animals per group; scale bar = 50 µm).

In addition, degenerative changes of neurons and neuropil and focal aggregation of microglial cells were recorded in the optic tectum for heavy-metal- and IMI-exposed animals with the lesion score (+++) (Table S2 in Supplementary Materials). The Hg + Cd group had the lesion score (+). The intensity of DAB anti-GFAP staining was significantly higher in IMI, Hg + Cd, and Hg + Cd + IMI (+++). An intermediate intensity score (++) was observed in the optic tectum of the group exposed to Hg, Cd, and IMI. Thus, GFAP expression is confirmed to increase with astrocytic emergence and activation; it is commonly adapted as a marker to study astrocyte response to various physiological and pathological conditions. Astrocytes play a significant role in nervous system development, synaptic transmission, and nerve tissue repair.

Researchers identified GFAP as a sensitive indicator of early response to neurodegenerative injury [114], and it can be detected in astrocytes even in the absence of neuronal death [115]. The severity of the injury correlates with the intensity of GFAP expression in reactive astrocytes [116,117]. In various central nervous system (CNS) disorders, activated astrocytes exhibit an increased GFAP expression due to elevated mRNA levels and increased cytoskeletal GFAP proteins. It is important to note that the elevation of GFAP levels alone is insufficient for classifying reactive astrocytes, suggesting that increased GFAP levels occur in response to pathological stimuli and regional astrocyte differences. Furthermore, initial GFAP expression in astrocytes can occur following physiological stimuli [116]. Astrocytes, as integral CNS cells, play crucial and diverse roles [118]. They are involved in metal uptake and sequestration, thereby preventing the accumulation of metals in the CNS at toxic concentrations [119,120]. Additionally, astrocytes protect neurons against various insults, including excess glutamate [121,122] and heavy metals, which are a primary source of neurotoxicity leading to neurodegenerative changes. The influx of heavy metals affects astroglial homeostatic and neuroprotective cascades, including the glutamate/GABA–glutamine complex, antioxidant systems, and energy metabolism. The astrocytes produce nerve growth factor and S100 protein, essential for neurite elongation and outgrowth [123,124]. Activated astrocytes are involved in angiogenesis, a crucial process in CNS development and repair that relies on physical interaction between astrocytes and endothelial cells. Notably, endothelial cells separated from astrocytes do not form capillaries [125,126].

In the control and HF groups, there was a noticeable difference in the reaction to anti-GFAP DAB staining. The HF group showed a more intense response. Additionally, the scoring indicated a result of (-) in the control group and a result of (+) in the HF group.

Figure 5 demonstrates the presence of the S100B protein in all experimental groups. The mesencephalic optic tectum predominantly contains S100B in nerve fibers rather than nerve cells. The protein showed immunoreactivity across all the fiber profiles, which run from the bottom to the top of the optic tectum perpendicular to its outer surface. Additionally, immunoreactivity in the medial and lateral regions of the cerebral valve was noticed. The inner surface of the optic tectum, adjacent to the tectal ventricle, contained S100B-positive ependymal and subependymal cells. The bulkier and rounder shapes of ependymal cells set them apart and make them easily identifiable.

In contrast, subependymal (glial) cells exhibited elongated radial processes that extended through the optic tectum and reached the surface of the pial membrane (*pia mater*). In the control group, the observed immunoreactivity was rated as (+), while in the HF group, it was (++). The mixed group displayed intense positivity (+++) in cytoplasmic and extracellular regions, whereas moderate positivity (++) was detected in groups exposed to individual or combined chemicals (IMI, Hg + Cd). Furthermore, the dorsal and lateral sections of the *torus longitudinalis* were lined with S100B-positive ependymal cells, and the nerve fibers forming the commissure in the ventral region of the *torus longitudinalis* were also S100B-positive. The presence of extracellular protein S100B might have a role in regulating tissue development and facilitating regeneration or repair processes.

In the assessment involving IHC staining for PCNA (Figure 5), a marker indicative of the nuclei of dividing cells, it was observed that the PCNA marker was evident in both the control group and the group exposed to HF (++). The neurotoxic effects reduced the PCNA staining in the IMI-exposed group, indicating a decrease in cell proliferation. An intense (+++) reaction was recorded in the Hg + Cd group, a moderately intense (++) reaction in the IMI and mixture groups, and a weakly positive (+) reaction in the Hg + Cd + IMI (refer to Table S2 in the Supplementary Materials). The toxicities of Cd and Hg may be mitigated, prompting the body to stimulate cell regeneration and potentially repair the damage.

In the cerebellum, IMI and Hg + Cd + IMI cause neuronal degeneration, spongiosis, and vacuolation with focal aggregation of glial cells and damage to granule neurons. A reduction in cell density, depletion of the nuclear area, and chromatin condensation in granular cells were also identified (Figure 6).

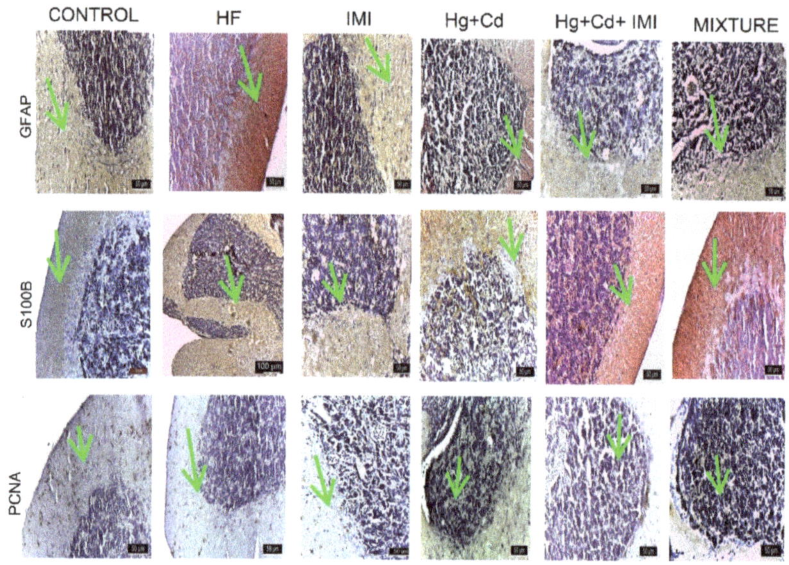

Figure 6. IHC expression of GFAP, S100, and PCNA markers in the cerebellum for the control and exposed groups, green arrow marks IHC positive cells (n = 5 animals per group; scale bar = 50 μm).

Vacuolation of grey and white matter, necrotic cells, and the presence of tissue edema that caused large lesions due to OS were observed during the analysis. Significant damage was documented in the cerebellar cortex after exposure to Hg + Cd, resulting in a neuronal loss lesion score of ++ (Table S3 in Supplementary Materials). The cerebellum showed slight vacuolization and degeneration of cells in the Purkinje cell layer with a score of (++). No changes were recorded in the control and HF groups, while in the mixture group, the lesion score was (++).

The intensity of DAB anti-GFAP staining in the cerebellum was as follows: significantly more intense, receiving a score of (+++) for the mixture group; intermediate intensity (++) for the Hg + Cd, Hg + Cd + IMI, and IMI-exposed groups; and reduced intensity for the control group (Table S3 in the Supplementary Materials). The distribution pattern of the S100B protein in the cerebellum differs from that of other central nervous segments. It is mainly localized in neurons and less in glial cells. The cerebellar body showed S100B protein in small neurons, localized primarily in the superficial layer. In addition, neurons forming the deep cerebellar nuclei were also immunoreactive for S100B. Purkinje neurons localized in the basal zone showed a strong reaction for S100B protein in both the perikaryon and the dendritic arbor. Intense reactions were observed in the group exposed to a ternary mixture of S100B, and reactions were rated as medium (++) in HF and Hg + Cd and low in the control group. The PCNA reaction was most intense (+++) in the Hg + Cd group, followed by a medium reaction (++) in the mixture, HF, and control groups, and low in the Hg + Cd + IMI group.

In the spinal cord, the intensity of DAB anti-GFAP staining was as follows: significantly more intense (+++) for the mixture group; of intermediate intensity (++) in Hg + Cd, Hg + Cd + IMI, and IMI-exposed groups, and reduced (+) in the control group, as shown in Table S4 in the Supplementary Materials. The distribution pattern of S100B protein in the cerebellum differs from that of other central nervous segments. It was localized mainly in neurons and less in glial cells. The reaction to PCNA was most intense (+++) in the Hg + Cd group, followed by a medium reaction (++) in the mixture, HF, and control group, and low (+) in the Hg + Cd + IMI group (Figure 7).

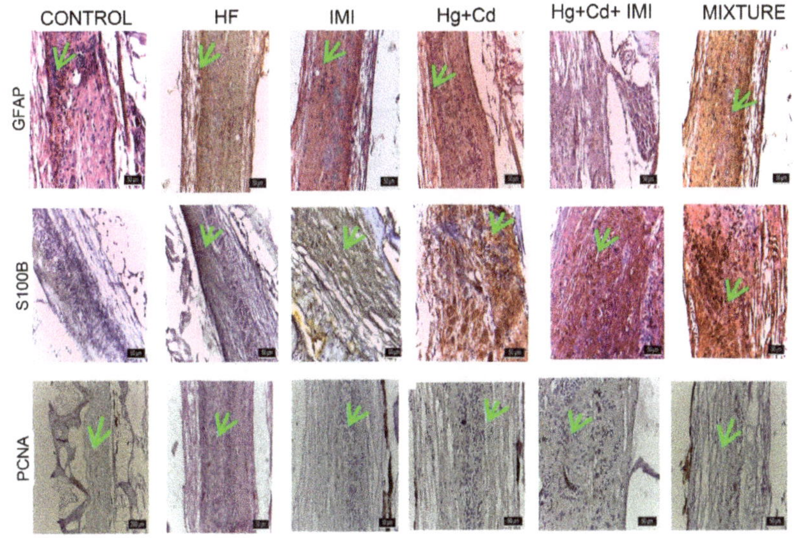

Figure 7. IHC expression of GFAP, S100, and PCNA markers in the spinal cord for the control and exposed groups, green arrow marks IHC positive cells (n = 5 animals per group; scale bar = 50 μm).

In the current investigation, the exposure of zebrafish to contaminants triggered a cascade of responses characterized by heightened oxidative and nervous system stress. This radical interference led to membrane damage, potential impairment, and morphological alterations in mitochondrial cristae, culminating in disrupted adenosine triphosphate (ATP) production and a subsequent decrease in intracellular ATP content [127]. The observed mitochondrial effects may have substantial implications for neurological function. Neurons have the highest energy requirements of all somatic cells. They are the largest consumers of oxygen (O_2)—the brain consumes O_2 at a rate ten times higher than any other tissue, making it much more vulnerable to oxidative damage [128,129]. In addition, cells have difficulty repairing the distal ends of dendrites and axons due to their distance from the cell body [129]. The heightened impact is associated with oxidative injury to mitochondria, which play a significant role in upholding the health of axons [130]. Although considerable efforts have been dedicated to the IMI toxicity on aquatic organisms, only a few data are on apoptosis and immunotoxicity [131]. IMI in common carp moderately induces apoptosis in the brain but also induces severe histopathological changes, inflammation, and OS in the gills, liver, and brain, in agreement with our results. The mechanisms of the activation of some biomarkers (iNOS, 8-OHdG, TNF-α) and expression of some apoptotic genes (caspase 3) during pesticide intoxication have been reported [132,133]. Cd also induces caspase-mediated apoptosis and necrosis in cortical neurons [127]. In addition, Cd exposure can cause a decrease in several critical brain enzymes, including AChE, acid phosphatases, alkaline phosphatase, ATPase, and catalase [134]. The observed neurotoxic effects of MeHg on zebrafish embryos could be because MeHg is known to disrupt microtubule formation [135], which could profoundly affect cell division. The toxic impact of MeHg could also be because it negatively affects mitochondrial function, thus decreasing the metabolic capabilities of cells. A study in adult zebrafish showed that exposure to 15 $\mu g \cdot L^{-1}$ of MeHg under ambient conditions caused uncoupling and mitochondrial oxidative phosphorylation in skeletal muscle [136].

The GFAP and S100B markers were selected for the protein homology studies between human, mouse, and zebrafish orthologs. GFAP (astrocyte intermediate filament), whose expression increases as astrocytes arise (astrocyte activation), is commonly adopted as a marker to study the reaction of astrocytes in various situations of physiological and pathological conditions, playing a vital role in nervous system development, synaptic transmission, and nerve tissue repair. In the present study, the expression of GFAP was considerably increased in the IMI-treated group and discrete in the control group. The observed changes may suggest heightened dopaminergic and serotonergic activity in the brain, along with initial behavioral shifts [137]. This study noted strong and extensive expression of S100B and moderate, discrete expression of GFAP in the zebrafish's central and peripheral nervous systems, aligning with previous research findings [138]. For instance, S100B is recognized for its direct interaction with Dbf2-related nuclear protein kinase (NDR kinase), inhibiting the recruitment of NDR kinase substrates [138–140]. S100B deactivates advanced glycation end products (RAGEs) by binding with vital fibroblast growth factor and its receptor [139,140]. Extracellular S100 proteins play a role in regulating the apoptosis, proliferation, differentiation, and migration of various cell types, such as monocytes, macrophages, neutrophils, lymphocytes, myoblasts, epithelial cells, endothelial cells, smooth muscle cells, neurons, and fibroblasts. Additionally, besides their function as Ca-binding proteins, S100 proteins were later identified as DAMP molecules [141] associated with cell death and tissue damage, triggering rapid inflammatory responses or producing biologically active molecules [142,143]. S100B protein's presence in the adult zebrafish nervous system contributes to intracellular Ca homeostasis regulation as a triggering or activating protein. It also exhibits neurotrophic activity, suppresses phosphorylation, induces apoptosis [144], and maintains cytoskeletal stability [145]. Furthermore, S100B protein is produced, stored, and released by astrocytes, tanycytes, oligodendrocytes, and radial cells and exerts paracrine and autocrine effects on neurons and glia. It has been shown that S100B expression defines a late developmental stage, after which the GFAP-expressing stage of development loses its potential as neuronal stem cells [146]. Since S100 protein continues to be deposited in these regions of adult zebrafish, and subventricular zone glial cells play an essential role in adult zebrafish neurogenesis, S100 protein could be involved in this function, which is maintained in zebrafish during their entire lifespan [147,148]. Using immunohistochemical staining for PCNA, a marker for the nuclei of cells undergoing cell division is expressed in developing zebrafish [149]. The expression of PCNA was significantly decreased in animals exposed to different concentrations of MeHg [150]. The neural toxicity of MeHg tested in this study caused a decrease in PCNA staining, suggesting a decline in cell proliferation. Smith et al. (2010) observed that in adult zebrafish, the telencephalic cell density was significantly decreased at all MeHg exposures during development [151]. These observations are similar to those reported by Yang et al. (2007), where MeHg exposure significantly impaired zebrafish development [152].

Higher production of antioxidants can act as a protective shield during oxidative damage [153–155]. OS has been implicated in the pathogenesis of neurotoxic mechanisms of several pesticides because the brain has a higher susceptibility to oxidative damage. OS occurs due to a mismatch between pro-oxidant and antioxidant systems. The resulting OS can cause oxidative damage to lipids, nucleic acids, and proteins [137]. The combination of heavy metals and IMI causes degenerative changes in the optic tectum and cerebellum and is somewhat reduced in the spinal cord. The use of GFAP, S100, and PCNA markers highlights the lesions in groups exposed to pesticides and heavy metals and their amelioration by reducing the effect of OS.

3. Materials and Methods

3.1. HF Composition and Analysis

3.1.1. HF Composition

The commercial HF was received directly from the producer. This product contains a mixture of 90.39% polyfloral honey and 9.61% other natural ingredients, such as royal jelly,

Spirulina *Arthrospira platensis*, extracts of *Arctium lappa* L., *Matricaria chamomilla*, *Eleutherococcus senticosus*, *Silybum marianum*, *Panax ginseng*, *Astragalus membranaceus*, *Boswellia serrata*, and essential oils from *Matricaria chamomilla*, *Salvia sclarea*, and *Salvia officinalis*. Adding natural ingredients to the polyfloral honey resulted in a notable alteration in its color, causing it to transition to a deep shade of blue (as depicted in Figure S1 in Supplementary Materials). Of the 12 constituents, the royal jelly [156], *Eleutherococcus senticosus* [157], *Arctium lappa* L. [158], *Matricaria chamomilla* [159,160], and *Silybum marianum* [161] are reported to exhibit neuroprotective effects and reduce oxidative damage. Polyfloral honey was obtained from the apiary located in the Bunesti-Viscri region (Romania) in the 2021 bee-keeping season. For transparency and to avoid any potential conflicts of interest concerning honey-based products, the brand name of the producer has been kept confidential. This product was chosen for this study to simulate a realistic scenario.

3.1.2. Physicochemical Analysis

The moisture content (%), total soluble substances (°Brix), and free acidity (mequiv$\cdot$kg^{-1}) for polyfloral honey without (control) or with additives (HF) were assessed following the guidelines provided by the Codex Alimentarius Commission—Revised Codex Standard for Honey Codex Stan 12-1981, Rev. 1 (1987), Rev. 2 (2001) [46]. Furthermore, the pH and conductivity (mS$\cdot$cm^{-1}) were measured using multiparameter analysis equipment from XS Instruments, Carpi, Italy.

3.1.3. Determination of Mineral Elements

The mineral analysis was conducted at the laboratory of the Faculty of Food Science and Engineering, Dunarea de Jos University of Galati, Romania. In preparation for the quantification analysis, minerals from polyfloral honey samples were extracted. This process involved treating approximately 0.5 g of honey samples (with and without additives) with a mixture of 5 mL HNO_3 Suprapur® of 65% concentration (certified Merck, Darmstadt, Germany) and 2 mL H_2O_2 EMSURE® of 30% concentration (certified Merck, Darmstadt, Germany), followed by digestion using a microwave-assisted pressure digestion system (TOPwave, Analytik Jena, Jena, Germany) The mineralized samples were then transferred into decontaminated polyethylene flasks (50 mL volume) and diluted with ultrapure water (18.2 M$\Omega\cdot$cm^{-2}, equipment LaboStar™ UV 4 Siemens Water Technologies, Barsbüttel, Germany). The treated samples were processed and then analyzed for the quantification of Calcium (Ca), potassium (K), magnesium (Mg), and sodium (Na) through the flame atomic absorption spectrometry technique (FL-AAS) using an air–acetylene–nitrous oxide flame. Furthermore, the determination of copper (Cu), iron (Fe), manganese (Mn), and zinc (Zn) was performed using the graphite furnace atomic absorption spectrometry technique (GF AAS). The ContrAA 700 equipment manufactured by Analytik Jena, Jena, Germany, was utilized to conduct these measurements. Calibration curves were established through the preparation and analysis of a reference standard solution, ICP multi-element standard solution IV with 23 elements in diluted HNO_3 (Certipur®, 1000 mg$\cdot$L^{-1}, Merck, Darmstadt, Germany). Each sample underwent triplicate analysis, and the metal ion content was determined at the mg$\cdot$g^{-1} level.

3.1.4. Determination of Total Phenolic Content (TPC), Total Flavonoid Content (TFC), and Total Carotenoid Content (TCC)

The TPC was determined according to the method proposed by Csakvari et al. (2021) [162], while the TFC and TAC were measured as outlined by Rababah et al. (2014) [163], for 4% solutions of honey in distilled water. The results for TPC and TFC were expressed in gallic acid equivalents (GAE) and quercetin equivalents (QE) per 100 g sample, respectively, using appropriate calibration curves with standard solutions (gallic acid monohydrate, Sigma-Aldrich Chemie GmbH, Taufkirchen, Germany).

3.1.5. Quantification of the Total Antioxidant Capacity (TAC)

The total antioxidant activity was determined by 1,1-diphenyl-2-picrylhydrazyl (DPPH; Sigma-Aldrich Chemie GmbH, Taufkirchen, Germany) and 2,2′-azino-bis-(3-ethylbenzothiazoline-6-sulfonic) acid (ABTS; Sigma-Aldrich Chemie GmbH, Taufkirchen, Germany) assays, as earlier reported by Bogdan et al. (2021) [164] and Csakvari et al. (2021) [162]. The results were expressed as mg gallic acid equivalents (GAE) per 100 g of sample and mg Trolox equivalents (TE) per 100 g of sample, respectively.

3.1.6. RF-HPLC Analysis of Honey Flavonoids and Other Phenolic Derivates

The identification and quantification of phenolic compounds were performed using an ultra-high-performance liquid chromatograph (Nexera X2, Shimadzu, Tokyo, Japan) equipped with a diode array detector (M30A, Shimadzu, Tokyo, Japan) and a Nucleosil 100-3 C18 reversed-phase column (4.0 mm column inner diameter × 125 mm column length, 3 μm particle size, Macherey-Nagel GmbH, Düren, Germany). The column temperature was maintained at 30 °C, and the flow rate was 0.5 mL per min. The solvents used for the chromatographic elution consisted of ultrapure water (18.2 MΩ·cm^{-2}, equipment Adrona Crystal EX, Adrona, Riga, Latvia) with 0.1% trifluoroacetic acid (A) (Sigma-Aldrich Chemie GmbH, Taufkirchen, Germany) and acetonitrile (B) (Sigma-Aldrich Chemie GmbH, Taufkirchen, Germany). The chromatographic elution program used was as follows: 5% B, then 42% B for 5 min, followed by 35% B, 5% B in 35 min. The HF was diluted 1:5 with 96% ethanol (CH_3CH_2OH, EMSURE®, Merck KGaA, Darmstadt, Germany) and then filtered using 25 mm syringe filters (Labbox Labware S.L., Barcelona, Spain). Then, the HF was injected at a volume of 10 μL and the spectra were acquired between 200 and 600 nm. The standards used were ascorbic acid, pyrogallol, gallic acid, riboflavin, rutin, caffeic acid, vanillic acid, syringic acid, *p*-coumaric acid, catechin, rosmarinic acid, ferulic acid, quercetin, and kaempferol (Sigma-Aldrich Chemie GmbH, Taufkirchen, Germany). Polyphenols were identified by comparing the retention times and UV-Vis spectra with the previously mentioned standards.

3.1.7. Antibacterial Assay

The screening of HF for antibacterial activity was performed by the disk diffusion approach. It was performed using six standard strains of microorganisms from Microbiologics (Saint Cloud, MN, USA), including three strains of Gram-negative bacteria [*Escherichia coli* (ATCC 8739), *Salmonella enteritidis* (ATCC 13076), *Pseudomonas aeruginosa* (ATCC 27853)] and three strains of Gram-positive bacteria [*Staphylococcus aureus* (ATCC 25923), *Enterococcus faecalis* (ATCC 19433), *Listeria monocytogenes* (ATCC 13932)]. The strains were first grown on Mueller–Hinton Agar (MHA, Scharlab, Barcelona, Spain) for 24 h at 37 °C in order to develop single colonies, and then one colony was inoculated in 10 mL of Mueller–Hinton Broth (MHB, Carl Roth, Karlsruhe, Germany) under shaking conditions (160 rpm) at 37 °C for 24 h. After incubation, an inoculum of 0.5 MacFarland turbidity standards (~2 × 10^8 CFU·mL^{-1}) of each strain was prepared, and 200 μL was surface-spread on MHA plates. Antibiotic assay discs (Whatman™, Cytiva, Marlborough, MA, USA) of 6 mm in diameter, saturated with 20 μL of 0.5 mg·mL^{-1}, 1 mg·mL^{-1}, and 1.5 mg·mL^{-1} HF solutions prepared with sterile distilled water, were placed on the surface of inoculated plates. As a reference, two different antibiotics (Bio-Rad, Marnes-la-Coquette, France) for each bacterial strain were used as presented in Table S1 (Supplementary Materials). The plates were incubated at 37 °C, and the zone of inhibition against the test microorganism was measured after 24 h. The test was carried out in triplicates.

3.2. Zebrafish Protocols and Analysis

3.2.1. Chemicals and Reagents Used for Exposure Protocols and Biochemical Analysis

A standard solution of Hg Certipur® (1000 mg·L^{-1}), a standard solution of Cd Certipur® (1000 mg·L^{-1}), nitric acid 65% Suprapur® (HNO_3, 100441), hydrogen peroxide 30% Perhydro® (H_2O_2, 107210), phosphate-buffered saline (PBS, P4417-50TAB), acetylthiocholine iodide (ATChI, A5751), Tris hydrochloride solution (Tris-HCl) solution (Trizma®, T2819), 5,5′-dithiobis(2-nitrobenzoic acid) (DTNB, 322123), Bovine Serum Albumin (BSA, A8022-10G), Bradford Reagent (B6916), Lipid Peroxidation Assay Kit (MDA, MAK085), Glutathione Peroxidase Cellular Activity Assay Kit (GPx, CGP1-1KT), and Superoxide Dismutase Determination Kit (SOD, 19160-1KT-F) were all purchased from Merck, Darmstadt, Germany. The IMI used in the experiment is the active compound of a known insecticide (IMI, 100 g·L^{-1}) bought from a local market. To ensure transparency and mitigate any potential conflict of interest related to the IMI, the manufacturer's identity has been kept confidential. This product was deliberately selected for this study to represent an accurate real-life scenario.

3.2.2. Zebrafish Maintenance and Exposure Protocols

Wild-type (AB strain) zebrafish (*Danio rerio*) of both sexes (6–8 months old, 0.38 ± 0.06 g) were obtained from a local supplier and acclimated in the experimental room for two weeks before the experiments. Zebrafish maintenance and experimental procedures were based on previous publications with slight modifications [21,51]. After acclimatization, the animals were randomly separated into the following six groups with 15 animals in each group, as shown in Figure 8: Group 1 (control)—fish were exposed to standard tank water and served as the control group; Group 2 (HF)—fish were exposed to honey enriched with additives at a concentration of 500 mg·L^{-1}; Group 3 (Hg + Cd)—fish were exposed to 15 µg·L^{-1} Hg and 5 µg·L^{-1} Cd; Group 4 (IMI)—fish were exposed to 0.5 mg·L^{-1} IMI; Group 5 (Hg + Cd + IMI)—fish were exposed to 15 µg·L^{-1} Hg, 5 µg·L^{-1} Cd, and 0.5 mg·L^{-1} IMI; Group 6 (mixture)—fish were exposed to 15 µg·L^{-1} Hg, 5 µg·L^{-1} Cd, and 0.5 mg·L^{-1} IMI, in combination with 500 mg·L^{-1} HF. According to a previous study [70], a density of approximately 1.59 g of fish per L of water was maintained. Two replicated experiments were carried out for the control and each treatment. During the acclimatization and the 21 days of exposure, the fish were fed twice a day with commercial fish food (TetraMin) at a level of approximately 1% of their body weight per day. Also, the test solutions were renewed every day to maintain the compounds' concentrations and water quality at stable levels. The concentrations of the heavy metals and pesticide were set based on previous studies [31,37–39,42,43,165], while the concentration of HF was selected based on its maximum solubility in water, to achieve a homogeneous and clear aqueous solution of HF.

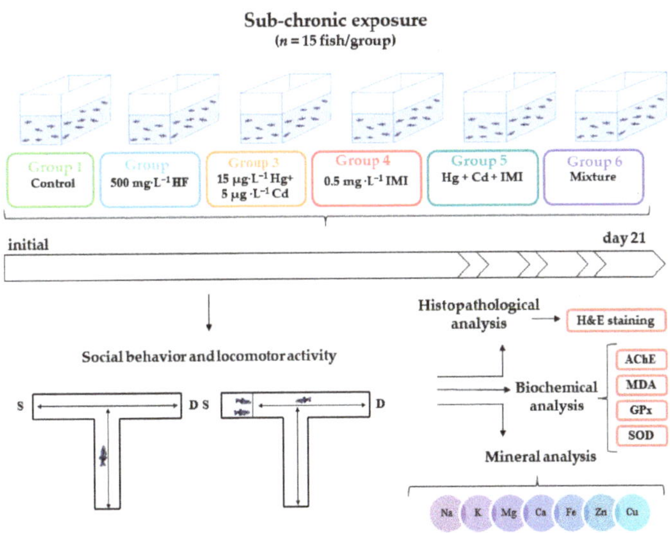

Figure 8. The schematic illustration of the experimental setup (*n* = 15 for each experimental group with two experimental replicates).

3.2.3. Behavioral Assessment

Initially and at the end of the 21 days of exposure, zebrafish from control and exposed groups were submitted to behavioral tests to evaluate animal locomotion and social responses (shoal preference). Both tests were performed in a multipurpose T maze (10 cm height × 50 cm length × 50 cm width) filled with water to a height of 5 cm. The fish were allowed to acclimate to the experimental tank for 30 s, after which behavior was video-recorded over a 4 min period.

The video files were recorded with a professional infrared light camera located above the T maze and analyzed with EthoVision® XT 11.5 software (Noldus Information Technology BV, Wageningen, The Netherlands). The software was previously programmed, calibrated, and optimized for the behavioral tests. The total distance traveled (cm) and mean speed ($cm \cdot s^{-1}$) were considered the main parameters of locomotion, whereas the immobility duration (s), characterized in zebrafish by the delay in the swimming activity, represented a parameter of anxiety-like behavior. For the social interaction test, 4 fish from the same experimental group were placed in a compartment in the left arm closed with a transparent slot. During the 4 min session following 30 s of acclimation, the fish could freely swim from the central zone to the other (left zone—social zone; right zone—non-social zone). The analysis of behavior changes was based on a comparison of responses of exposed zebrafish either to responses of unexposed zebrafish (control group) or to responses measured during the pre-exposure period (baseline behavior assessment).

3.2.4. Biochemical Assays

Following the completion of treatment, the zebrafish brains of each group were dissected and homogenized with a KIMBLE® Dounce tissue grinder (Sigma-Aldrich, Saint Louis, MO, USA) by adding 6-fold ice of 0.05 M Tris-HCl with a pH of 8.0, according to the approach of Richetti et al. (2011) [166]. Then, the supernatant was used for the determination of AChE activity, according to a previously reported protocol [22].

For the determination of SOD, GPx activities, and MDA levels, the fish bodies were homogenized with 10 volumes of 0.1 M ice-cold PBS with a pH of 7.4 and then centrifuged (microcentrifuge Biocen 22 R, ORTO ALRESA, Madrid, Spain) for 15 min at 5500 rpm (4 °C). The enzyme activities and MDA levels were measured using test kits according to the manufacturer's instructions. All measurements were normalized by the determined protein concentration in each sample according to the Bradford method with BSA as a standard.

3.2.5. Mineral Analysis of Zebrafish Whole Body

The concentrations of Cu, Zn, Fe, Ca, Mg, K, and Na were determined following a previously described protocol [22,28,167] and using the same reagents and equipment described in Section 3.1.3. Five fish specimens were selected from each experimental group and subjected to the mineralization process after being washed with ultrapure water.

The validation of the method was performed using a certified reference material (fish muscle, ERM-BB422), which was prepared by applying the same protocol used for experimental samples. The reference material has been certified by the Institute for Reference Materials and Measurements (IRMM) of the European Commission's Joint Research Centre (JRC). All concentrations are reported as $\mu g \cdot g^{-1}$ wet weight.

3.2.6. Histological Analysis and Immunohistochemistry Activity for Brain and Spinal Cord Tissues

After each fish was euthanized using ice-cold water, specimens were fixed in 10% neutral buffered formalin (HT501128, Sigma-Aldrich Chemie GmbH, Taufkirchen, Germany) for 1 h and 48 h in Bouin's solution (HT10132, Sigma-Aldrich Chemie GmbH, Taufkirchen, Germany) and then dehydrated in different ethanol solutions (Ethanol absolute for analysis, EMSURE®, Merck, Darmstadt, Germany), embedded in paraffin, and sectioned at 5 µm thickness using a rotary microtome (Cut 6062, SLEE medical GmbH, Mainz, Germany).

Five microscope slides were prepared from each paraffin block and subjected to staining following the standard hematoxylin and eosin (H&E) staining protocol (Sigma-Aldrich, Saint Louis, MO, USA). Immunohistochemistry (IHC) staining was conducted using a series of specific antibodies as markers. Three key regions were thoroughly discussed and analyzed (optic tectum, cerebellum, and spinal cord) with three markers, proliferating cell nuclear antigen (PCNA), S100 beta (S100B), and glial fibrillary acidic protein (GFAP), respectively. These antibodies included GFAP from Synaptic Systems GmbH (Catalog No. 173002, Göttingen, Germany), S100B from Signalway Antibody (Catalog No. C48942, Greenbelt, MA, USA), and PCNA from GeneTex (catalog No. GTX124496, Irvine, CA, USA). For each fish, three slides were subjected to the following procedure: They were initially dewaxed and then microwaved at 95 °C for 10 min in a citrate buffer, pH 6.0, 10×, Antigen Retriever (C9999, Sigma-Aldrich, Saint Louis, MO, USA). Afterward, the slides were allowed to cool for 20 min and subsequently rinsed twice in PBS (P4417, Sigma-Aldrich, Saint Louis, MO, USA) for 5 min each. Next, they were incubated overnight at 4 °C in a humid chamber with primary antibodies. The primary antibodies were diluted to a ratio of 1:1000 for GFAP, 1:100 for S100B, and 1:250 for PCNA. On the following day, the slides were washed three times with PBS for 5 min each and then incubated with secondary antibodies (Goat Anti-Rabbit HRP, ab205718, Abcam, Cambridge, UK) diluted to the same ratio as primary antibodies. Immunohistochemistry sections were developed using a 3,30-diaminobenzidine Substrate Kit (DAB, ab64238, Abcam, Cambridge, UK) and subsequently counterstained with hematoxylin (Merck KGaA, Darmstadt, Germany). Subsequently, they were examined and photographed using an optical microscope (Leica DM750, Leica Microsystems CMS GmbH, Wetzlar, Germany). In addition, the IHC lesion severity scores were calculated as follows: (−) = absent or rarely observed labeling, no lesions; (+) = mild–low lesions (numbers of cells positively labeled up to 10%), (++) = moderate–medium lesions (numbers of cells positively labeled between 10 and 50%), (+++) = severe, major lesions (more than 50% of cells positively labeled) [168].

3.3. Statistical Analysis

The data are represented by mean ± standard deviation (SD), and the statistical difference was analyzed either by two-way ANOVA followed by a Tukey's HSD post hoc test for multiple comparisons between groups or a *t*-test using a statistical analysis software, GraphPad Prism (GraphPad Prism Software, San Diego, CA, USA), version 10.3.1 for Windows. The SigmaPlot software (Systat Software Inc., Erkrath, Germany), version 15 for Windows, was used for the normality test (Shapiro–Wilk), equal variance test (Brown–Forsythe), and the power of the test, with $\alpha = 0.050$. Moreover, a priori sample size calculation was carried out on various studies [44,169–172] and assuming a one-way between-subjects ANOVA for independent groups to detect a medium to large effect with power exceeding 80% with $\alpha = 0.05$ by using G*Power 3.1.9.7 software (University of Düsseldorf, Germany).

4. Conclusions

In our study, we observed that the combined exposure of zebrafish to Cd, Hg, and IMI resulted in a more pronounced impact than any individual treatment alone. Thus, co-exposure to these two heavy metals and IMI over 21 days led to a significant decline in main locomotion parameters and an increase in immobile duration. Exposure to these contaminants adversely affected the social behavior of the fish, with the most substantial reduction in time spent near social stimuli recorded in the group subjected to the ternary mixture. The presence of heavy metals and a pesticide in the habitat of zebrafish resulted in a noticeable decrease in AChE activity within their brains. This decline was accompanied by increased levels of MDA and heightened activity of antioxidant enzymes, such as SOD and GPx, in the fish's tissues. In addition, the group exposed to the mixture of Cd, Hg, and IMI exhibited excess Ca levels, while elevated Mg levels were noted in the Hg + Cd and mixture groups. In contrast, significant deficiencies in Fe and K were observed in the Hg + Cd and mixture groups. Furthermore, the mixture treatment resulted in degenerative changes in the optic tectum and cerebellum, with relatively milder effects observed in the spinal cord. The results showed that HF co-administration can attenuate the contaminant-induced social and locomotor activity impairments, changes in OS parameters and AChE activity, and brain injury. These supplements hold potential utility within the fish industry for alleviating the adverse effects of environmental contaminants infiltrating aquaculture systems through various pathways. Future research endeavors are warranted to elucidate the precise mechanisms underlying both the toxicological effects induced by concurrent exposure to heavy metals and pesticides and the observed protective effects of health supplements. Furthermore, establishing a dose–response relationship for these supplements remains crucial for advancing our understanding.

Supplementary Materials: The following supporting information can be downloaded at: https://www.mdpi.com/article/10.3390/ijms252111730/s1.

Author Contributions: Conceptualization, T.I., R.J. and E.P.; methodology, T.I., R.J. and D.I.; software, R.J. and D.I.; validation, R.J., C.-M.G., T.I., A.L., C.M. (Cristian Moisa), L.C., C.M. (Claudia Muresan), D.M.C., G.M., F.D.L., C.S., G.-A.D., I.-A.S. and E.P.; formal analysis, R.J., A.L., C.M. (Cristian Moisa), L.C., C.M. (Claudia Muresan), D.M.C., G.M., C.S. and I.-A.S.; investigation R.J. and E.P.; resources, T.I.; data curation, R.J. and E.P.; writing—original draft preparation, R.J., E.P., C.-M.G., T.I., A.L., C.M. (Cristian Moisa), L.C., C.M. (Claudia Muresan), D.M.C., G.M., F.D.L., C.S. and I.-A.S.; writing—review and editing, G.S., C.S., D.I., R.J., E.P. and M.N.; visualization, R.J., E.P., D.I. and M.N.; supervision, T.I., A.C., M.N. and C.-M.G.; project administration, T.I. and A.C.; funding acquisition, T.I. All authors have read and agreed to the published version of the manuscript.

Funding: This work was supported by a grant from the Ministry of Research, Innovation and Digitization, CNCS—UEFISCDI, project number PN-III-P4-PCE-2021-0639, within PNCDI III. SC Synergy Plant Products SRL provided support in the form of materials.

Institutional Review Board Statement: The animal study protocol was approved by the Ethics Committee of the Faculty of Biology, Alexandru Ioan Cuza University of Iasi (2992/05.10.2021).

Informed Consent Statement: Not applicable.

Data Availability Statement: Data are contained within the article or Supplementary Materials.

Conflicts of Interest: The authors declare no conflicts of interest.

References

1. Tomczyk, M.; Miłek, M.; Sidor, E.; Kapusta, I.; Litwińczuk, W.; Puchalski, C.; Dżugan, M. The Effect of Adding the Leaves and Fruits of *Morus alba* to Rape Honey on Its Antioxidant Properties, Polyphenolic Profile, and Amylase Activity. *Molecules* **2019**, *25*, 84. [CrossRef] [PubMed]
2. Dżugan, M.; Sowa, P.; Kwaśniewska, M.; Wesołowska, M.; Czernicka, M. Physicochemical Parameters and Antioxidant Activity of Bee Honey Enriched with Herbs. *Plant Foods Hum. Nutr.* **2017**, *72*, 74–81. [CrossRef] [PubMed]
3. Grabek-Lejko, D.; Miłek, M.; Sidor, E.; Puchalski, C.; Dżugan, M. Antiviral and Antibacterial Effect of Honey Enriched with *Rubus* spp. as a Functional Food with Enhanced Antioxidant Properties. *Molecules* **2022**, *27*, 4859. [CrossRef] [PubMed]
4. Guldas, M.; Gurbuz, O.; Cakmak, I.; Yildiz, E.; Sen, H. Effects of honey enrichment with Spirulina platensis on phenolics, bioaccessibility, antioxidant capacity and fatty acids. *LWT* **2022**, *153*, 112461. [CrossRef]
5. Miłek, M.; Ciszkowicz, E.; Sidor, E.; Hęclik, J.; Lecka-Szlachta, K.; Dżugan, M. The Antioxidant, Antibacterial and Anti-Biofilm Properties of Rapeseed Creamed Honey Enriched with Selected Plant Superfoods. *Antibiotics* **2023**, *12*, 235. [CrossRef]
6. Miłek, M.; Grabek-Lejko, D.; Stępień, K.; Sidor, E.; Mołoń, M.; Dżugan, M. The Enrichment of Honey with *Aronia melanocarpa* Fruits Enhances Its in vitro and in vivo Antioxidant Potential and Intensifies its Antibacterial and Antiviral Properties. *Food Funct.* **2021**, *12*, 8920–8931. [CrossRef]
7. Pereira, C.; Barreira, J.C.M.; Calhelha, R.C.; Lopes, M.; Queiroz, M.J.R.P.; Vilas-Boas, M.; Barros, L.; Ferreira, I.C.F.R. Is Honey Able to Potentiate the Antioxidant and Cytotoxic Properties of Medicinal Plants Consumed as Infusions for Hepatoprotective Effects? *Food Funct.* **2015**, *6*, 1435–1442. [CrossRef]
8. Sowa, P.; Tarapatskyy, M.; Puchalski, C.; Jarecki, W.; Dżugan, M. A Novel Honey-Based Product Enriched with Coumarin from *Melilotus* flowers. *J. Food Meas. Charact.* **2019**, *13*, 1748–1754. [CrossRef]
9. Żebracka, A.; Winiarska-Mieczan, A.; Nowakowicz-Dębek, B.; Banach, M.; Drabik, A.; Pulit-Prociak, J.; Chmielowiec-Korzeniowska, A. Assessment of the Microbiological Quality and Bactericidal Properties of Flavoured Honey. *Med. Weter.* **2022**, *78*, 556–562. [CrossRef]

10. Hussien, H.M.; Abdou, H.M.; Yousef, M.I. Cypermethrin Induced Damage in Genomic DNA and Histopathological Changes in Brain and Haematotoxicity in Rats: The Protective Effect of Sesame Oil. *Brain Res. Bull.* **2013**, *92*, 76–83. [CrossRef]
11. Tunali-Akbay, T.; Sener, G.; Salvarli, H.; Sehirli, O.; Yarat, A. Protective Effects of *Ginkgo biloba* Extract Against Mercury(II)-induced Cardiovascular Oxidative Damage in Rats. *Phytother. Res.* **2007**, *21*, 26–31. [CrossRef] [PubMed]
12. Banaee, M.; Impellitteri, F.; Samani, H.E.-Z.; Piccione, G.; Faggio, C. Dietary *Arthrospira platensis* in Rainbow Trout (*Oncorhynchus mykiss*): A Means to Reduce Threats Caused by $CdCl_2$ Exposure? *Toxics* **2022**, *10*, 731. [CrossRef] [PubMed]
13. El-Demerdash, F.M.; Jebur, A.B.; Nasr, H.M.; Hamid, H.M. Modulatory Effect of *Turnera diffusa* Against Testicular Toxicity Induced by Fenitrothion and/or Hexavalent Chromium in Rats. *Environ. Toxicol.* **2019**, *34*, 330–339. [CrossRef] [PubMed]
14. González, S.; López-Roldán, R.; Cortina, J.-L. Presence and Biological Effects of Emerging Contaminants in Llobregat River Basin: A Review. *Environ. Pollut.* **2012**, *161*, 83–92. [CrossRef]
15. Machado, C.S.; Fregonesi, B.M.; Alves, R.I.S.; Tonani, K.A.A.; Sierra, J.; Martinis, B.S.; Celere, B.S.; Mari, M.; Schuhmacher, M.; Nadal, M.; et al. Health Risks of Environmental Exposure to Metals and Herbicides in the Pardo River, Brazil. *Environ. Sci. Pollut. Res.* **2017**, *24*, 20160–20172. [CrossRef]
16. Salvadó, V.; Quintana, X.D.; Hidalgo, M. Monitoring of Nutrients, Pesticides, and Metals in Waters, Sediments, and Fish of a Wetland. *Arch. Environ. Contam. Toxicol.* **2006**, *51*, 377–386. [CrossRef]
17. Olutona, G.O.; Livingstone, S.T. Detection of Organochlorine Pesticide (OCPs) Residues and Trace Metals in Some Selected Malt Drinks in Nigeria. *Beverages* **2018**, *4*, 65. [CrossRef]
18. Marković, M.; Cupać, S.; Durović, R.; Milinović, J.; Kljajić, P. Assessment of Heavy Metal and Pesticide Levels in Soil and Plant Products From Agricultural Area of Belgrade, Serbia. *Arch. Environ. Contam. Toxicol.* **2010**, *58*, 341–351. [CrossRef]
19. He, W.; Guo, W.; Qian, Y.; Zhang, S.; Ren, D.; Liu, S. Synergistic Hepatotoxicity by Cadmium and Chlorpyrifos: Disordered Hepatic Lipid Homeostasis. *Mol. Med. Rep.* **2015**, *12*, 303–308. [CrossRef]
20. Chen, L.; Qu, G.; Sun, X.; Zhang, S.; Wang, L.; Sang, N.; Du, Y.; Liu, J.; Liu, S. Characterization of the Interaction between Cadmium and Chlorpyrifos with Integrative Techniques in Incurring Synergistic Hepatoxicity. *PLoS ONE* **2013**, *8*, e59553. [CrossRef]
21. Jijie, R.; Solcan, G.; Nicoara, M.N.; Micu, D.; Strungaru, S.-A. Antagonistic effects in zebrafish (*Danio rerio*) behavior and oxidative stress induced by toxic metals and deltamethrin acute exposure. *Sci. Total Environ.* **2019**, *698*, 134299. [CrossRef] [PubMed]
22. Paduraru, E.; Flocea, E.I.; Lazado, C.C.; Simionov, I.A.; Nicoara, M.; Ciobica, A.; Faggio, C.; Jijie, R. Vitamin C Mitigates Oxidative Stress and Behavioral Impairments Induced by Deltamethrin and Lead Toxicity in Zebrafish. *Int. J. Mol. Sci.* **2021**, *22*, 12714. [CrossRef] [PubMed]
23. Ku, T.; Yan, W.; Jia, W.; Yun, Y.; Zhu, N.; Li, G.; Sang, N. Characterization of Synergistic Embryotoxicity of Nickel and Buprofezin in Zebrafish. *Environ. Sci. Technol.* **2015**, *49*, 4600–4608. [CrossRef]
24. Lajmanovich, R.C.; Peltzer, P.M.; Attademo, A.M.; Martinuzzi, C.S.; Simoniello, M.F.; Colussi, C.L.; Boccioni, A.P.C.; Sigrist, M. First Evaluation of Novel Potential Synergistic Effects of Glyphosate and Arsenic Mixture on *Rhinella arenarum* (Anura: Bufonidae) Tadpoles. *Heliyon* **2019**, *5*, e02601. [CrossRef]
25. Liu, J.; Sun, L.; Zhang, H.; Shi, M.; Dahlgren, R.A.; Wang, X.; Wang, H. Response Mechanisms to Joint Exposure of Triclosan and Its Chlorinated Derivatives on Zebrafish (*Danio rerio*) Behavior. *Chemosphere* **2018**, *193*, 820–832. [CrossRef] [PubMed]
26. Yang, Y.; Yu, Y.; Zhou, R.; Yang, Y.; Bu, Y. The Effect of Combined Exposure of Zinc and Nickel on the Development of Zebrafish. *J. Appl. Toxicol.* **2021**, *41*, 1765–1778. [CrossRef] [PubMed]
27. Lu, K.; Qiao, R.; An, H.; Zhang, Y. Influence of Microplastics on the Accumulation and Chronic Toxic Effects of Cadmium in Zebrafish (*Danio rerio*). *Chemosphere* **2018**, *202*, 514–520. [CrossRef]
28. Jijie, R.; Paduraru, E.; Simionov, I.-A.; Faggio, C.; Ciobica, A.; Nicoara, M. Effects of Single and Combined Ciprofloxacin and Lead Treatments on Zebrafish Behavior, Oxidative Stress, and Elements Content. *Int. J. Mol. Sci.* **2023**, *24*, 4952. [CrossRef]
29. Di Paola, D.; Iaria, C.; Capparucci, F.; Cordaro, M.; Crupi, R.; Siracusa, R.; D'Amico, R.; Fusco, R.; Impellizzeri, D.; Cuzzocrea, S.; et al. Aflatoxin B1 Toxicity in Zebrafish Larva (Danio rerio): Protective Role of Hericium erinaceus. *Toxins* **2021**, *13*, 710. [CrossRef]
30. Costa, B.P.D.; Moura, L.A.; Pinto, S.A.G.; Lima-Maximino, M.G.; Maximino, C. Zebrafish Models in Neural and Behavioral Toxicology across the Life Stages. *Fishes* **2020**, *5*, 23. [CrossRef]
31. Xia, Y.; Zhu, J.; Xu, Y.; Zhang, H.; Zou, F.; Meng, X. Effects of Ecologically Relevant Concentrations of Cadmium on Locomotor Activity and Microbiota in Zebrafish. *Chemosphere* **2020**, *257*, 127220. [CrossRef] [PubMed]
32. Tian, J.; Hu, J.; He, W.; Zhou, L.; Huang, Y. Parental Exposure to Cadmium Chloride Causes Developmental Toxicity and Thyroid Endocrine Disruption in Zebrafish Offspring. *Comp. Biochem. Physiol. C Toxicol. Pharmacol.* **2020**, *234*, 108782. [CrossRef] [PubMed]
33. Patel, U.N.; Patel, U.D.; Khadayata, A.V.; Vaja, R.K.; Modi, C.M.; Patel, H.B. Long-term Exposure of the Binary Mixture of Cadmium and Mercury Damages the Developed Ovary of Adult Zebrafish. *Environ. Sci. Pollut. Res.* **2022**, *29*, 44928–44938. [CrossRef] [PubMed]
34. Li, P.; Li, Z.-H. Environmental Co-exposure to TBT and Cd Caused Neurotoxicity and Thyroid Endocrine Disruption in Zebrafish, a Three-generation Study in a Simulated Environment. *Environ. Pollut.* **2020**, *259*, 113868. [CrossRef]
35. Lacave, J.M.; Bilbao, E.; Gilliland, D.; Mura, F.; Dini, L.; Cajaraville, M.P.; Orbea, A. Bioaccumulation, Cellular and Molecular Effects in Adult Zebrafish After Exposure to Cadmium Sulphide Nanoparticles and to Ionic Cadmium. *Chemosphere* **2020**, *238*, 124588. [CrossRef]

36. Abu Bakar, N.; Sata, N.S.M.; Ramlan, N.F.; Ibrahim, W.N.W.; Zulkifli, S.Z.; Abdullah, C.A.C.; Ahmad, S.; Amal, M.N. Evaluation of the Neurotoxic Effects of Chronic Embryonic Exposure with Inorganic Mercury on Motor and Anxiety-like Responses in Zebrafish (*Danio rerio*) Larvae. *Neurotoxicol. Teratol.* **2017**, *59*, 53–61. [CrossRef]
37. Strungaru, S.A.; Robea, M.A.; Plavan, G.; Todirascu-Ciornea, E.; Ciobica, A.; Nicoara, M. Acute Exposure to Methylmercury Chloride Induces Fast Changes in Swimming Performance, Cognitive Processes and Oxidative Stress of Zebrafish (*Danio rerio*) as Reference Model for Fish Community. *J. Trace Elem. Med. Biol.* **2018**, *47*, 115–123. [CrossRef]
38. Macirella, R.; Brunelli, E. Morphofunctional Alterations in Zebrafish (*Danio rerio*) Gills after Exposure to Mercury Chloride. *Int. J. Mol. Sci.* **2017**, *18*, 824. [CrossRef] [PubMed]
39. Xie, D.; Chen, Q.; Gong, S.; An, J.; Li, Y.; Lian, X.; Liu, Z.; Shen, Y.; Giesy, J.P. Exposure of Zebrafish to Environmentally Relevant Concentrations of Mercury During Early Life Stages Impairs Subsequent Reproduction in Adults but can be Recovered in Offspring. *Aquat. Toxicol.* **2020**, *229*, 105655. [CrossRef]
40. Sun, Y.; Li, Y.; Liu, Z.; Chen, Q. Environmentally Relevant Concentrations of Mercury Exposure Alter Thyroid Hormone Levels and Gene Expression in the Hypothalamic-pituitary-thyroid Axis of Zebrafish Larvae. *Fish Physiol. Biochem.* **2018**, *44*, 1175–1183. [CrossRef]
41. Chung, K.-T.; Chen, L.-W.; Tseng, H.-W.; Wu, C.-H. Neonicotinoid Imidacloprid Affects the Social Behavior of Adult Zebrafish by Damaging Telencephalon Neurons through Oxidation Stress, Inflammation, and Apoptosis. *Life* **2023**, *13*, 1418. [CrossRef] [PubMed]
42. Luo, T.; Wang, X.; Jin, Y. Low Concentrations of Imidacloprid Exposure Induced Gut Toxicity in Adult Zebrafish (*Danio rerio*). *Comp. Biochem. Physiol. C Toxicol. Pharmacol.* **2021**, *241*, 108972. [CrossRef] [PubMed]
43. Luo, T.; Weng, Y.; Huang, Z.; Zhao, Y.; Jin, Y. Combined Hepatotoxicity of Imidacloprid and Microplastics in Adult Zebrafish: Endpoints at Gene Transcription. *Comp. Biochem. Physiol. C Toxicol. Pharmacol.* **2021**, *246*, 109043. [CrossRef] [PubMed]
44. Hou, J.; Zhang, L.; Xu, W.; Liu, Z.; Yu, J.; Yu, R.; Chen, L. Glycometabolic Disorder Induced by Chronic Exposure to Low-concentration Imidacloprid in Zebrafish. *Sci. Total Environ.* **2024**, *937*, 173421. [CrossRef] [PubMed]
45. Alvarez-Suarez, J.M.; Giampieri, F.; Brenciani, A.; Mazzoni, L.; Gasparrini, M.; González-Paramás, A.M.; Santos-Buelga, C.; Morroni, G.; Simoni, S.; Forbes-Hernández, T.Y.; et al. *Apis mellifera* vs *Melipona beecheii* Cuban Polifloral Honeys: A Comparison Based on Their Physicochemical Parameters, Chemical Composition and Biological Properties. *LWT* **2018**, *87*, 272–279. [CrossRef]
46. *Codex Stan 12-1981*; Revised Codex Standard for Honey Codex Stan 12-1981, Rev. 1 (1987), Rev. (2001). FAO: Rome, Italy, 2022.
47. The European Parliament and The Council of The European Union. European Parliament Directive 2014/63/EU of The European Parliament and of the Council of 15 May 2014 amending Council Directive 2001/110/EC relating to honey. *Off. J. Eur. Union* **2014**, *L164*, 1–5.
48. Sakač, M.B.; Jovanov, P.T.; Marić, A.Z.; Pezo, L.L.; Kevrešan, Ž.S.; Novaković, A.R.; Nedeljković, N.M. Physicochemical Properties and Mineral Content of Honey Samples from Vojvodina (Republic of Serbia). *Food Chem.* **2019**, *276*, 15–21. [CrossRef]
49. Shamsudin, S.; Selamat, J.; Sanny, M.; Abdul Razak, S.B.; Jambari, N.N.; Khatib, A. A Comparative Characterization of Physicochemical and Antioxidants Properties of Processed *Heterotrigona itama* Honey from Different Origins and Classification by Chemometrics Analysis. *Molecules* **2019**, *24*, 3898. [CrossRef]
50. Albu, A.; Radu-Rusu, C.-G.; Pop, I.M.; Frunza, G.; Nacu, G. Quality Assessment of Raw Honey Issued from Eastern Romania. *Agriculture* **2021**, *11*, 247. [CrossRef]
51. da Silva, P.M.; Gauche, C.; Gonzaga, L.V.; Costa, A.C.O.; Fett, R. Honey: Chemical Composition, Stability and Authenticity. *Food Chem.* **2016**, *196*, 309–323. [CrossRef]
52. Živkov Baloš, M.; Popov, N.; Jakšić, S.; Mihaljev, Ž.; Pelić, M.; Ratajac, R.; Ljubojević Pelić, D. Sunflower Honey—Evaluation of Quality and Stability during Storage. *Foods* **2023**, *12*, 2585. [CrossRef] [PubMed]
53. Can, Z.; Yildiz, O.; Sahin, H.; Akyuz Turumtay, E.; Silici, S.; Kolayli, S. An investigation of Turkish honeys: Their physico-chemical properties, antioxidant capacities and phenolic profiles. *Food Chem.* **2015**, *180*, 133–141. [CrossRef] [PubMed]
54. Nascimento, K.S.d.; Gasparotto Sattler, J.A.; Lauer Macedo, L.F.; Serna González, C.V.; Pereira de Melo, I.L.; da Silva Araújo, E.; Granato, D.; Sattler, A.; de Almeida-Muradian, L.B. Phenolic Compounds, Antioxidant Capacity and Physicochemical Properties of Brazilian *Apis mellifera* Honeys. *LWT* **2018**, *91*, 85–94. [CrossRef]
55. Kunat-Budzyńska, M.; Rysiak, A.; Wiater, A.; Grąz, M.; Andrejko, M.; Budzyński, M.; Bryś, M.S.; Sudziński, M.; Tomczyk, M.; Gancarz, M.; et al. Chemical Composition and Antimicrobial Activity of New Honey Varietals. *Int. J. Environ. Res. Public Health* **2023**, *20*, 2458. [CrossRef]
56. Pătruică, S.; Alexa, E.; Obiștioiu, D.; Cocan, I.; Radulov, I.; Berbecea, A.; Lazăr, R.N.; Simiz, E.; Vicar, N.M.; Hulea, A.; et al. Chemical Composition, Antioxidant and Antimicrobial Activity of Some Types of Honey from Banat Region, Romania. *Molecules* **2022**, *27*, 4179. [CrossRef]
57. Boussaid, A.; Chouaibi, M.; Rezig, L.; Hellal, R.; Donsì, F.; Ferrari, G.; Hamdi, S. Physicochemical and Bioactive Properties of Six Honey Samples From Various Floral Origins From Tunisia. *Arab. J. Chem.* **2018**, *11*, 265–274. [CrossRef]
58. Conti, M.E. Lazio Region (Central Italy) Honeys: A Survey of Mineral Content and Typical Quality Parameters. *Food Control* **2000**, *11*, 459–463. [CrossRef]
59. Wilczyńska, A.; Newerli-Guz, J.; Szweda, P. Influence of the Addition of Selected Spices on Sensory Quality and Biological Activity of Honey. *J. Food Qual.* **2017**, *2017*, 6963904. [CrossRef]

60. Đorđević, S.; Nedić, N.; Pavlović, A.; Milojković-Opsenica, D.; Tešić, Ž.; Gašić, U. Honey with Added Value—Enriched with Rutin and Quercetin from *Sophora* Flower. *J. Herb. Med.* **2022**, *34*, 100580. [CrossRef]
61. Alzahrani, H.A.; Alsabehi, R.; Boukraâ, L.; Abdellah, F.; Bellik, Y.; Bakhotmah, B.A. Antibacterial and Antioxidant Potency of Floral Honeys from Different Botanical and Geographical Origins. *Molecules* **2012**, *17*, 100540–100549. [CrossRef]
62. Weston, R.J.; Mitchell, K.R.; Allen, K.L. Antibacterial Phenolic Components of New Zealand Manuka Honey. *Food Chem.* **1999**, *64*, 295–301. [CrossRef]
63. Cheung, Y.; Meenu, M.; Yu, X.; Xu, B. Phenolic Acids and Flavonoids Profiles of Commercial Honey from Different Floral Sources and Geographic Sources. *Int. J. Food Prop.* **2019**, *22*, 290–308. [CrossRef]
64. Mandal, M.D.; Mandal, S. Honey: Its Medicinal Property and Antibacterial Activity. *Asian Pac. J. Trop. Biomed.* **2011**, *1*, 154–160. [CrossRef] [PubMed]
65. Albaridi, N.A. Antibacterial Potency of Honey. *Int. J. Microbiol.* **2019**, *2019*, 2464507. [CrossRef] [PubMed]
66. Almasaudi, S. The Antibacterial Activities of Honey. *Saudi J. Biol. Sci.* **2021**, *28*, 2188–2196. [CrossRef]
67. Bui Thi, N.H.; Nguyen Thi, N.A.; Audira, G.; Siregar, P.; Liang, S.T.; Huang, J.C.; Hsiao, C.D. Chronic Exposure to Low Concentration Lead Chloride-Induced Anxiety and Loss of Aggression and Memory in Zebrafish. *Int. J. Mol. Sci.* **2020**, *21*, 1844. [CrossRef]
68. Hu, S.; Han, J.; Yang, L.; Li, S.; Guo, Y.; Zhou, B.; Wu, H.; Hu, S.; Han, J.; Yang, L.; et al. Impact of Co-Exposure to Titanium Dioxide Nanoparticles and Pb on Zebrafish Embryos. *Chemosphere* **2019**, *233*, 579–589. [CrossRef]
69. Zhang, W.; Fan, R.; Luo, S.; Ying, l.; Jin, Y.; Li, Y.; Xiong, M.; Yuan, X.; Jia, L.; Chen, Y. Antagonistic Effects and Mechanisms of Carbendazim and Chlorpyrifos on the Neurobehavior of Larval Zebrafish. *Chemosphere* **2022**, *293*, 133522. [CrossRef]
70. Guerra, L.J.; do Amaral, A.M.B.; de Quadros, V.A.; da Luz Fiuza, T.; Rosemberg, D.B.; Prestes, O.D.; Zanella, R.; Clasen, B.; Loro, V.L. Biochemical and Behavioral Responses in Zebrafish Exposed to Imidacloprid Oxidative Damage and Antioxidant Responses. *Arch. Environ. Contam. Toxicol.* **2021**, *81*, 255–264. [CrossRef]
71. Faria, M.; Prats, E.; Gómez-Canela, C.; Hsu, C.-Y.; Arick, M.A.; Bedrossiantz, J.; Orozco, M.; Garcia-Reyero, N.; Ziv, T.; Ben-Lulu, S.; et al. Therapeutic Potential of N-acetylcysteine in Acrylamide Acute Neurotoxicity in Adult Zebrafish. *Sci. Rep.* **2019**, *9*, 16467. [CrossRef]
72. Abdulmajeed, W.I.; Sulieman, H.B.; Zubayr, M.O.; Imam, A.; Amin, A.; Biliaminu, S.A.; Oyewole, L.A.; Owoyele, B.V. Honey Prevents Neurobehavioural Deficit and Oxidative Stress Induced by Lead Acetate Exposure in Male Wistar Rats—A Preliminary Study. *Metab. Brain Dis.* **2016**, *31*, 37–44. [CrossRef] [PubMed]
73. Robea, M.A.; Jijie, R.; Nicoara, M.; Plavan, G.; Ciobica, A.S.; Solcan, C.; Audira, G.; Hsiao, C.-D.; Strungaru, S.-A. Vitamin C Attenuates Oxidative Stress and Behavioral Abnormalities Triggered by Fipronil and Pyriproxyfen Insecticide Chronic Exposure on Zebrafish Juvenile. *Antioxidants* **2020**, *9*, 944. [CrossRef] [PubMed]
74. Robea, M.A.; Ilie, O.D.; Nicoara, M.N.; Solcan, G.; Romila, L.E.; Ureche, D.; Ciobica, A. Vitamin B12 Ameliorates Pesticide-Induced Sociability Impairment in Zebrafish (*Danio rerio*): A Prospective Controlled Intervention Study. *Animals* **2024**, *14*, 405. [CrossRef] [PubMed]
75. Zimmermann, F.F.; Gaspary, K.V.; Leite, C.E.; De Paula Cognato, G.; Bonan, C.D. Embryological Exposure to Valproic Acid Induces Social Interaction Deficits in Zebrafish (*Danio rerio*): A Developmental Behavior Analysis. *Neurotoxicol. Teratol.* **2015**, *52*, 36–41. [CrossRef]
76. Robea, M.A.; Ciobica, A.; Curpan, A.S.; Plavan, G.; Strungaru, S.; Lefter, R.; Nicoara, M. Preliminary Results Regarding Sleep in a Zebrafish Model of Autism Spectrum Disorder. *Brain Sci.* **2021**, *11*, 556. [CrossRef]
77. Chen, J.; Tanguay, R.L.; Simonich, M.; Nie, S.; Zhao, Y.; Li, L.; Bai, C.; Dong, Q.; Huang, C.; Lin, K. TBBPA Chronic Exposure Produces Sex-specific Neurobehavioral and Social Interaction Changes in Adult Zebrafish. *Neurotoxicol. Teratol.* **2016**, *56*, 9–15. [CrossRef]
78. Fu, C.-W.; Horng, J.-L.; Tong, S.-K.; Cherng, B.-W.; Liao, B.-K.; Lin, L.-Y.; Chou, M.-Y. Exposure to Silver impairs Learning and Social Behaviors in Adult Zebrafish. *J. Hazard. Mater.* **2021**, *403*, 124031. [CrossRef]
79. Naderi, M.; Puar, P.; JavadiEsfahani, R.; Kwong, R.W.M. Early Developmental Exposure to Bisphenol A and Bisphenol S Disrupts Socio-cognitive Function, Isotocin Equilibrium, and Excitation-Inhibition Balance in Developing Zebrafish. *Neurotoxicology* **2022**, *88*, 144–154. [CrossRef]
80. Hong, X.; Zhao, G.; Zhou, Y.; Chen, R.; Li, J.; Zha, J. Risks to Aquatic Environments Posed by 14 Pharmaceuticals as Illustrated by Their Effects on Zebrafish Behaviour. *Sci. Total Environ.* **2021**, *771*, 145450. [CrossRef]
81. Karri, V.; Schuhmacher, M.; Kumar, V. Heavy Metals (Pb, Cd, As and MeHg) as Risk Factors for Cognitive Dysfunction: A General Review of Metal Mixture Mechanism in Brain. *Environ. Toxicol. Pharmacol.* **2016**, *48*, 203–213. [CrossRef]
82. Erejuwa, O.O.; Sulaiman, S.A.; Ab Wahab, M.S. Honey: A Novel Antioxidant. *Molecules* **2012**, *17*, 4400–4423. [CrossRef]
83. Ahmed, S.; Sulaiman, S.A.; Baig, A.A.; Ibrahim, M.; Liaqat, S.; Fatima, S.; Jabeen, S.; Shamim, N.; Othman, N.H. Honey as a Potential Natural Antioxidant Medicine: An Insight into Its Molecular Mechanisms of Action. *Oxidative Med. Cell. Longev.* **2018**, *2018*, 8367846. [CrossRef] [PubMed]
84. Ranneh, Y.; Akim, A.M.; Hamid, H.A.; Khazaai, H.; Fadel, A.; Zakaria, Z.A.; Albujja, M.; Bakar, M.F.A. Honey and Its Nutritional and Anti-inflammatory Value. *BMC Complement. Med. Ther.* **2021**, *21*, 30. [CrossRef] [PubMed]
85. Khalil, M.I.; Sulaiman, S.A. The Potential Role of Honey and Its Polyphenols in Preventing Heart Diseases: A Review. *Afr. J. Tradit. Complement. Altern. Med.* **2010**, *7*, 315–321. [CrossRef]

86. Olas, B. Honey and Its Phenolic Compounds as an Effective Natural Medicine for Cardiovascular Diseases in Humans? *Nutrients* **2020**, *12*, 283. [CrossRef]
87. Bt Hj Idrus, R.; Sainik, N.; Nordin, A.; Saim, A.B.; Sulaiman, N. Cardioprotective Effects of Honey and Its Constituent: An Evidence-Based Review of Laboratory Studies and Clinical Trials. *Int. J. Environ. Res. Public Health* **2020**, *17*, 3613. [CrossRef]
88. Kumar, S.; Verma, M.; Hajam, Y.A.; Kumar, R. Honey Infused with Herbs: A Boon to Cure Pathological Diseases. *Heliyon* **2024**, *10*, e23302. [CrossRef]
89. Cruz, F.F.; Leite, C.E.; Pereira, T.C.; Bogo, M.R.; Bonan, C.D.; Battastini, A.M.; Campos, M.M.; Morrone, F.B. Assessment of Mercury Chloride-induced Toxicity and the Relevance of P2X7 Receptor Activation in Zebrafish Larvae. *Comp. Biochem. Physiol. C Toxicol. Pharmacol.* **2013**, *158*, 159–164. [CrossRef] [PubMed]
90. Jin, Y.; Liu, Z.; Liu, F.; Ye, Y.; Peng, T.; Fu, Z. Embryonic Exposure to Cadmium (II) and Chromium (VI) Induce Behavioral Alterations, Oxidative Stress and Immunotoxicity in Zebrafish (*Danio rerio*). *Neurotoxicol. Teratol.* **2015**, *48*, 9–17. [CrossRef]
91. Ge, W.; Yan, S.; Wang, J.; Zhu, L.; Chen, A.; Wang, J. Oxidative Stress and DNA Damage Induced by Imidacloprid in Zebrafish (*Danio rerio*). *J. Agric. Food Chem.* **2015**, *63*, 1856–1862. [CrossRef]
92. Banni, M.; Chouchene, L.; Said, K.; Kerkeni, A.; Messaoudi, I. Mechanisms Underlying the Protective Effect of Zinc and Selenium Against Cadmium-induced Oxidative Stress in Zebrafish *Danio rerio*. *Biometals* **2011**, *24*, 981–992. [CrossRef] [PubMed]
93. Green, A.J.; Planchart, A. The Neurological Toxicity of Heavy Metals: A Fish Perspective. *Comp. Biochem. Physiol. C Toxicol. Pharmacol.* **2018**, *208*, 12–19. [CrossRef] [PubMed]
94. Zhang, J.G.; Ma, D.D.; Xiong, Q.; Qiu, S.Q.; Huang, G.Y.; Shi, W.J.; Ying, G.G. Imidacloprid and Thiamethoxam Affect Synaptic Transmission in Zebrafish. *Ecotoxicol. Environ. Saf.* **2021**, *227*, 112917. [CrossRef] [PubMed]
95. Capriello, T.; Grimaldi, M.C.; Cofone, R.; D'Aniello, S.; Ferrandino, I. Effects of Aluminium and Cadmium on Hatching and Swimming Ability in Developing Zebrafish. *Chemosphere* **2019**, *222*, 243–249. [CrossRef]
96. Rosales-Pérez, K.E.; Elizalde-Velázquez, G.A.; Gómez-Oliván, L.M.; Orozco-Hernández, J.M.; Cardoso-Vera, J.D.; Heredia-García, G.; Islas-Flores, H.; García-Medina, S.; Galar-Martínez, M. Brain Damage Induced by Contaminants Released in a Hospital from Mexico: Evaluation of Swimming Behavior, Oxidative Stress, and Acetylcholinesterase in Zebrafish (*Danio rerio*). *Chemosphere* **2022**, *294*, 133791. [CrossRef]
97. Azman, K.F.; Zakaria, R.; AbdAziz, C.; Othman, Z.; Al-Rahbi, B. Tualang honey improves memory performance and decreases depressive-like behavior in rats exposed to loud noise stress. *Noise Health* **2015**, *17*, 83–89. [CrossRef]
98. Arabmoazzen, S.; Sarkaki, A.; Saki, G.; Mirshekar, M.A. Antidiabetic Effect of Honey Feeding in Noise Induced Hyperglycemic Rat: Involvement of Oxidative Stress. *Iran. J. Basic Med. Sci.* **2015**, *18*, 745–751.
99. Qaid, E.Y.A.; Zakaria, R.; Mohd Yusof, N.A.; Sulaiman, S.F.; Shafin, N.; Othman, Z.; Ahmad, A.H.; Abd Aziz, C.B.; Muthuraju, S. Tualang Honey Ameliorates Hypoxia-induced Memory Deficits by Reducing Neuronal Damage in the Hippocampus of Adult Male Sprague Dawley Rats. *Turk. J. Pharm. Sci.* **2020**, *17*, 555–564. [CrossRef]
100. Merciai, R.; Guasch, H.; Kumar, A.; Sabater, S.; García-Berthou, E. Trace Metal Concentration and Fish Size: Variation Among Fish Species in a Mediterranean River. *Ecotoxicol. Environ. Saf.* **2014**, *107*, 154–161. [CrossRef]
101. Solayman, M.; Islam, M.A.; Paul, S.; Ali, Y.; Khalil, M.I.; Alam, N.; Gan, S.H. Physicochemical Properties, Minerals, Trace Elements, and Heavy Metals in Honey of Different Origins: A Comprehensive Review. *Compr. Rev. Food Sci. Food Saf.* **2016**, *15*, 219–233. [CrossRef]
102. Hassan, A.T.; Kwong, R.W.M. The Neurophysiological Effects of Iron in Early Life Stages of Zebrafish. *Environ. Pollut.* **2020**, *267*, 115625. [CrossRef] [PubMed]
103. Li, S.; Cao, C.; Shi, H.; Yang, S.; Qi, L.; Zhao, X.; Sun, C. Effect of Quercetin Against Mixture of Four Organophosphate pesticides Induced Nephrotoxicity in Rats. *Xenobiotica* **2016**, *46*, 225–233. [CrossRef] [PubMed]
104. Dale, K.; Rasinger, J.D.; Thorstensen, K.L.; Penglase, S.; Ellingsen, S. Vitamin E Reduces Endosulfan-induced Toxic Effects on Morphology and Behavior in Early Development of Zebrafish (*Danio rerio*). *Food Chem. Toxicol.* **2017**, *101*, 84–93. [CrossRef] [PubMed]
105. Zeng, X.; Du, Z.; Ding, X.; Jiang, W. Protective Effects of Dietary Flavonoids Against pesticide-induced Toxicity: A Review. *Trends Food Sci. Technol.* **2021**, *109*, 271–279. [CrossRef]
106. Priyadarsini, K.I. The Chemistry of Curcumin: From Extraction to Therapeutic Agent. *Molecules* **2014**, *19*, 20091–20112. [CrossRef]
107. Pivovarov, A.S.; Calahorro, F.; Walker, R.J. Na(+)/K(+)-pump and Neurotransmitter Membrane Receptors. *Invertebr. Neurosci.* **2018**, *19*, 1. [CrossRef]
108. Glasdam, S.M.; Glasdam, S.; Peters, G.H. The Importance of Magnesium in the Human Body: A Systematic Literature Review. *Adv. Clin. Chem.* **2016**, *73*, 169–193. [CrossRef]
109. Araujo, G.F.; Soares, L.O.S.; Junior, S.F.S.; Barreto de Carvalho, L.V.; Rocha, R.C.C.; Saint'Pierre, T.; Hauser-Davis, R.A.; Correia, F.V.; Saggioro, E.M. Oxidative Stress and Metal Homeostasis Alterations in *Danio rerio* (zebrafish) under Single and Combined Carbamazepine, Acetamiprid and Cadmium Exposures. *Aquat. Toxicol.* **2022**, *245*, 106122. [CrossRef]
110. Alsop, D.; Wood, C.M. Metal Uptake and Acute Toxicity in Zebrafish: Common Mechanisms Across Multiple Metals. *Aquat. Toxicol.* **2011**, *105*, 385–393. [CrossRef]
111. Shaw, P.; Mondal, P.; Dey Bhowmik, A.; Bandyopadhyay, A.; Sudarshan, M.; Chakraborty, A.; Chattopadhyay, A. Environmentally Relevant Hexavalent Chromium Disrupts Elemental Homeostasis and Induces Apoptosis in Zebrafish Liver. *Bull. Environ. Contam. Toxicol.* **2022**, *108*, 716–724. [CrossRef]

112. Siblerud, R.; Mutter, J.; Moore, E.; Naumann, J.; Walach, H. A Hypothesis and Evidence That Mercury May be an Etiological Factor in Alzheimer's Disease. *Int. J. Environ. Res. Public Health* **2019**, *16*, 5152. [CrossRef] [PubMed]
113. Devarapogu, R.; Asupatri, U.R. Effects of Zinc Supplementation in Mitigating the Harmful Effects of Chronic Cadmium Exposure in a Zebrafish Model. *Environ. Toxicol. Pharmacol.* **2023**, *100*, 104158. [CrossRef] [PubMed]
114. Hol, E.M.; Pekny, M. Glial Fibrillary Acidic Protein (GFAP) and the Astrocyte Intermediate Filament System in Diseases of the Central Nervous System. *Curr. Opin. Cell Biol.* **2015**, *32*, 121–130. [CrossRef] [PubMed]
115. Brenner, M.; Kisseberth, W.C.; Su, Y.; Besnard, F.; Messing, A. GFAP Promoter Directs Astrocyte-specific Expression in Transgenic Mice. *J. Neurosci.* **1994**, *14*, 1030–1037. [CrossRef] [PubMed]
116. Escartin, C.; Galea, E.; Lakatos, A.; O'Callaghan, J.P.; Petzold, G.C.; Serrano-Pozo, A.; Steinhäuser, C.; Volterra, A.; Carmignoto, G.; Agarwal, A.; et al. Reactive Astrocyte Nomenclature, Definitions, and Future Directions. *Nat. Neurosci.* **2021**, *24*, 312–325. [CrossRef]
117. Ben Haim, L.; Rowitch, D.H. Functional Diversity of Astrocytes in Neural Circuit Regulation. *Nat. Rev. Neurosci.* **2017**, *18*, 31–41. [CrossRef]
118. Gradisnik, L.; Velnar, T. Astrocytes in the Central Nervous System and their Functions in Health and Disease: A Review. *World J. Clin. Cases* **2023**, *11*, 3385–3394. [CrossRef]
119. Oberheim, N.A.; Goldman, S.A.; Nedergaard, M. Heterogeneity of Astrocytic Form and Function. *Methods Mol. Biol.* **2012**, *814*, 23–45. [CrossRef]
120. Grupp, L.; Wolburg, H.; Mack, A.F. Astroglial Structures in the Zebrafish Brain. *J. Comp. Neurol.* **2010**, *518*, 4277–4287. [CrossRef]
121. Kettenmann, H.; Verkhratsky, A. Neuroglia—Living Nerve Glue. *Fortschr. Neurol. Psychiatr.* **2011**, *79*, 588–597. [CrossRef]
122. Hamby, M.E.; Sofroniew, M.V. Reactive Astrocytes as Therapeutic Targets for CNS Disorders. *Neurotherapeutics* **2010**, *7*, 494–506. [CrossRef] [PubMed]
123. Floden, A.M.; Combs, C.K. Microglia Repetitively Isolated from in vitro Mixed Glial Cultures Retain their Initial Phenotype. *J. Neurosci. Methods* **2007**, *164*, 218–224. [CrossRef] [PubMed]
124. Fellin, T.; Ellenbogen, J.M.; De Pittà, M.; Ben-Jacob, E.; Halassa, M.M. Astrocyte regulation of sleep circuits: Experimental and modeling perspectives. *Front. Comput. Neurosci.* **2012**, *6*, 65. [CrossRef] [PubMed]
125. Souza, D.G.; Bellaver, B.; Souza, D.O.; Quincozes-Santos, A. Characterization of adult rat astrocyte cultures. *PLoS ONE* **2013**, *8*, e60282. [CrossRef] [PubMed]
126. He, Y.; Taylor, N.; Bhattacharya, A. Isolation and Culture of Astrocytes from Postnatal and Adult Mouse Brains. *Methods Mol. Biol.* **2019**, *1938*, 37–47. [CrossRef] [PubMed]
127. López, E.; Arce, C.; Oset-Gasque, M.J.; Cañadas, S.; González, M.P. Cadmium Induces Reactive Oxygen Species Generation and Lipid Peroxidation in Cortical Neurons in Culture. *Free Radic. Biol. Med.* **2006**, *40*, 940–951. [CrossRef]
128. Bolam, J.P.; Pissadaki, E.K. Living on the Edge with Too Many Mouths to Feed: Why Dopamine Neurons Die. *Mov. Disord.* **2012**, *27*, 1478–1483. [CrossRef]
129. Rayman, M.P. Selenium and Human Health. *Lancet* **2012**, *379*, 1256–1268. [CrossRef]
130. O'Donnell, K.C.; Vargas, M.E.; Sagasti, A. WldS and PGC-1α Regulate Mitochondrial Transport and Oxidation State After Axonal Injury. *J. Neurosci.* **2013**, *33*, 14778–14790. [CrossRef]
131. Tisler, T.; Jemec, A.; Mozetic, B.; Trebse, P. Hazard Identification of Imidacloprid to Aquatic Environment. *Chemosphere* **2009**, *76*, 907–914. [CrossRef]
132. Duzguner, V.; Erdogan, S. Chronic exposure to imidacloprid induces inflammation and oxidative stress in the liver & central nervous system of rats. *Pestic. Biochem. Physiol.* **2012**, *104*, 58–64. [CrossRef]
133. Abdel-Rahman Mohamed, A.; Mohamed, W.A.M.; Khater, S.I. Imidacloprid Induces Various Toxicological Effects Related to the Expression of 3β-HSD, NR5A1, and OGG1 Genes in Mature and Immature Rats. *Environ. Pollut.* **2017**, *221*, 15–25. [CrossRef] [PubMed]
134. Antonio, M.T.; Peinado, V.; González, J.C.; Leret, M.L. Effects of Maternal Cadmium Administration on Development of Monoaminergic, GABAergic and Glutamatergic Systems. *Environ. Toxicol. Pharmacol.* **2010**, *29*, 87–90. [CrossRef]
135. Philbert, M.A.; Billingsley, M.L.; Reuhl, K.R. Mechanisms of Injury in the Central Nervous System. *Toxicol. Pathol.* **2000**, *28*, 43–53. [CrossRef] [PubMed]
136. Cambier, S.; Bénard, G.; Mesmer-Dudons, N.; Gonzalez, P.; Rossignol, R.; Brèthes, D.; Bourdineaud, J.P. At Environmental Doses, Dietary Methylmercury Inhibits Mitochondrial Energy Metabolism in Skeletal Muscles of the Zebra fish (*Danio rerio*). *Int. J. Biochem. Cell. Biol.* **2009**, *41*, 791–799. [CrossRef]
137. Borges, A.C.; Da Silva Montes, C.; Barbosa, L.A.; Ferreira, M.A.P.; Berrêdo, J.F.; Martins Rocha, R. Integrated Use of Histological and Ultrastructural Biomarkers for Assessing Mercury Pollution in Piranhas (*Serrasalmus rhombeus*) from the Amazon Mining Region. *Chemosphere* **2018**, *202*, 788–796. [CrossRef]
138. Bhattacharya, S.; Large, E.; Heizmann, C.W.; Hemmings, B.; Chazin, W.J. Structure of the Ca^{2+}/S100B/NDR Kinase Peptide Complex: Insights Into S100 Target Specificity and Activation of the Kinase. *Biochemistry* **2003**, *42*, 14416–14426. [CrossRef]
139. Leclerc, E.; Fritz, G.; Weibel, M.; Heizmann, C.W.; Galichet, A. S100B and S100A6 Differentially Modulate Cell Survival by Interacting with Distinct RAGE (Receptor for Advanced Glycation End Products) Immunoglobulin Domains. *J. Biol. Chem.* **2007**, *282*, 31317–31331. [CrossRef]

140. Sorci, G.; Riuzzi, F.; Arcuri, C.; Giambanco, I.; Donato, R. Amphoterin Stimulates Myogenesis and Counteracts the Antimyogenic Factors Basic Fibroblast Growth Factor and S100B via RAGE Binding. *Mol. Cell. Biol.* **2004**, *24*, 4880–4894. [CrossRef]
141. Vogl, T.; Tenbrock, K.; Ludwig, S.; Leukert, N.; Ehrhardt, C.; van Zoelen, M.A.; Nacken, W.; Foell, D.; van der Poll, T.; Sorg, C.; et al. Mrp8 and Mrp14 are Endogenous Activators of Toll-like Receptor 4, Promoting Lethal, Endotoxin-induced Shock. *Nat. Med.* **2007**, *13*, 1042–1049. [CrossRef]
142. Chan, J.K.; Roth, J.; Oppenheim, J.J.; Tracey, K.J.; Vogl, T.; Feldmann, M.; Horwood, N.; Nanchahal, J. Alarmins: Awaiting a Clinical Response. *J. Clin. Investig.* **2012**, *122*, 2711–2719. [CrossRef] [PubMed]
143. Newton, K.; Dixit, V.M. Signaling in Innate Immunity and Inflammation. *Cold Spring Harb. Perspect. Biol.* **2012**, *4*, a006049. [CrossRef] [PubMed]
144. Donato, R. Intracellular and Extracellular Roles of S100 Proteins. *Microsc. Res. Tech.* **2003**, *60*, 540–551. [CrossRef]
145. Sorci, G.; Agneletti, A.L.; Donato, R. Effects of S100A1 and S100B on Microtubule Stability. An in vitro Study Using Triton-cytoskeletons from Astrocyte and Myoblast Cell Lines. *Neuroscience* **2000**, *99*, 773–783. [CrossRef] [PubMed]
146. Raponi, E.; Agenes, F.; Delphin, C.; Assard, N.; Baudier, J.; Legraverend, C.; Deloulme, J.C. S100B Expression Defines a State in Which GFAP-expressing Cells Lose Their Neural Stem Cell Potential and Acquire a More Mature Developmental Stage. *Glia* **2007**, *55*, 165–177. [CrossRef]
147. Grandel, H.; Kaslin, J.; Ganz, J.; Wenzel, I.; Brand, M. Neural Stem Cells and Neurogenesis in the Adult Zebrafish Brain: Origin, Proliferation Dynamics, Migration and Cell Fate. *Dev. Biol.* **2006**, *295*, 263–277. [CrossRef]
148. Ninkovic, J.; Götz, M. Signaling in Adult Neurogenesis: From Stem Cell Niche to Neuronal Networks. *Curr. Opin. Neurobiol.* **2007**, *17*, 338–344. [CrossRef]
149. Mahler, J.; Driever, W. Expression of the Zebrafish Intermediate Neurofilament Nestin in the Developing Nervous System and in Neural Proliferation Zones at Postembryonic Stages. *BMC Dev. Biol.* **2007**, *7*, 89. [CrossRef]
150. Hassan, S.A.; Moussa, E.A.; Abbott, L.C. The Effect of Methylmercury Exposure on Early Central Nervous System Development in the Zebrafish (*Danio rerio*) embryo. *J. Appl. Toxicol.* **2012**, *32*, 707–713. [CrossRef]
151. Smith, L.E.; Carvan, M.J.; Dellinger, J.A.; Ghorai, J.K.; White, D.B.; Williams, F.E.; Weber, D.N. Developmental Selenomethionine and Methylmercury Exposures Affect Zebrafish Learning. *Neurotoxicol. Teratol.* **2010**, *32*, 246–255. [CrossRef]
152. Yang, L.; Kemadjou, J.R.; Zinsmeister, C.; Bauer, M.F.; Legradi, J.; Müller, F.; Pankratz, M.J.; Jäkel, J.; Strähle, U. Transcriptional Profiling Reveals Barcode-like Toxicogenomic Responses in the Zebrafish Embryo. *Genome Biol.* **2007**, *8*, R227. [CrossRef] [PubMed]
153. Hamed, H.S.; El-Sayed, Y.S. Antioxidant Activities of *Moringa oleifera* Leaf Extract Against Pendimethalin-induced Oxidative Stress and Genotoxicity in Nile tilapia, *Oreochromis niloticus* (L.). *Fish Physiol. Biochem.* **2019**, *45*, 71–82. [CrossRef] [PubMed]
154. Mallya, R.; Chatterjee, P.K.; Vinodini, N.A.; Chatterjee, P.; Mithra, P. *Moringa oleifera* Leaf Extract: Beneficial Effects on Cadmium Induced Toxicities—A Review. *J. Clin. Diagn. Res.* **2017**, *11*, CE01–CE04. [CrossRef] [PubMed]
155. Sreelatha, S.; Padma, P.R. Antioxidant Activity and Total phenolic Content of *Moringa oleifera* Leaves in Two Stages of Maturity. *Plant Foods Hum. Nutr.* **2009**, *64*, 303–311. [CrossRef]
156. Guardia de Souza, E.S.T.; do Val de Paulo, M.E.F.; da Silva, J.R.M.; da Silva Alves, A.; Britto, L.R.G.; Xavier, G.F.; Lopes Sandoval, M.R. Oral Treatment with Royal Jelly Improves Memory and Presents Neuroprotective Effects on icv-STZ Rat Model of Sporadic Alzheimer's Disease. *Heliyon* **2020**, *6*, e03281. [CrossRef]
157. Li, X.T.; Zhou, J.C.; Zhou, Y.; Ren, Y.S.; Huang, Y.H.; Wang, S.M.; Tan, L.; Yang, Z.Y.; Ge, Y.W. Pharmacological Effects of *Eleutherococcus senticosus* on the Neurological Disorders. *Phytother. Res.* **2022**, *36*, 3490–3504. [CrossRef]
158. Yang, Y.; Gao, H.; Liu, W.; Liu, X.; Jiang, X.; Li, X.; Wu, Q.; Xu, Z.; Zhao, Q. *Arctium lappa* L. Roots Ameliorates Cerebral Ischemia Through Inhibiting Neuronal Apoptosis and Suppressing AMPK/mTOR-Mediated Autophagy. *Phytomedicine* **2021**, *85*, 153526. [CrossRef]
159. Sayyar, Z.; Yazdinezhad, A.; Hassan, M.; Anarkooli, I.J. Protective Effect of *Matricaria chamomilla* Ethanolic Extract on Hippocampal Neuron Damage in Rats Exposed to Formaldehyde. *Oxid. Med. Cell. Longev.* **2018**, *2018*, 6414317. [CrossRef] [PubMed]
160. Asgharzade, S.; Rabiei, Z.; Rafieian-kopaei, M. Effects of *Matricaria chamomilla* Extract on Motor Coordination Impairment Induced by Scopolamine in Rats. *Asian Pac. J. Trop. Biomed.* **2015**, *5*, 829–833. [CrossRef]
161. Wang, X.; Zhang, Z.; Wu, S.C. Health Benefits of *Silybum marianum*: Phytochemistry, Pharmacology, and Applications. *J. Agric. Food Chem.* **2020**, *68*, 11644–11664. [CrossRef]
162. Csakvari, A.C.; Moisa, C.; Radu, D.G.; Olariu, L.M.; Lupitu, A.I.; Panda, A.O.; Pop, G.; Chambre, D.; Socoliuc, V.; Copolovici, L.; et al. Green Synthesis, Characterization, and Antibacterial Properties of Silver Nanoparticles Obtained by Using Diverse Varieties of *Cannabis sativa* Leaf Extracts. *Molecules* **2021**, *26*, 4041. [CrossRef]
163. Rababah, T.M.; Al-Omoush, M.; Brewer, S.; Alhamad, M.; Yang, W.; Alrababah, M.; Al-Ghzawi, A.A.M.; Al-u'datt, M.; Ereifej, K.; Alsheyab, F.; et al. Total Phenol, Antioxidant Activity, Flavonoids, Anthocyanins and Color of Honey as Affected by Floral Origin Found in the Arid and Semiarid Mediterranean Areas. *J. Food Process. Preserv.* **2014**, *38*, 1119–1128. [CrossRef]
164. Bogdan, M.A.; Bungau, S.; Tit, D.M.; Zaha, D.C.; Nechifor, A.C.; Behl, T.; Chambre, D.; Lupitu, A.I.; Copolovici, L.; Copolovici, D.M. Chemical Profile, Antioxidant Capacity, and Antimicrobial Activity of Essential Oils Extracted from Three Different Varieties (Moldoveanca 4, Vis Magic 10, and Alba 7) of *Lavandula angustifolia*. *Molecules* **2021**, *26*, 4381. [CrossRef]
165. Wu, S.; Li, X.; Liu, X.; Yang, G.; An, X.; Wang, Q.; Wang, Y. Joint Toxic Effects of Triazophos and Imidacloprid on Zebrafish (*Danio rerio*). *Environ. Pollut.* **2018**, *235*, 470–481. [CrossRef]

166. Richetti, S.K.; Rosemberg, D.B.; Ventura-Lima, J.; Monserrat, J.M.; Bogo, M.R.; Bonan, C.D. Acetylcholinesterase Activity and Antioxidant Capacity of Zebrafish Brain is Altered by Heavy Metal Exposure. *Neurotoxicology* **2011**, *32*, 116–122. [CrossRef]
167. Plavan, G.; Jitar, O.; Teodosiu, C.; Nicoara, M.; Micu, D.; Strungaru, S.A. Toxic Metals in Tissues of Fishes from the Black Sea and Associated Human Health Risk Exposure. *Environ. Sci. Pollut. Res.* **2017**, *24*, 7776–7787. [CrossRef]
168. Lee, J.W.; Kim, J.-W.; De Riu, N.; Moniello, G.; Hung, S.S.O. Histopathological Alterations of Juvenile Green (*Acipenser medirostris*) and White Sturgeon (*Acipenser transmontanus*) Exposed to Graded Levels of Dietary Methylmercury. *Aquat. Toxicol.* **2012**, *109*, 90–99. [CrossRef]
169. Félix, L.; Carreira, P.; Peixoto, F. Effects of Chronic Exposure of Naturally Weathered Microplastics on Oxidative Stress Level, Behaviour, and Mitochondrial Function of Adult Zebrafish (*Danio rerio*). *Chemosphere* **2023**, *310*, 136895. [CrossRef]
170. Boopathi, S.; Haridevamuthu, B.; Mendonca, E.; Gandhi, A.; Priya, P.S.; Alkahtani, S.; Al-Johani, N.S.; Arokiyaraj, S.; Guru, A.; Arockiaraj, J.; et al. Combined Effects of a High-fat Diet and Polyethylene Microplastic Exposure Induce Impaired Lipid Metabolism and Locomotor Behavior in Larvae and Adult Zebrafish. *Sci. Total Environ.* **2023**, *902*, 165988. [CrossRef]
171. Rodrigues Garcia, T.; Freire, P.T.C.; da Silva, A.W.; Ferreira, M.K.A.; Rebouças, E.L.; Mendes, F.R.S.; Marinho, E.M.; Marinho, M.M.; Teixeira, A.M.R.; Marinho, E.S.; et al. Anxiolytic and Anticonvulsant Effect of Ibuprofen Derivative Through GABAergic Neuromodulation in Adult Zebrafish. *J. Biomol. Struct. Dyn.* **2023**, *41*, 12055–12062. [CrossRef]
172. Maffioli, E.; Nonnis, S.; Negri, A.; Fontana, M.; Frabetti, F.; Rossi, A.R.; Tedeschi, G.; Toni, M. Environmental Temperature Variation Affects Brain Lipid Composition in Adult Zebrafish (*Danio rerio*). *Int. J. Mol. Sci.* **2024**, *25*, 9629. [CrossRef] [PubMed]

International Journal of *Molecular Sciences*

MDPI

Article

Do Microplastics Have Neurological Implications in Relation to Schizophrenia Zebrafish Models? A Brain Immunohistochemistry, Neurotoxicity Assessment, and Oxidative Stress Analysis

Alexandra Savuca [1,2,†], Alexandrina-Stefania Curpan [2], Luminita Diana Hritcu [3,†], Teodora Maria Buzenchi Proca [4], Ioana-Miruna Balmus [5], Petru Fabian Lungu [2], Roxana Jijie [6], Mircea Nicusor Nicoara [1,7,*], Alin Stelian Ciobica [7,8,9,10], Gheorghe Solcan [4] and Carmen Solcan [4]

1 Doctoral School of Geosciences, Faculty of Geography and Geology, "Alexandru Ioan Cuza" University of Iasi, Carol I Avenue, 20A, 700505 Iași, Romania; alexandra.savuca@yahoo.com

2 Doctoral School of Biology, Faculty of Biology, "Alexandru Ioan Cuza" University of Iasi, Carol I Avenue, 20A, 700505 Iași, Romania; andracurpan@yahoo.com (A.-S.C.); lungufabian123@gmail.com (P.F.L.)

3 Internal Medicine Clinic, University of Life Sciences "Ion Ionescu de la Brad", Mihail Sadoveanu Street, No. 3, 700490 Iasi, Romania; lumidih@yahoo.com

4 Faculty of Veterinary Medicine, University of Life Sciences "Ion Ionescu de la Brad", Mihail Sadoveanu Street, No. 3, 700490 Iasi, Romania; buzenchi.teodora@yahoo.com (T.M.B.P.); gsolcan@uaiasi.ro (G.S.); csolcan@uaiasi.ro (C.S.)

5 Department of Exact Sciences and Natural Sciences, Institute of Interdisciplinary Research, "Alexandru Ioan Cuza" University of Iasi, Carol I Avenue, 20A, 700505 Iași, Romania; balmus.ioanamiruna@yahoo.com

6 Research Center on Advanced Materials and Technologies, Department of Exact and Natural Sciences, Institute of Interdisciplinary Research, "Alexandru Ioan Cuza" University of Iasi, Carol I Avenue, 20A, 700505 Iași, Romania; roxanajijie@yahoo.com

7 Department of Biology, Faculty of Biology, "Alexandru Ioan Cuza" University of Iasi, Carol I Avenue, 20A, 700505 Iași, Romania; alin.ciobica@uaic.ro

8 Academy of Romanian Scientists, 3 Ilfov, 050044 Bucharest, Romania

9 Center of Biomedical Research, Romanian Academy, Iasi Branch, Teodor Codrescu 2, 700481 Iasi, Romania

10 Preclinical Department, Apollonia University, 700511 Iasi, Romania

* Correspondence: mirmag@uaic.ro

† These authors contributed equally to this work.

Abstract: The effects of exposure to environmental pollutants on neurological processes are of increasing concern due to their potential to induce oxidative stress and neurotoxicity. Considering that many industries are currently using different types of plastics as raw materials, packaging, or distribution pipes, microplastics (MPs) have become one of the biggest threats to the environment and human health. These consequences have led to the need to raise the awareness regarding MPs negative neurological effects and implication in neuropsychiatric pathologies, such as schizophrenia. The study aims to use three zebrafish models of schizophrenia obtained by exposure to ketamine (Ket), methionine (Met), and their combination to investigate the effects of MP exposure on various nervous system structures and the possible interactions with oxidative stress. The results showed that MPs can interact with ketamine and methionine, increasing the severity and frequency of optic tectum lesions, while co-exposure (MP+Met+Ket) resulted in attenuated effects. Regarding oxidative status, we found that all exposure formulations led to oxidative stress, changes in antioxidant defense mechanisms, or compensatory responses to oxidative damage. Met exposure induced structural changes such as necrosis and edema, while paradoxically activating periventricular cell proliferation. Taken together, these findings highlight the complex interplay between environmental pollutants and neurotoxicants in modulating neurotoxicity.

Keywords: microplastics; zebrafish; oxidative stress; immunohistochemistry; schizophrenia

1. Introduction

Schizophrenia (SZ) is a chronic psychiatric disorder with a global prevalence of approximately 0.4% [1]. This disorder is generally characterized by the presence of positive symptoms (delusions, hallucinations), cognitive symptoms (disorganized speech and thought processes) [2], negative symptoms (reduced emotional expression, inability to start or continue tasks, anhedonia, alogia) [2,3], or a combination of the above. The presence of negative symptoms is correlated with a significant segment of the comorbidities associated with SZ [2]. The onset of SZ usually occurs early during neurogenesis, well before cognitive development, meaning that the development of the social skills needed in adulthood could be impaired [4,5]. The poor prognosis of SZ is associated with early onset, the male sex, occurrence of negative symptoms, a family history of SZ, social isolation, adverse childhood experiences, and a lack of affection [6].

Zebrafish has become a reliable species for modeling schizophrenia-like symptoms, anxiety, depression, and neurodegenerative disorders [7], especially through pharmaceutical manipulation [8]. In the context of the present study, one of the most popular approaches for modeling schizophrenia is the administration of ketamine [9] for mimicking the positive symptoms [10], methionine for the negative symptoms [11], and their combined effect for a more robust model that includes the cognitive dysfunction as well [12].

Diet plays a crucial role in psychiatric disorders due to its impact on brain function and overall mental health; it is worth mentioning that this is an often-missed component in SZ management. The proper function of the brain requires a variety of nutrients, including vitamins, minerals, omega-3 fatty acids, and amino acids [13,14]. Thus, it was shown that deficiencies of certain nutrients could contribute to the development or exacerbation of psychiatric symptoms. Also, while chronic inflammation was associated with several psychiatric disorders, it was recently reported that diets rich in processed food, refined sugars, and unhealthy fats could promote inflammation [15]. On the other hand, diets mainly based on fruits, vegetables, whole grains, and omega-3 fatty acid intake could exhibit anti-inflammatory effects [15].

Emerging research has suggested a strong connection between the brain and the gut, known as the brain–gut axis. The gut microbiota, consisting of trillions of microorganisms living in the gastrointestinal tract, was demonstrated to modulate several brain functions and behavior via the production of neurotransmitters and immune regulators [16,17].

A growing concern of modern society is related to the widespread presence of PM in various environments, even reaching food sources [18]. MPs can contaminate food sources via air, water, and soil contamination, as well as via packaging and processing [19,20]. While the effects of MP-contaminated food are not currently fully described, there are some concerns about the potential impact of MPs on human health [21]. Additionally, MPs were recently detected in human's samples such as thrombi [22] or, even worse, in semen samples [23]. This is an even clearer indication of the fact that people are extraordinarily exposed to this type of pollutant. MPs can be mixed with additives and other pollutants that were previously linked to inflammatory response when entering a living body [24], or cyanobacteria, if we consider seafood, where recent studies indicate that co-exposure of Microcystin-LR, an natural toxin produced by cyanobacteria, lead to an increased MDA level along with hepatic histopathological injuries [25] Thus, the intake of MP-contaminated food could indirectly affect psychiatric health by influencing gut microbiota composition and function [20,26]. MPs have been shown to induce oxidative stress, leading to cellular damage and inflammation [27–29]. Oxidative stress is an important pathophysiological component of psychiatric disorders, having the potential to exacerbate or aggravate some of the symptoms [30–33], as well as to actively contribute to the development of several neuropsychiatric diseases (i.e., SZ, Alzheimer's disease, Parkinson's disease, and major depressive disorder) [34–39].

According to several studies conducted in zebrafish models, MPs can pass through the blood–brain barrier [38], accumulating at high levels in nervous system structures and leading to a considerable increase in oxidative imbalance and apoptosis [39,40]. In addition,

SZ patients have been shown to exhibit increased levels of lipid peroxidation and protein oxidation in the blood, cerebrospinal fluid, and tissues, along with altered antioxidants and reduced glutathione [40–42].

In the present study, we experimentally induced schizophrenia-like symptoms in zebrafish using methionine, ketamine, and a combination of methionine and ketamine, along with exposure to microplastics. Thus, we aimed to study the contribution of contaminated food in the development of oxidative imbalance, with special attention to the systemic and tissue status in the context of well-established models of SZ in zebrafish based on methionine (Met) and ketamine administration (Ket), as previously described by our group [12].

2. Results

2.1. Oxidative Stress Levels

The analysis of biochemical data revealed several effects of MPs and pharmacological agents. We observed that the administration of Ket, Met, and their combination did not lead to significantly increased SOD specific activity compared to the control group. On the other hand, only MP administration in MP+Ket led to a decrease in SOD specific activity compared to the Ket group ($p = 0.018$) (Figure 1A, Table 1). Other significant variations in the specific enzymatic activity of SOD were recorded for Met vs. MP+Ket ($p = 0.046$), Met+Ket vs. MP+Ket ($p = 0.049$), and MP+Ket vs. MP+Met ($p = 0.005$).

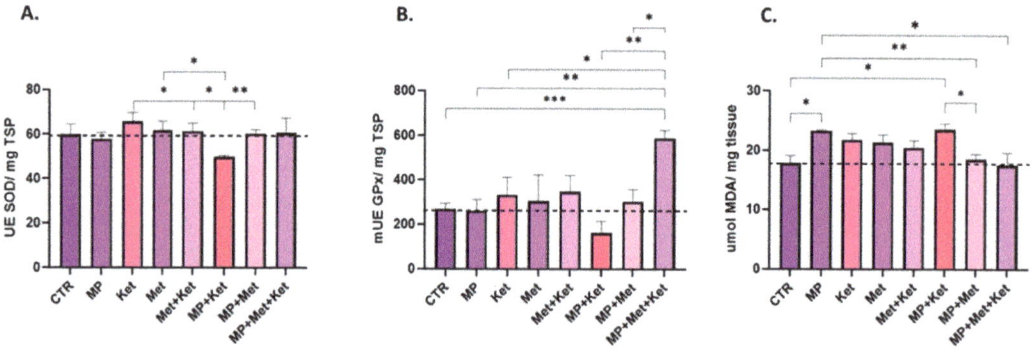

Figure 1. The graphic representation of the oxidative stress markers levels. (**A**)—SOD specific activity (UE SOD/mg TSP), (**B**)—GPx specific activity (mUE GPx/mg TSP), (**C**)—relative MDA content (μmol MDA/mg tissue). All data are expressed as means ± SEM (n = 5/per group, * $p < 0.05$; ** $p < 0.01$; *** $p < 0.001$, two-tailed student t test). TSP = total soluble protein content.

Regarding GPx specific activity (Figure 1B, Table 1), we observed significant differences between the following groups: CTR vs. MP+Met+Ket ($p < 0.001$), MP vs. MP+Met+Ket ($p = 0.005$), Ket vs. MP+Met+Ket ($p = 0.041$), MP+Ket vs. MP+Met+Ket ($p < 0.01$), and MP+Met vs. MP+Met+Ket ($p = 0.01$), MP+Ket vs. MP+ Met+ KET ($p = 0.003$). When strictly compared with the control group, all groups had higher levels of GPx except MP and MP+Ket, indicating different degrees of antioxidant defense mechanisms and a compensatory mechanism.

We found significant variations in the relative content of MDA (Figure 1C) when compared with the control group in two treatments, namely MP ($p = 0.006$) and MP+Ket ($p = 0.0211$), indicating increased oxidative stress. Moreover, significant differences occurred for MP vs. MP+Met, with a significant decrease in the MDA content ($p = 0.002$), as well as for MP vs. MP+Met+Ket ($p = 0.022$). Other significant differences were found also in the following combinations: Ket vs. MP+Met ($p = 0.028$) and MP+Ket vs. MP+Met ($p = 0.012$) (Table 1).

Table 1. Oxidative stress marker variation, depending on the administered treatment.

	Experimental Group			Oxidative Stress Parameters ‡		
	MP	Ket	Met	SOD	MDA	GPx
1	+	−	−	↓	↑ **	−
2	−	+	−	↑	↑	↑
3	−	−	+	↑	↑	↑
4	−	+	+	↑	↑	↑
5	+	+	−	↓	↑ *	↓
6	+	−	+	−	↓	↑
7	+	+	+	−	−	↑ ***

+ substance administered; − substance not administered; ‡ the control group was used as reference; */**/*** the power of significance between the treatment and the control groups (* $p < 0.05$; ** $p < 0.01$; *** $p < 0.001$).

2.2. Immunohistology Results

Several histopathological changes were observed in the different layers of the optic tectum in zebrafish following exposure to microplastics, ketamine, methionine, and their combination (Figure 2, Table 2). The severity and frequency of lesions in these layers were more pronounced when combining MP+Ket and MP+Met and more attenuated when combining MP+Met+Ket and Met+Ket.

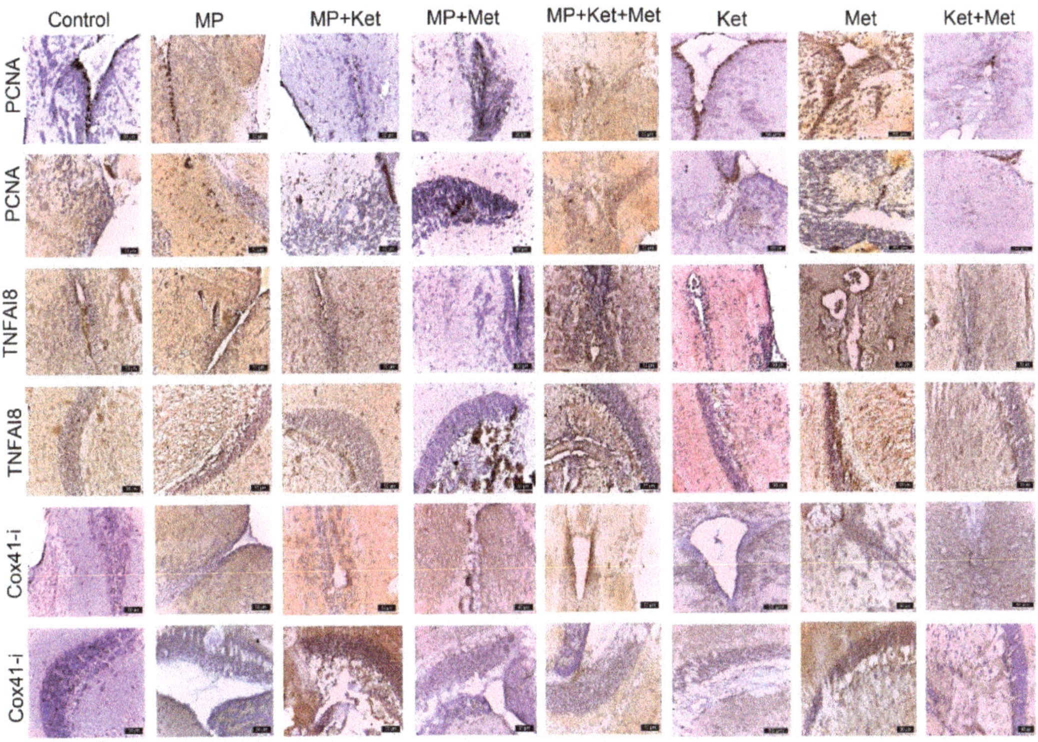

Figure 2. Different expression of *PCNA*, *TNFAI8*, and *Cox41-i* in different experimental and control groups. In rows 1, 3, and 5 the periventricular expression is observed, and in rows 2, 4, and 6, the expression is observed at the level of the optic tectum (bar = 50 μm).

Table 2. Score of expression of PCNA, TNFAI8, and *Cox41-i* in different experimental and control groups (number of positive immunolabeled cells/field at a magnitude of ×400).

Antibody	Control	MP	MP+Ket	MP+Met	MP+Met+Ket	Ket	Met	Met+Ket
PCNA SPV	++	++	+	+++	++	+++	+++	+
PCNA TO	++	+++	++	++	++	+	+	+
TNFAI8 SPV	+	+++	++++	++	++++	++	++	++
TNFAI8 TO	+	+++	++	++	+++	++	++++	++
Cox41-i SPV	++	+++	+++	+++	++++	++++	++	++
Cox 41-i TO	+	+++	++++	+++	++	++++	++	++

SPV = periventricular space; TO = optic tectum; + = 1–15 positive cells; ++ = 15–30 positive cells; +++ = 30–60 positive cells; ++++ = 60–100 positive cells.

The optical tectum showed marked structural differences in fish from all experimental groups. Areas of granular cell necrosis and detachment in the SPV area, granular cell spongiosis in SFGC, small areas of necrosis in the outer layer, connective tissue degeneration, edema, and elevated and degenerated external squamous cells were observed. These changes were more pronounced in individuals exposed to the combination of microplastics and ketamine and included detachment and necrosis in the granular cells in the SPV area and necrosis in mononuclear cells, as well as the high detachment and degeneration of neuronal cells in the SO and MS areas.

The pathology was also pronounced in fish exposed to ketamine and methionine, and observations included the spongiosis of granular cells in the SPV area and vacuolation in granular cells in the SAC layer due to degeneration. A large number of blood capillaries in the SFGC area and a reduction in the SO and MS layers due to neuronal cell degeneration were also observed. In the periventricular area, ectasia of the vessels, a variable proliferation of periventricular stem cells, and edema were observed.

PCNA expression varies by group. The number of positively labeled cells was recorded, in descending order, in the following groups: MP, MP+Met, Met, Ket, MP+Ket, MP+Met+Ket, and Met+Ket. In the control group, positive PCNA cells are present, in moderate amounts, at the level of the regeneration areas.

MPs and Met have an important role in stimulating the multiplication of progenitor stem cells or radial glial cells (RGCs) in the periventricular zone, optic tectum, and valvula cerebeli. In all experimental groups, both MPs, Met, and Ket alone and combined produce many changes that are preceded by stimulation of neurogenesis. Thus, the presence of positive PCNA cells can be explained, both at the level of the regeneration niches and at those in migration, in groups with MP, MP+Ket, and MP+Met (Figure 2).

In the groups receiving ketamine, PCNA positive cells were observed in the two areas of the periventricular and cerebellar valve, but a smaller number of migratory cells was also noted. It is possible that ketamine treatment not only alters the expression pattern of *PCNA* but also the time of migration/differentiation of cerebellar cell types in adults, contributing to a state of general alteration of CNS cell proliferation.

Regarding methionine, it caused structural changes in all groups that were exposed to the agents, either alone or in combination, and generated areas of necrosis in the ZPV, edema, and blood vessels and also an activation of periventricular neurogenesis with *PCNA* positive cells. *PCNA* expression varied by group. In the SPV area, the number of positively labeled cells was recorded, in descending order, as follows: MP+Met, Met, Ket, MP, MP+Met+Ket, control, Met+Ket, and MP+Ket (Table 2). In the TO, the number of positively labeled cells was recorded, in descending order, as follows: MP, MP+Met, Met+Ket, and MP+Met+Ket and there was a more reduced positivity in Ket, Met, and

Met+Ket (Table 1). In the control group, 1–30 positive *PCNA* cells/field were present at the level of the regeneration areas.

TNFαI8 registers the most intense expression in the periventricular area and in the optical tectum and, subsequently, varies by group. In the SPV area, the number of positively labeled cells was recorded, in descending order, as follows: MP+Met, MP+Met+Ket, MP, Met+Ket, Met, Ket, and control. In TO, the number of positively labeled cells for TNFAI8 expression was recorded in descending order, as follows: Met, MP, MP+Met+Ket, Ket, Met+Ket, MP+Met, MP+Ket, and control (Table 2). The presence of *TNFαI8* expression in the nervous system of fish from experimental groups denotes the presence of an inflammatory process generated by MP, Ket, Met, and their combination (Figure 2).

The expression of *Cox41-i* was also recorded. In the SPV area, the number of positively labeled cells for *Cox41-I* was recorded, in descending order, as follows: MP+Met+Ket, KET, MP, MP+Ket, MP+Met, Met, Met+Ket, and control (Table 2). In the TO, the number of positively labeled cells for *Cox41-i* was recorded, in descending order, as follows: MP+Ket, Ket, MP+Met, MP, MP+Met+Ket, Ket, Met+Ket, and control (Table 2).

The presence of a different *Cox41-i* expression in cells in the SPV zone and the TO denotes different degrees of ATP generation, dependent on the expression of hypoxia induction factor (Figure 2). The *Cox41-i* overexpression in some groups indicates increased energy requirements to restore the cellular metabolism of nerve cells as a result of the oxidative stress and neurotoxicity produced by types of various exposure.

BDNF 2 expression in the SPV area was recorded, in descending order, as follows: control, MP+Met +Ket, MP+Met, MP, Ket, Met, MP+Ket, and Met+Ket. In the TO, the number of positively labeled cells was recorded, in descending order, as follows: MP+Met, Met, MP+Met+Ket, control, MP, MP+Ket, Ket, and Met+Ket (Figure 3, Table 3).

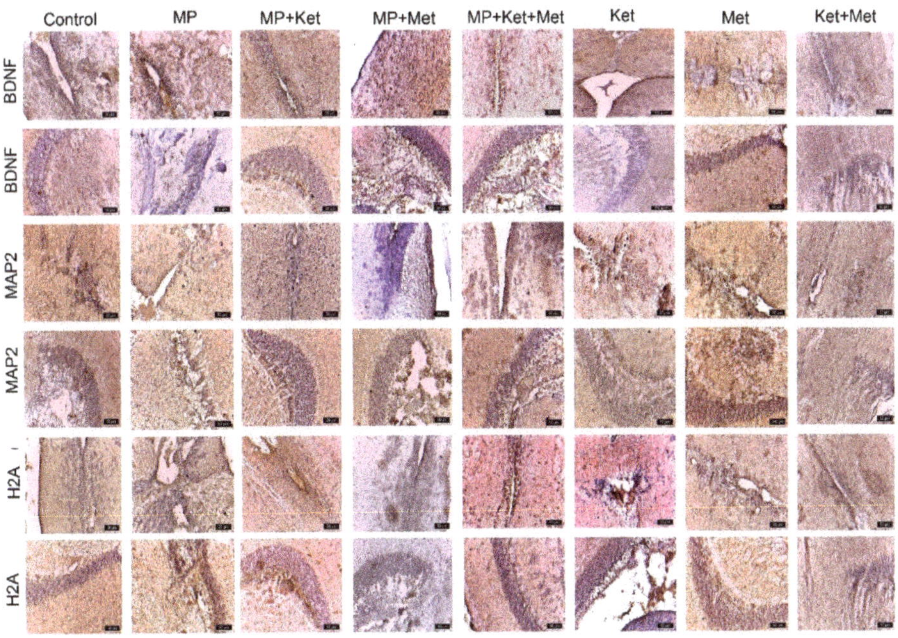

Figure 3. Different expression of *BDNF*, *MAP2*, and *H2A* in different experimental groups and controls. In rows 1, 3, and 5, the periventricular expression is observed, and in rows 2, 4, and 6, the expression is observed at the level of the optic tectum (bar = 50 μm).

MAP2 expression in the SPV area was recorded, in descending order, in the following groups: Met, MP+Met+Ket, Ket, control, MP, Met +Ket, MP+Met, and MP+Ket. In the

TO, *MAP2* expression was recorded, in descending order, as follows: Met, MP+Met+Ket, MP+Ket, Ket, MP+Met, control, MP, and Met+Ket (Figure 3, Table 3).

Table 3. Score for the expression of BDNF, MAP2, and H2A in different experimental groups and the control (number of positive immunolabeled cells/field at a magnitude of ×400).

Antibody	Control	MP	MP+Ket	MP+Met	MP+Met+Ket	Ket	Met	Met+Ket
BDNF SPV	++++	+++	++	+++	++++	+++	+++	++
BDNF TO	++	++	++	++	++	+	+++	+
MAP2 SPV	+++	+++	++	+++	++++	+++	++++	+++
MAP2 TO	++	++	+++	++	+++	+++	++++	+
H2A SPV	++	+++	+++	++	++++	+++	+++	++++
H2A TO	++	+++	+++	++	+++	++	+++	+++

SPV = periventricular space; TO = optic tectum; + =1–15 positive cells; ++ = 15–30 positive cells; +++ = 30–60 positive cells; ++++ = 60–100 positive cells.

H2A expression in the SPV area was recorded, in descending order, as follows: MP+Met+Ket, Met+Ket, MP+Ket, Met, Ket, MP, MP+Met, and control. In the TO, *H2A* expression was recorded, in descending order, as follows: MP+Ket, Met, Met+Ket, MP+Met+Ket, MP, Ket, MP+Met, and control (Figure 3, Table 3).

The expression of *S100* is intense in all groups (Figure 4, Table 4). Because the *S100* protein continues to be expressed in these regions of adult zebrafish and by subventricular zone glial cells, the *S100* protein plays an important role in the neurogenesis of adult zebrafish. In the SPV area, *S100* expression was recorded, in descending order, as follows: MP, MP+Ket, MP+Met+Ket, Ket, MP+Ket, Met, Met+Ket, and control. In the TO, *S100* expression was recorded, in descending order, as follows: MP+Met, Ket, MP, Met+Ket, MP+Ket, Met, MP+Met+Ket, and control (Figure 4, Table 4).

Table 4. Score of expression of expression of S100, GFAP, and Tub2 in different experimental and control groups (number of positive immunolabeled cells/field at a magnitude of ×400).

Antibody	Control	MP	MP+Ket	MP+Met	MP+Met+Ket	Ket	Met	Met+Ket
S100 SPV	+++	++++	++++	+++	++++	+++	+++	+++
S100 TO	+++	++++	+++	++++	+++	++++	+++	+++
GFAP SPV	+	++++	+++	++++	+++	++++	++	+++
GFAP TO	+	++++	++++	++	+	++	++	++
Tub2 SPV	+++	++	+++	+++	+++	++	+	+++
Tub2 TO	++	++	++	+++	++	++	+	++

SPV = periventricular space; TO = optic tectum; + = 1–15 positive cells; ++ = 15–30 positive cells; +++ = 30–60 positive cells; ++++ = 60–100 positive cells.

The expression of *GFAP* varies, including having the lowest intensity of expression in the control group and the most intense expression in the MP group. In the SPV area, *GFAP* expression was recorded, in descending order, as follows: MP, MP+Ket, MP+Met, Ket, MP+Met+Ket, Met+Ket, Met, and control. In the TO, the expression of GFAP descended in the following order: MP+Ket, MP, Ket, Met, Met+Ket, MP+Met, MP+Met+Ket, and control (Figure 4, Table 4). The changes produced by MP, Met, and Ket alone and combined caused a reaction in astrocytes but without producing scar areas as in mammals.

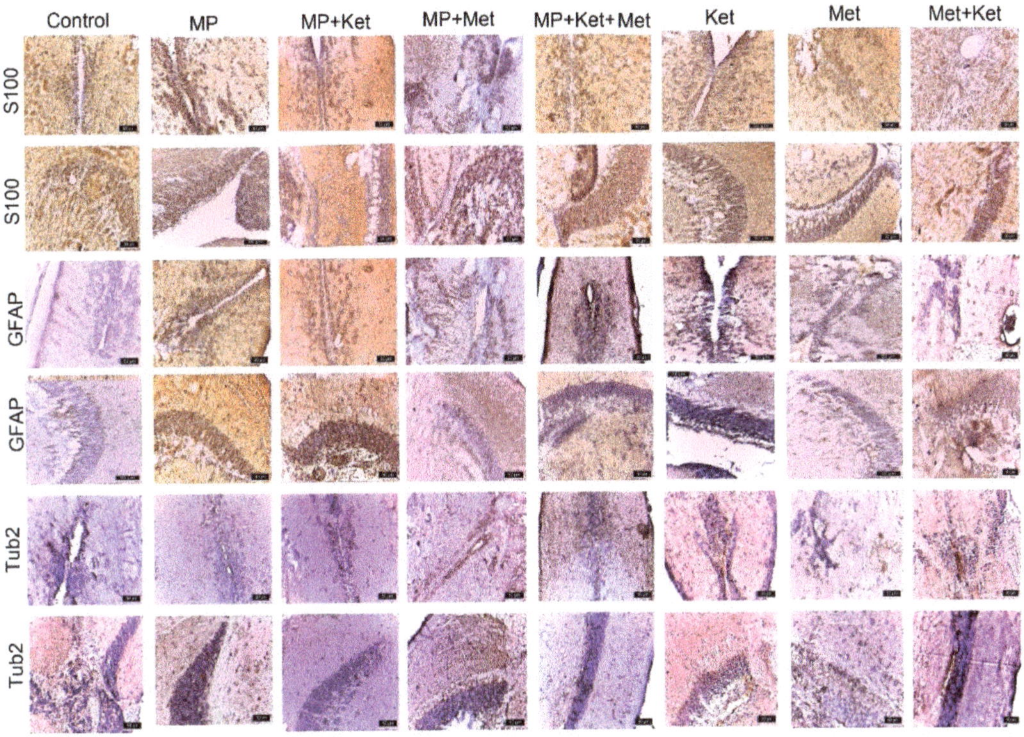

Figure 4. Different expression of *S100*, *GFAP*, and *Tub2* in different experimental and control groups. In rows 1, 3, and 5, the periventricular expression is observed, and in rows 2, 4, and 6, the expression is observed at the level of the optic tectum (bar = 50 μm).

Tub2 expression was observed in periventricular cells in all experimental groups but was lower in the control. In the SPV area, the expression of *Tub2* is more intense in the MP+Ket group, descending in the following order: MP+Ket, MP+Met, Met+Ket, MP, MP+Met+Ket, Ket, control, and Met. In the TO area, *Tub2* expression was recorded, in descending order, as follows: MP+Met, MP+Met+Ket, MP, MP+Ket, Met+Ket, Ket, control, and Met (Figure 4, Table 4).

3. Discussion

This study demonstrated effects on oxidative stress and immunohistochemistry (IHC) for microplastics alone and in three Schizophrenia models, focusing on an already vulnerable population category; this approach represents an important step forward to fill the gaps in this field of molecular toxicology.

In terms of oxidative stress, SOD is an antioxidant enzyme that plays a crucial role in neutralizing superoxide radicals. High levels of SOD activity suggest an adaptive mechanism to counteract increased oxidative stress and protect cells from potential damage [43]. On the other hand, decreased SOD activity could suggest a drop in this enzyme's ability to neutralize superoxide radicals, which could lead to the increased vulnerability of cells to oxidative stress. These effects could be seen in locations when the balance between the production of reactive oxygen species and the antioxidant defense performance is disrupted [43]. Previous studies have shown that 28-day chronic exposure to polypropylene, both alone and co-exposed with triclosan, could lead to antioxidant defense weakening via decreasing the neutralizing activity of SOD [44,45]. In these situations, increased MDA production was often observed [45]. High MDA levels could be associated with increased oxidative stress and could at least explain the occurrence of certain pathological processes, including inflammation, cardiovascular disease, and neurodegenerative disorders [46].

In a study designed to assess the transgenerational effects of methionine exposure, it was found that MDA levels show decreases over the control group, whereas SOD activity increased in the f1 generation [47]. Notably, a significant decrease in SOD activity was observed after cessation of Met exposure [48]. With regards to the effects of Ket administration, the previous studies indicated that the levels of SOD and MDA were increased in a dose-dependent manner even after Ket administration discontinuation [49]. In these contexts, it was suggested that both Met and Ket could alter the oxidative balance, influence the antioxidant enzymatic defense performance, and lead to the occurrence of cellular damage, in a similar manner to the results of our current study.

GPx is another member of the antioxidant enzymatic defense system that is implicated in cellular protection against the harmful effects of ROS overproduction. Its main action is to catalyze the neutralization of hydrogen peroxide. However, in some situations, GPx can also use other peroxides as enzymatic substrates, leading to their neutralization (including lipid peroxides) [50]. Our results showed that the exposure to MPs alone did not lead to altered GPx antioxidant activity; this was in contrast to [29], which showed that 21-day exposure to a cocktail of different microplastics (1 mg L^{-1}) led to significantly lower GPx activity as compared to the control group. On the other hand, we observed decreased GPx activity when MP and Ket were co-administered, meaning a reduced ability to neutralize ROS and increased cellular susceptibility to oxidative damage.

In contrast to other studies that reported no changes in GPx activity when Met and Ket were independently administered [51,52], we found that the co-administration of MP, Met, and Ket led to significantly increased GPx activity. As the effects of Met and Ket on the dynamics of GPx activity were similar to those in the already mentioned studies [51,52], the different patterns of variation we observed for the MP+Met+Ket co-administration could suggest that the fish metabolism deployed counteractive measures against oxidative stress by modulating defense mechanisms. This response is often observed in various conditions, including inflammation, the immune response to pathogens, or upon exposure to environmental stressors [53].

Plastic pollution is a critical worldwide environmental concern and comprises different sizes of plastic particles, categorized by the scientific community as macroplastics, microplastics, and nanoplastics [54]. MPs are currently defined as small plastic fragments of various shapes and sizes (ranging between 100 nm and 5 mm) and are considered to be an ecological danger that requires the utmost attention due to their presence in aquatic ecosystems and their toxicity for all living organisms [55]. Moreover, the association of MPs with other substances, such as pharmaceutical products and byproducts, has become of great interest to researchers [56–59].

Multiple studies have shown that MPs, alone or in combination with heavy metals, could exhibit detrimental effects on the neurological functions of fish, such as acetylcholinesterase activity inhibition [60,61] or neurotransmitter activity alterations [28,62–64]. Moreover, studies have reported on the negative effects of MPs in zebrafish during different developmental stages, especially early neurogenesis impairment, which could lead to persistent and severe alterations of morphology and locomotion, as well as feeding and predator avoidance behavior [61,65–67]. Furthermore, the multi-generational maternal and paternal transfer of nanoplastics was previously documented in the *Danio rerio* species [68].

Regarding the potential of MPs to modulate ROS production within the brains of fish, several studies have reported that this effect could be observed regardless of the size of the MPs (nano, micro, and macro) [69–71]. In this context, the over-production of ROS within the tissues of the central nervous system was previously associated with decreased cognitive performances and weakened antioxidant defense [72].

Previous studies on the effects of MPs revealed that a concentration of 100 mg L^{-1} of MP in tilapia and carp is sufficient to produce several histopathological changes, including necrosis, spongiosis, degeneration, and oedema [73,74]. Similar effects were reported in goldfish (*Carassius auratus*) exposed to pure MP [75] and zebrafish larvae (*Danio rerio*) exposed to pure low-density polyethylene fragments [76]. Exposure to polypropylene MP in concentrations of 300 mg L^{-1} and 600 mg L^{-1} lead to downregulated pathways and genes associated with cell proliferation regulation and DNA damage repair mechanisms in the liver, highlighting the need for future studies regarding PP toxicity [77]. Savari et al. in 2020 analyzed the brain tissues of *Epinephelus coioides* and reported hyperemia, hemorrhage, karyosis, tissue necrosis, hyper chromatin, vacuolation, endothelial hypertrophy, dropsy degeneration, and granular ectopic accumulation following the exposure to methylmercury [78]. These findings suggest that MP exposure could lead to similar effects to those observed in the case of heavy metal intoxication.

The optic tectum is a bilobate structure located in the dorsal part of the midbrain. This region of the central nervous system (CNS) mainly acts as a visual center and consists of six different layers with neurons of different shapes and sizes. In contrast to mammals, adult zebrafish have a high number of neurogenic niches distributed throughout the brain with a high capacity for regeneration without the formation of scars or injuries that contribute to visual information processing [79,80]. Also in zebrafish, these highly proliferative areas are widespread and can be detected in all subdivisions of the brain, including the telencephalon, diencephalon, midbrain, and metencephalon [81]. Significant levels of proliferation could be found, especially in the thalamus and specifically in the regions surrounding the habenula, near the optic tectum, along with its subdivisions [80,81]. Adolf et al. (2006) have shown via IHC staining with proliferating cell nuclear antigen (*PCNA*) that the telencephalon contains two different types of neural precursor cells: (1) slow-cycling ones, distributed along the ventricular surface, and (2) fast-cycling ones, mainly organized in a sub palatial group [82]. Slow-cycling progenitors have been identified as radial glial cells (RGCs), while rapidly multiplying progenitors have been described as neuroblasts [81].

Cell proliferation and neurogenesis are two processes susceptible to environmental pollution that have negative consequences for the body. Some studies mention that PCNA-positive cells and cell proliferation themselves decrease in zebrafish brains after exposure at very young ages to Cu and Cu+MP, consistent with reduced *PCNA* gene expression [61]. This inhibition of proliferation and the reduction in *PCNA* expression was recorded in the Ket+Met group, suggesting that this combination could be responsible for oxidative stress, changes in protein phosphorylation, inhibition of G1 to S transition during mitosis, and dysregulation of intracellular calcium ion (Ca^{2+}) homeostasis, as previously documented [82].

It has been shown that between 6 months and 2 years of age, the pool of neural stem cells and progenitor cells gradually depletes, which impacts neuronal regeneration, especially in adults [83]. Thus, despite the observation of cyclic cells in the brains of 7-month-old zebrafish via IHC, the lack of detection of *PCNA* via WB agrees with the findings in the literature, indicating a decrease in the population of NSC cells and cyclic cells in the CNS of zebrafish that were older than 6 months, which makes quantification more difficult.

Our findings showed that Ket administration altered autophagy, as well as proliferation. Researchers have suggested that Ket-induced neurotoxicity could be correlated with increased levels of ROS [84]. In addition, Ket increased ROS production alongside the cell death and differential expression of oxidative-stress-related genes in human neurons in a time- and dose-dependent manner [85]. It is recognized that frequent Ket administration reduces the expression of autophagy proteins, such as LC3, in rat models of traumatic brain injury, probably by activating the mTOR signaling pathway [86], and indeed, Ket has been described as activating mTOR signaling in mammals [84], which is a key regulator of cell growth and proliferation. These results support the disruption of neuronal proliferative and autophagic processes by Ket, as reported in other species and described in humans.

BDNF could act as an antioxidant factor, as it is known to increase the activity levels of some antioxidant enzymes (i.e., SOD and GPx) [87]. The presence of intense *BDNF* expression in the experimental groups demonstrated the ability to activate the antioxidant enzyme system to restore damage caused by MP, Met, Ket, and their combination. The presence of *MAP2* in dendritic extensions emphasizes their role in stabilizing microtubules and is characteristic of differentiated neurons. In addition, immunostaining revealed that *H2A.z* was ubiquitously expressed in the developing cortex and was predominantly localized to the nucleus in the VZ/SVZ (indicating *H2A.z* expression in neuronal progenitor cells), suggesting that *H2A.z* is required for cortical neurogenesis. Thus, our study provides further potential evidence to identify *H2A.z* as a risk gene for neurodevelopmental disorders.

Some significant observations of the current study are the fact that the PCNA and BDNF expression in the MP groups increased, which suggests the presence of DNA lesions and a possible cognitive decline, whereas in terms of inflammation, it was present in all groups besides the control. Moreover, *MAP2* expression has been associated with schizophrenia, which in the context of our study supports the animal models used. In terms of the connection between the immunohistology findings and oxidative stress, *Cox41-i* expression is responsible for the production of superoxide species, whereas tumor necrosis factor interleukin-8 (*TNFIL-8*) is responsible for inducing oxidative stress by producing reactive oxygen species (ROS) and pro-inflammatory cytokines.

Broadly speaking, the results of our current study contribute to filling the knowledge gap within the topic of the neurotoxicity risks of MPs to aquatic organisms. The biochemical and immunohistological responses are consistent with the neurotoxic effects observed in fish treated with MP, Ket, Met, and their combinations. Moreover, alterations in fish exposed to microplastics are consistent with those observed in fish treated with other environmental toxicants, such as heavy metals and metal nanoparticles. It was also observed that the alteration of the evaluated biomarkers was dependent on the exposure design (with special regards to the co-exposure groups), which could lead to further criteria for biological responses to MP exposure, especially in the context of Schizophrenia disorder models. The effect of MP size may be related to the ability to penetrate cells and tissues and their potential to induce molecular damage due to their physical and chemical properties.

4. Materials and Methods

4.1. Ethical Statement

All experimental procedures were approved by the Ethics Committee of the Faculty of Biology, "Alexandru Ioan Cuza" University of Iasi (no. 2533/08.09.2022) and the Ethics Committee of the Faculty of Veterinary Medicine, University of Agricultural Sciences and Veterinary Medicine "Ion Ionescu de la Brad", Iasi (no. 165/26.01.2022) and were conducted in agreement with the European Directive and Romanian Legislation (2010/63/EU and 43/2014) regarding the protection of animals used for scientific or experimental purposes.

4.2. Animals and Housing

A total of 80 wild-type adult zebrafish obtained from an authorized local source were housed according to the literature guidelines (28 °C, 10 L tanks with oxygen pumps, light/dark cycle of 12:12 h, and fed once a day with commercial food for fish). To eliminate any possibility of interference with our experiments and animal welfare, we tested the water parameters daily, measuring nitrites, nitrates, water hardness, alkalinity, pH, and chloride.

4.3. Experimental Models of Schizophrenia and MPs Pollution

After the acclimatization period (14 days), the batch of zebrafish was randomly assigned to eight experimental groups (n = 10/group), the control group (CTR), ketamine group (Ket), methionine group (Met), ketamine+methionine group (Met+Ket), microplastics control (MP), ketamine + MP group (MP+Ket), methionine + MP group (MP+Met), and the co-exposure group (MP+Met+Ket).

Since the current report is part of a more complex study, the experimental design was performed according to the previously published protocol [12]. Briefly, Ket and Met were chronically administrated (5 min exposure to 0.1% Ket/day in separate individual tanks for 5 consecutive days; the tank water containing extemporaneously prepared 6.0 mM Met, replaced every 24 h for 7 consecutive days), while MPs (polypropylene microfibers, <2 mm size) were administered at a concentration of 2 mg L^{-1} at the same time as the food, in an amount required based on the weight of the fish, for 7 consecutive days [88]) for a period of 7 days. The co-exposure protocol consisted of delayed Ket administration (48 h after MP and Met administration).

4.4. Oxidative Stress Analysis

After the exposure protocol was performed, all the animals were euthanized according to European guidelines for standard euthanasia procedures by immersion in cold water for at least 10 min or until opercular movements ceased. The whole bodies were rinsed with distilled water and underwent mechanical processing. The homogenate was mixed with phosphate-buffered saline (PBS) solution and centrifuged at 3500 rpm for 15 min. The fresh supernatant was collected, aliquoted, and used to determine the enzymatic activities of superoxide dismutase (SOD) and glutathione peroxidase (GPx) and the malondialdehyde (MDA) content.

Superoxide dismutase (SOD) activity was determined via an indirect method based on Dojindo's tetrazolium salt (WST-1) reaction with a superoxide anion according to the manufacturer's analytical kit protocols (SOD Assay Kit, Sigma, Roedermark, Germany).

Likewise, GPx activity was determined according to the manufacturer's protocols (GPx Cellular Activity Assay Kit CGP-1, Sigma, Germany) via an indirect method of dynamic observation of substrate consumption, where the rate of NADPH consumption during the considered time unit was used as an indicator of GPx activity.

Malondialdehyde (MDA), which is a measure of lipid peroxidation, was spectrophotometrically measured at 532 nm in a UV–VIS spectrum (Analytik Jena Specord 200, Germany). The MDA content was determined using the thiobarbituric acid reactive substances (TBARS) method, according to the previously described protocols [89,90]. The concentrations of MDA were expressed as relative content per mg of tissue (μmol MDA/mg tissue).

The total soluble protein (TSP) content was assessed based on the Bradford method using a commercial kit according to the manufacturer protocols (Rapid Protein Quantification Kit, Sigma, Germany). The results obtained were calculated against a bovine serum albumin etalon curve and were used to express the antioxidant enzyme activity as specific activity (EU/mg TSP) [89,90]

4.5. Statistical Analysis

All the numerical analyses were performed using Graph Pad Prism software version 9 (San Diego, CA, USA). Firstly, the normality and distribution of the data were determined using the Shapiro–Wilk test. Since the normality test was passed, a two-tailed Student's t test was applied for comparisons between the groups. The data are expressed as the average ± standard error of the mean (SEM), and a $p < 0.05$ was considered to be statistically significant.

4.6. Immunohistology Assays

The fish heads were collected shortly after the fish were euthanized, fixed with Bouin for 48 h and decalcified with EDTA 15% for 7 days, followed by dehydration with ethylic alcohol 70–99.8% and clarification with xylene and fixing in paraffin. The samples were then sectioned at 4 μm, stained with hematoxylin plus eosin (H&E), and immunohistochemically (IHC) colored with the following antibodies: *PCNA*, *TNFAI8*, *Cox41*, *BDNF*, *Map2*, *H2A*, *S100*, *GFAP*, and *Tub2* (Table 5). The sectioning was performed longitudinally.

For each antibody, the number of immunolabeled cells was counted on five fields at magnitude of ×400. The score for immunopositivity was established as follows: + = 1–15 positive cells; ++ = 15–30 positive cells; +++ = 30–60 positive cells; ++++ = 60–100 positive cells.

Table 5. Primary and secondary antibodies alongside the originating species and the applied dilution.

Primary Antibody	Species for Primary Antibody	Primary Antibody Dilution	Species for Secondary Antibody	Secondary Antibody Dilution
PCNA	Rabbit	1:250	Goat anti rabbit	1:250
TNFAI8	Rabbit	1:150	Goat anti rabbit	1:150
Cox41	Rabbit	1:250	Goat anti rabbit	1:250
BDNF	Mouse	1:50	Goat anti mouse	1:50
Map2	Mouse	1:100	Goat anti mouse	1:100
H2A	Rabbit	1:100	Goat anti rabbit	1:100
S100	Rabbit	1:100	Goat anti rabbit	1:100
GFAP	Rabbit	1:1000	Goat anti rabbit	1:10,000
Tub2	Mouse	1:100	Goat anti mouse	1:100

5. Conclusions

Considering the impact that microplastics pollution may have on health and overall wellbeing, the present study proposed an experimental approach measuring the effect of MPs on certain a psychiatric condition, namely schizophrenia. This approach represents an important step forward as it focuses on an already vulnerable category of people and the importance of a clean diet. The presence of MPs alone can alter the measured biomarkers, indicating the potential toxicity of microplastics even without additional contaminants. The severity and frequency of lesions in the optic tectum are more pronounced when MPs are combined with ketamine, as well as with methionine. MPs and methionine seem to play a significant role in stimulating the multiplication of progenitor stem cells or radial glial cells in areas such as the periventricular region, optic tectum, and cerebellar valve. Moreover, methionine exposure leads to structural changes, including creating areas of necrosis in the periventricular zone, across all exposed groups, whether alone or in combination. Despite these detrimental effects, methionine also activated periventricular neurogenesis, as evidenced by the presence of PCNA-positive cells. Moreover, we observed that MPs could have protective effects against oxidative stress, as their combined exposure with Met and Met+Ket led to a successful second antioxidant line of defense alongside reduced lipid peroxidation. Overall, the findings suggest complex interactions between microplastics, ketamine, and methionine in the context of neurogenesis and optic tectum lesions. Future studies will be necessary for a more complete understanding of the mechanisms of action.

Author Contributions: A.S., A.-S.C., L.D.H. and T.M.B.P.—conceptualization, investigation, analysis, visualization, software, and writing—original draft preparation; I.-M.B., P.F.L. and R.J.—investigation and analysis; M.N.N., A.S.C., G.S. and C.S. —writing—review and editing, supervision, and project administration. All authors have read and agreed to the published version of the manuscript.

Funding: This research received no external funding.

Institutional Review Board Statement: The animal study protocol was approved by the Ethics Committee of the Faculty of Biology, "Alexandru Ioan Cuza" University of Iasi (no. 2533/08.09.2022) and the Ethics Committee of the Faculty of Veterinary Medicine, University of Agricultural Sciences and Veterinary Medicine "Ion Ionescu de la Brad", Iasi (no. 165/26.01.2022) and were conducted in agreement with the European Directive and Romanian Legislation (2010/63/EU and 43/2014) regarding the protection of animals used for scientific or experimental purposes.

Informed Consent Statement: Not applicable.

Data Availability Statement: Data are contained within the article.

Conflicts of Interest: The authors declare no conflicts of interest.

References

1. Winship, I.R.; Dursun, S.M.; Baker, G.B.; Balista, P.A.; Kandratavicius, L.; Maia-de-Oliveira, J.P.; Hallak, J.; Howland, J.G. An Overview of Animal Models Related to Schizophrenia. *Can. J. Psychiatry* **2019**, *64*, 5–17. [CrossRef] [PubMed]
2. Andreasen, N.C.; Olsen, S. Negative v Positive Schizophrenia. Definition and Validation. *Arch. Gen. Psychiatry* **1982**, *39*, 789–794. [CrossRef] [PubMed]
3. Stefan, M.; Travis, M.; Murray, R. *An Atlas of Schizophrenia*; The Parthenon Publishing Group: London, UK, 2002; ISBN 1859799745.
4. Jablensky, A. The Diagnostic Concept of Schizophrenia: Its History, Evolution, and Future Prospects. *Dialogues Clin. Neurosci.* **2010**, *12*, 271–287. [CrossRef] [PubMed]
5. Guo, J.Y.; Ragland, J.D.; Carter, C.S. Memory and Cognition in Schizophrenia. *Mol. Psychiatry* **2019**, *24*, 633–642. [CrossRef] [PubMed]
6. Rajji, T.K.; Ismail, Z.; Mulsant, B.H. Age at Onset and Cognition in Schizophrenia: Meta-Analysis. *Br. J. Psychiatry* **2009**, *195*, 286–293. [CrossRef] [PubMed]
7. Kalueff, A.V.; Stewart, A.M.; Gerlai, R. Zebrafish as an Emerging Model for Studying Complex Brain Disorders. *Trends Pharm. Sci.* **2014**, *35*, 63–75. [CrossRef] [PubMed]
8. Choi, T.-Y.; Choi, T.-I.; Lee, Y.-R.; Choe, S.-K.; Kim, C.-H. Zebrafish as an Animal Model for Biomedical Research. *Exp. Mol. Med.* **2021**, *53*, 310–317. [CrossRef] [PubMed]
9. Gawel, K.; Banono, N.S.; Michalak, A.; Esguerra, C.V. A Critical Review of Zebrafish Schizophrenia Models: Time for Validation? *Neurosci. Biobehav. Rev.* **2019**, *107*, 6–22. [CrossRef] [PubMed]
10. Białoń, M.; Wąsik, A. Advantages and Limitations of Animal Schizophrenia Models. *Int. J. Mol. Sci.* **2022**, *23*, 5968. [CrossRef]
11. Wang, L.; Alachkar, A.; Sanathara, N.; Belluzzi, J.D.; Wang, Z.; Civelli, O. A Methionine-Induced Animal Model of Schizophrenia: Face and Predictive Validity. *Int. J. Neuropsychopharmacol.* **2015**, *18*, pyv054. [CrossRef]
12. Curpan, A.-S.; Savuca, A.; Hritcu, L.D.; Solcan, C.; Nicoara, M.N.; Luca, A.-C.; Ciobica, A.-S. A New Approach to Explore the Correlation between Declarative Memory and Anxiety in Animal Models of Schizophrenia and Microplastic Pollution. *Behav. Brain Res.* **2024**, *458*, 114742. [CrossRef] [PubMed]
13. Marx, W.; Moseley, G.; Berk, M.; Jacka, F. Nutritional Psychiatry: The Present State of the Evidence. *Proc. Nutr. Soc.* **2017**, *76*, 427–436. [CrossRef] [PubMed]
14. Onaolapo, O.J.; Onaolapo, A.Y. Nutrition, Nutritional Deficiencies, and Schizophrenia: An Association Worthy of Constant Reassessment. *World J. Clin. Cases* **2021**, *9*, 8295–8311. [CrossRef] [PubMed]
15. Cha, H.Y.; Yang, S.J. Anti-Inflammatory Diets and Schizophrenia. *Clin. Nutr. Res.* **2020**, *9*, 241. [CrossRef] [PubMed]
16. Berding, K.; Vlckova, K.; Marx, W.; Schellekens, H.; Stanton, C.; Clarke, G.; Jacka, F.; Dinan, T.G.; Cryan, J.F. Diet and the Microbiota–Gut–Brain Axis: Sowing the Seeds of Good Mental Health. *Adv. Nutr.* **2021**, *12*, 1239–1285. [CrossRef] [PubMed]
17. Sasso, J.M.; Ammar, R.M.; Tenchov, R.; Lemmel, S.; Kelber, O.; Grieswelle, M.; Zhou, Q.A. Gut Microbiome–Brain Alliance: A Landscape View into Mental and Gastrointestinal Health and Disorders. *ACS Chem. Neurosci.* **2023**, *14*, 1717–1763. [CrossRef]
18. Yuan, Z.; Nag, R.; Cummins, E. Human Health Concerns Regarding Microplastics in the Aquatic Environment—From Marine to Food Systems. *Sci. Total Environ.* **2022**, *823*, 153730. [CrossRef] [PubMed]
19. Van Raamsdonk, L.W.D.; Van Der Zande, M.; Koelmans, A.A.; Hoogenboom, P.L.A.; Peters, R.J.B.; Groot, M.J.; Peijnenburg, M.A.C.; Weesepoel, Y.J.A. Current Insights into Monitoring, Bioaccumulation, and Potential Health Effects of Microplastics Present in the Food Chain. *Foods* **2020**, *9*, 72. [CrossRef]
20. Al Mamun, A.; Prasetya, T.A.E.; Dewi, I.R.; Ahmad, M. Microplastics in Human Food Chains: Food Becoming a Threat to Health Safety. *Sci. Total Environ.* **2023**, *858*, 159834. [CrossRef]
21. Prata, J.C.; da Costa, J.P.; Lopes, I.; Duarte, A.C.; Rocha-Santos, T. Environmental Exposure to Microplastics: An Overview on Possible Human Health Effects. *Sci. Total Environ.* **2020**, *702*, 134455. [CrossRef]
22. Wu, D.; Feng, Y.; Wang, R.; Jiang, J.; Guan, Q.; Yang, X.; Wei, H.; Xia, Y.; Luo, Y. Pigment Microparticles and Microplastics Found in Human Thrombi Based on Raman Spectral Evidence. *J. Adv. Res.* **2023**, *49*, 141–150. [CrossRef] [PubMed]
23. Montano, L.; Giorgini, E.; Notarstefano, V.; Notari, T.; Ricciardi, M.; Piscopo, M.; Motta, O. Raman Microspectroscopy Evidence of Microplastics in Human Semen. *Sci. Total Environ.* **2023**, *901*, 165922. [CrossRef] [PubMed]
24. Yu, Y.; Kumar, M.; Bolan, S.; Padhye, L.P.; Bolan, N.; Li, S.; Wang, L.; Hou, D.; Li, Y. Various Additive Release from Microplastics and Their Toxicity in Aquatic Environments. *Environ. Pollut.* **2024**, *343*, 123219. [CrossRef] [PubMed]
25. Lin, W.; Hu, F.; Liu, F.; Liao, L.; Ling, L.; Li, L.; Yang, J.; Yang, P. Microcystin-LR and Polystyrene Microplastics Jointly Lead to Hepatic Histopathological Damage and Antioxidant Dysfunction in Male Zebrafish. *Environ. Pollut.* **2024**, *347*, 123789. [CrossRef] [PubMed]
26. Zaheer, J.; Kim, H.; Ko, I.O.; Jo, E.-K.; Choi, E.-J.; Lee, H.-J.; Shim, I.; Woo, H.; Choi, J.; Kim, G.-H.; et al. Pre/Post-Natal Exposure to Microplastic as a Potential Risk Factor for Autism Spectrum Disorder. *Environ. Int.* **2022**, *161*, 107121. [CrossRef] [PubMed]

27. Qiao, R.; Sheng, C.; Lu, Y.; Zhang, Y.; Ren, H.; Lemos, B. Microplastics Induce Intestinal Inflammation, Oxidative Stress, and Disorders of Metabolome and Microbiome in Zebrafish. *Sci. Total Environ.* **2019**, *662*, 246–253. [CrossRef]
28. Ding, P.; Xiang, C.; Li, X.; Chen, H.; Shi, X.; Li, X.; Huang, C.; Yu, Y.; Qi, J.; Li, A.J.; et al. Photoaged Microplastics Induce Neurotoxicity via Oxidative Stress and Abnormal Neurotransmission in Zebrafish Larvae (*Danio rerio*). *Sci. Total Environ.* **2023**, *881*, 163480. [CrossRef]
29. Félix, L.; Carreira, P.; Peixoto, F. Effects of Chronic Exposure of Naturally Weathered Microplastics on Oxidative Stress Level, Behaviour, and Mitochondrial Function of Adult Zebrafish (*Danio rerio*). *Chemosphere* **2023**, *310*, 136895. [CrossRef]
30. Pizzino, G.; Irrera, N.; Cucinotta, M.; Pallio, G.; Mannino, F.; Arcoraci, V.; Squadrito, F.; Altavilla, D.; Bitto, A. Oxidative Stress: Harms and Benefits for Human Health. *Oxid. Med. Cell Longev.* **2017**, *2017*, 8416763. [CrossRef]
31. Bošković, M.; Vovk, T.; Plesničar, B.K.; Grabnar, I. Oxidative Stress in Schizophrenia. *Curr. Neuropharmacol.* **2011**, *9*, 301. [CrossRef]
32. Huang, W.J.; Zhang, X.; Chen, W.W. Role of Oxidative Stress in Alzheimer's Disease. *Biomed. Rep.* **2016**, *4*, 519. [CrossRef] [PubMed]
33. Dias, V.; Junn, E.; Mouradian, M.M. The Role of Oxidative Stress in Parkinson's Disease. *J Park. Dis.* **2013**, *3*, 461. [CrossRef] [PubMed]
34. Padurariu, M.; Ciobica, A.; Lefter, R.; Serban, I.L.; Stefanescu, C.; Chirita, R. The Oxidative Stress Hypothesis in Alzheimer's Disease. *Psychiatr. Danub.* **2013**, *25*, 401–409.
35. Ciobica, A.; Padurariu, M.; Dobrin, I.; Stefanescu, C.; Dobrin, R. Oxidative Stress in Schizophrenia—Focusing on the Main Markers. *Psychiatr. Danub.* **2011**, *23*, 237–245. [PubMed]
36. Stefanescu, C.; Ciobica, A. The Relevance of Oxidative Stress Status in First Episode and Recurrent Depression. *J. Affect Disord.* **2012**, *143*, 34–38. [CrossRef]
37. Robea, M.-A.; Balmus, I.-M.; Ciobica, A.; Strungaru, S.; Plavan, G.; Gorgan, L.D.; Savuca, A.; Nicoara, M. Parkinson's Disease-Induced Zebrafish Models: Focussing on Oxidative Stress Implications and Sleep Processes. *Oxid. Med. Cell Longev.* **2020**, *2020*, 1370837. [CrossRef] [PubMed]
38. Balmus, I.M.; Ciobica, A.; Antioch, I.; Dobrin, R.; Timofte, D. Oxidative Stress Implications in the Affective Disorders: Main Biomarkers, Animal Models Relevance, Genetic Perspectives, and Antioxidant Approaches. *Oxid. Med. Cell Longev.* **2016**, *2016*, 3975101. [CrossRef] [PubMed]
39. Robea, M.-A.; Săvucă, A.; Strungaru, Ș.-A.; Lenzi, C.; Grasso, C.; Plăvan, G.; Ciobică, A.; Nicoară, M. The General Biological Relevance of the Oxidative Stress Status in Replicating Some Neuropsychiatric and Digestive-Related Manifestations in Zebrafish. *Rom Biotechnol. Lett.* **2019**, *24*, 571–579. [CrossRef]
40. Dietrich-Muszalska, A.; Kontek, B. Lipid Peroxidation in Patients with Schizophrenia. *Psychiatry Clin. Neurosci.* **2010**, *64*, 469–475. [CrossRef]
41. Piatoikina, A.S.; Lyakhova, A.A.; Semennov, I.V.; Zhilyaeva, T.V.; Kostina, O.V.; Zhukova, E.S.; Shcherbatyuk, T.G.; Kasyanov, E.D.; Blagonravova, A.S.; Mazo, G.E. Association of Antioxidant Deficiency and the Level of Products of Protein and Lipid Peroxidation in Patients with the First Episode of Schizophrenia. *J. Mol. Neurosci.* **2022**, *72*, 217–225. [CrossRef]
42. Mabrouk, H.; Houas, I.; Mechria, H.; Mechri, A.; Douki, W.; Gaha, L.; Najjar, M.F. Oxidative Stress Markers in Schizophrenic Patients. *Immuno-Anal. Biol. Spécialisée* **2013**, *28*, 51–56. [CrossRef]
43. Younus, H. Therapeutic Potentials of Superoxide Dismutase. *Int. J. Health Sci. (Qassim)* **2018**, *12*, 88–93.
44. Alfhili, M.A.; Lee, M.-H. Triclosan: An Update on Biochemical and Molecular Mechanisms. *Oxid. Med. Cell Longev.* **2019**, *2019*, 1607304. [CrossRef]
45. Sheng, C.; Zhang, S.; Zhang, Y. The Influence of Different Polymer Types of Microplastics on Adsorption, Accumulation, and Toxicity of Triclosan in Zebrafish. *J. Hazard Mater.* **2021**, *402*, 123733. [CrossRef] [PubMed]
46. Del Rio, D.; Stewart, A.J.; Pellegrini, N. A Review of Recent Studies on Malondialdehyde as Toxic Molecule and Biological Marker of Oxidative Stress. *Nutr. Metab. Cardiovasc. Dis.* **2005**, *15*, 316–328. [CrossRef] [PubMed]
47. Zanandrea, R.; Wiprich, M.T.; Altenhofen, S.; Rubensam, G.; dos Santos, T.M.; Wyse, A.T.S.; Bonan, C.D. Paternal Exposure to Excessive Methionine Altered Behavior and Neurochemical Activities in Zebrafish Offspring. *Amino Acids* **2021**, *53*, 1153–1167. [CrossRef]
48. Zanandrea, R.; Wiprich, M.T.; Altenhofen, S.; Rubensam, G.; dos Santos, T.M.; Wyse, A.T.S.; Bonan, C.D. Withdrawal Effects Following Methionine Exposure in Adult Zebrafish. *Mol. Neurobiol.* **2020**, *57*, 3485–3497. [CrossRef]
49. Liao, P.-H.; Yang, W.-K.; Yang, C.-H.; Lin, C.-H.; Hwang, C.-C.; Chen, P.-J. Illicit Drug Ketamine Induces Adverse Effects from Behavioral Alterations and Oxidative Stress to P53-Regulated Apoptosis in Medaka Fish under Environmentally Relevant Exposures. *Environ. Pollut.* **2018**, *237*, 1062–1071. [CrossRef] [PubMed]
50. Balmus, I.-M.; Alexa, A.I.; Ciuntu, R.-E.; Danielescu, C.; Stoica, B.; Cojocaru, S.I.; Ciobica, A.; Cantemir, A. Oxidative Stress Markers Dynamics in Keratoconus Patients' Tears before and after Corneal Collagen Crosslinking Procedure. *Exp. Eye Res.* **2020**, *190*, 107897. [CrossRef]
51. Félix, L.M.; Vidal, A.M.; Serafim, C.; Valentim, A.M.; Antunes, L.M.; Monteiro, S.M.; Matos, M.; Coimbra, A.M. Ketamine Induction of P53-Dependent Apoptosis and Oxidative Stress in Zebrafish (*Danio rerio*) Embryos. *Chemosphere* **2018**, *201*, 730–739. [CrossRef]

52. Wu, P.; Tang, L.; Jiang, W.; Hu, K.; Liu, Y.; Jiang, J.; Kuang, S.; Tang, L.; Tang, W.; Zhang, Y.; et al. The Relationship between Dietary Methionine and Growth, Digestion, Absorption, and Antioxidant Status in Intestinal and Hepatopancreatic Tissues of Sub-Adult Grass Carp (*Ctenopharyngodon idella*). *J. Anim. Sci. Biotechnol.* **2017**, *8*, 63. [CrossRef] [PubMed]
53. Lubos, E.; Loscalzo, J.; Handy, D.E. Glutathione Peroxidase-1 in Health and Disease: From Molecular Mechanisms to Therapeutic Opportunities. *Antioxid. Redox Signal* **2011**, *15*, 1957. [CrossRef]
54. Sorensen, R.M.; Jovanović, B. From Nanoplastic to Microplastic: A Bibliometric Analysis on the Presence of Plastic Particles in the Environment. *Mar. Pollut. Bull.* **2021**, *163*, 111926. [CrossRef]
55. Ding, R.; Tong, L.; Zhang, W. Microplastics in Freshwater Environments: Sources, Fates and Toxicity. *Water Air Soil Pollut.* **2021**, *232*, 181. [CrossRef]
56. Stapleton, M.J.; Hai, F.I. Microplastics as an Emerging Contaminant of Concern to Our Environment: A Brief Overview of the Sources and Implications. *Bioengineered* **2023**, *14*, 2244754. [CrossRef] [PubMed]
57. Albendín, M.G.; Aranda, V.; Coello, M.D.; González-Gómez, C.; Rodríguez-Barroso, R.; Quiroga, J.M.; Arellano, J.M. Pharmaceutical Products and Pesticides Toxicity Associated with Microplastics (Polyvinyl Chloride) in Artemia Salina. *Int. J. Env. Res. Public Health* **2021**, *18*, 10773. [CrossRef]
58. Wang, H.; Xu, J.; Yuan, Y.; Wang, Z.; Zhang, W.; Li, J. The Exploration of Joint Toxicity and Associated Mechanisms of Primary Microplastics and Methamphetamine in Zebrafish Larvae. *Toxics* **2024**, *12*, 64. [CrossRef]
59. Han, X.; Fu, L.; Yu, J.; Li, K.; Deng, Z.; Shu, R.; Wang, D.; You, J.; Zeng, E.Y. Effects of Erythromycin on Biofilm Formation and Resistance Mutation of Escherichia Coli on Pristine and UV-Aged Polystyrene Microplastics. *Water Res.* **2024**, *256*, 121628. [CrossRef] [PubMed]
60. Barboza, L.G.A.; Vieira, L.R.; Branco, V.; Figueiredo, N.; Carvalho, F.; Carvalho, C.; Guilhermino, L. Microplastics Cause Neurotoxicity, Oxidative Damage and Energy-Related Changes and Interact with the Bioaccumulation of Mercury in the European Seabass, Dicentrarchus Labrax (Linnaeus, 1758). *Aquat. Toxicol.* **2018**, *195*, 49–57. [CrossRef]
61. Santos, D.; Félix, L.; Luzio, A.; Parra, S.; Bellas, J.; Monteiro, S.M. Single and Combined Acute and Subchronic Toxic Effects of Microplastics and Copper in Zebrafish (*Danio rerio*) Early Life Stages. *Chemosphere* **2021**, *277*, 130262. [CrossRef]
62. Chen, Q.; Gundlach, M.; Yang, S.; Jiang, J.; Velki, M.; Yin, D.; Hollert, H. Quantitative Investigation of the Mechanisms of Microplastics and Nanoplastics toward Zebrafish Larvae Locomotor Activity. *Sci. Total Environ.* **2017**, *584–585*, 1022–1031. [CrossRef] [PubMed]
63. Deng, Y.; Zhang, Y.; Lemos, B.; Ren, H. Tissue Accumulation of Microplastics in Mice and Biomarker Responses Suggest Widespread Health Risks of Exposure. *Sci. Rep.* **2017**, *7*, 46687. [CrossRef] [PubMed]
64. Prüst, M.; Meijer, J.; Westerink, R.H.S. The Plastic Brain: Neurotoxicity of Micro- and Nanoplastics. *Part Fibre Toxicol.* **2020**, *17*, 24. [CrossRef] [PubMed]
65. Manuel, P.; Almeida, M.; Martins, M.; Oliveira, M. Effects of Nanoplastics on Zebrafish Embryo-Larval Stages: A Case Study with Polystyrene (PS) and Polymethylmethacrylate (PMMA) Particles. *Environ. Res.* **2022**, *213*, 113584. [CrossRef] [PubMed]
66. Duan, Z.; Duan, X.; Zhao, S.; Wang, X.; Wang, J.; Liu, Y.; Peng, Y.; Gong, Z.; Wang, L. Barrier Function of Zebrafish Embryonic Chorions against Microplastics and Nanoplastics and Its Impact on Embryo Development. *J. Hazard Mater.* **2020**, *395*, 122621. [CrossRef] [PubMed]
67. Prata, J.C.; Venâncio, C.; Girão, A.V.; da Costa, J.P.; Lopes, I.; Duarte, A.C.; Rocha-Santos, T. Effects of Virgin and Weathered Polystyrene and Polypropylene Microplastics on Raphidocelis Subcapitata and Embryos of *Danio rerio* under Environmental Concentrations. *Sci. Total Environ.* **2022**, *816*, 151642. [CrossRef] [PubMed]
68. Pitt, J.A.; Trevisan, R.; Massarsky, A.; Kozal, J.S.; Levin, E.D.; di Giulio, R.T. Maternal Transfer of Nanoplastics to Offspring in Zebrafish (*Danio rerio*): A Case Study with Nanopolystyrene. *Sci. Total Environ.* **2018**, *643*, 324–334. [CrossRef]
69. Chen, Q.; Yin, D.; Jia, Y.; Schiwy, S.; Legradi, J.; Yang, S.; Hollert, H. Enhanced Uptake of BPA in the Presence of Nanoplastics Can Lead to Neurotoxic Effects in Adult Zebrafish. *Sci. Total Environ.* **2017**, *609*, 1312–1321. [CrossRef] [PubMed]
70. Sökmen, T.Ö.; Sulukan, E.; Türkoğlu, M.; Baran, A.; Özkaraca, M.; Ceyhun, S.B. Polystyrene Nanoplastics (20 Nm) Are Able to Bioaccumulate and Cause Oxidative DNA Damages in the Brain Tissue of Zebrafish Embryo (*Danio rerio*). *Neurotoxicology* **2020**, *77*, 51–59. [CrossRef]
71. Jung, B.-K.; Han, S.-W.; Park, S.-H.; Bae, J.-S.; Choi, J.; Ryu, K.-Y. Neurotoxic Potential of Polystyrene Nanoplastics in Primary Cells Originating from Mouse Brain. *Neurotoxicology* **2020**, *81*, 189–196. [CrossRef]
72. Massaad, C.A.; Klann, E. Reactive Oxygen Species in the Regulation of Synaptic Plasticity and Memory. *Antioxid. Redox Signal* **2011**, *14*, 2013–2054. [CrossRef]
73. Hamed, M.; Martyniuk, C.J.; Naguib, M.; Lee, J.-S.; Sayed, A.E.-D.H. Neurotoxic Effects of Different Sizes of Plastics (Nano, Micro, and Macro) on Juvenile Common Carp (*Cyprinus carpio*). *Front. Mol. Neurosci.* **2022**, *15*, 1028364. [CrossRef]
74. Hamed, M.; Soliman, H.A.M.; Badrey, A.E.A.; Osman, A.G.M. Microplastics Induced Histopathological Lesions in Some Tissues of Tilapia (*Oreochromis niloticus*) Early Juveniles. *Tissue Cell* **2021**, *71*, 101512. [CrossRef] [PubMed]
75. Romano, N.; Renukdas, N.; Fischer, H.; Shrivastava, J.; Baruah, K.; Egnew, N.; Sinha, A.K. Differential Modulation of Oxidative Stress, Antioxidant Defense, Histomorphology, Ion-Regulation and Growth Marker Gene Expression in Goldfish (*Carassius auratus*) Following Exposure to Different Dose of Virgin Microplastics. *Comp. Biochem. Physiol. Part C Toxicol. Pharmacol.* **2020**, *238*, 108862. [CrossRef] [PubMed]

76. Karami, A.; Groman, D.B.; Wilson, S.P.; Ismail, P.; Neela, V.K. Biomarker Responses in Zebrafish (*Danio rerio*) Larvae Exposed to Pristine Low-Density Polyethylene Fragments. *Environ. Pollut.* **2017**, *223*, 466–475. [CrossRef] [PubMed]
77. Tian, R.; Guan, M.; Chen, L.; Wan, Y.; He, L.; Zhao, Z.; Gao, T.; Zong, L.; Chang, J.; Zhang, J. Mechanism Insights into the Histopathological Changes of Polypropylene Microplastics Induced Gut and Liver in Zebrafish. *Ecotoxicol. Environ. Saf.* **2024**, *280*, 116537. [CrossRef]
78. Savari, S.; Safahieh, A.; Archangi, B.; Savari, A.; Abdi, R. The Histopathological Effect of Methylmercury on the Brain in Orange Spotted Grouper (*Epinephelus coioides*) in Zangi Creek and Laboratory. *Iran. J. Fish. Sci.* **2020**, *19*, 457–470. [CrossRef]
79. Boulanger-Weill, J.; Sumbre, G. Functional Integration of Newborn Neurons in the Zebrafish Optic Tectum. *Front Cell Dev. Biol.* **2019**, *7*, 57. [CrossRef]
80. Pellegrini, E.; Mouriec, K.; Anglade, I.; Menuet, A.; Le Page, Y.; Gueguen, M.; Marmignon, M.; Brion, F.; Pakdel, F.; Kah, O. Identification of Aromatase-positive Radial Glial Cells as Progenitor Cells in the Ventricular Layer of the Forebrain in Zebrafish. *J. Comp. Neurol.* **2007**, *501*, 150–167. [CrossRef]
81. Ghaddar, B.; Lübke, L.; Couret, D.; Rastegar, S.; Diotel, N. Cellular Mechanisms Participating in Brain Repair of Adult Zebrafish and Mammals after Injury. *Cells* **2021**, *10*, 391. [CrossRef]
82. Adolf, B.; Chapouton, P.; Lam, C.S.; Topp, S.; Tannhäuser, B.; Strähle, U.; Götz, M.; Bally-Cuif, L. Conserved and Acquired Features of Adult Neurogenesis in the Zebrafish Telencephalon. *Dev. Biol.* **2006**, *295*, 278–293. [CrossRef] [PubMed]
83. Kaslin, J.; Kroehne, V.; Ganz, J.; Hans, S.; Brand, M. Correction: Distinct Roles of Neuroepithelial-like and Radial Glia-like Progenitor Cells in Cerebellar Regeneration. *Development* **2017**, *144*, 3388. [CrossRef] [PubMed]
84. Choudhury, D.; Autry, A.E.; Tolias, K.F.; Krishnan, V. Ketamine: Neuroprotective or Neurotoxic? *Front. Neurosci.* **2021**, *15*, 672526. [CrossRef] [PubMed]
85. Bosnjak, J.Z.; Yan, Y.; Canfield, S.; Muravyeva, M.Y.; Kikuchi, C.; Wells, C.W.; Corbett, J.A.; Bai, X. Ketamine Induces Toxicity in Human Neurons Differentiated from Embryonic Stem Cells via Mitochondrial Apoptosis Pathway. *Curr. Drug Saf.* **2012**, *7*, 106–119. [CrossRef] [PubMed]
86. Wang, C.-Q.; Ye, Y.; Chen, F.; Han, W.-C.; Sun, J.-M.; Lu, X.; Guo, R.; Cao, K.; Zheng, M.-J.; Liao, L.-C. Posttraumatic Administration of a Sub-Anesthetic Dose of Ketamine Exerts Neuroprotection via Attenuating Inflammation and Autophagy. *Neuroscience* **2017**, *343*, 30–38. [CrossRef] [PubMed]
87. Correia, A.S.; Cardoso, A.; Vale, N. BDNF Unveiled: Exploring Its Role in Major Depression Disorder Serotonergic Imbalance and Associated Stress Conditions. *Pharmaceutics* **2023**, *15*, 2081. [CrossRef] [PubMed]
88. Avdesh, A.; Chen, M.; Martin-Iverson, M.T.; Mondal, A.; Ong, D.; Rainey-Smith, S.; Taddei, K.; Lardelli, M.; Groth, D.M.; Verdile, G.; et al. Regular Care and Maintenance of a Zebrafish (*Danio rerio*) Laboratory: An Introduction. *J. Vis. Exp.* **2012**, *18*, e4196. [CrossRef]
89. Balmus, I.-M.; Lefter, R.; Ciobica, A.; Cojocaru, S.; Guenne, S.; Timofte, D.; Stanciu, C.; Trifan, A.; Hritcu, L. Preliminary Biochemical Description of Brain Oxidative Stress Status in Irritable Bowel Syndrome Contention-Stress Rat Model. *Medicina* **2019**, *55*, 776. [CrossRef]
90. Bild, W.; Hritcu, L.; Stefanescu, C.; Ciobica, A. Inhibition of Central Angiotensin II Enhances Memory Function and Reduces Oxidative Stress Status in Rat Hippocampus. *Prog. Neuropsychopharmacol. Biol. Psychiatry* **2013**, *43*, 79–88. [CrossRef]

International Journal of
Molecular Sciences

MDPI

Article

Mild Oxidative Stress Induced by Sodium Arsenite Reduces Lipocalin-2 Expression Levels in Cortical Glial Cells

Ye-Jin Cho, So-Hyun Park and Kwon-Yul Ryu *

Department of Life Science, University of Seoul, Seoul 02504, Republic of Korea; fifi1022@naver.com (Y.-J.C.); sohyun1231@naver.com (S.-H.P.)
* Correspondence: kyryu@uos.ac.kr

Abstract: Astrocytes and microglia, the most abundant glial cells in the central nervous system, are involved in maintaining homeostasis in the brain microenvironment and in the progression of various neurological disorders. Lipocalin-2 (LCN2) is a small secretory protein that can be transcriptionally upregulated via nuclear factor kappa B (NF-κB) signaling. It is synthesized and secreted by glial cells, resulting in either the restoration of damaged neural tissues or the induction of neuronal apoptosis in a context-dependent manner. It has recently been reported that when glial cells are under lipopolysaccharide-induced inflammatory stress, either reduced production or accelerated degradation of LCN2 can alleviate neurotoxicity. However, the regulatory mechanisms of LCN2 in glial cells are not yet fully understood. In this study, we used primary astroglial-enriched cells which produce LCN2 and found that the production of LCN2 could be reduced by sodium arsenite treatment. Surprisingly, the reduced LCN2 production was not due to the suppression of NF-κB signaling. Mild oxidative stress induced by sodium arsenite treatment activated antioxidant responses and downregulated *Lcn2* expression without reducing the viability of astroglial-enriched cells. Intriguingly, reduced LCN2 production could not be achieved by simple activation of the nuclear factor erythroid-2-related factor 2 (Nrf2)–Kelch-like ECH-associated protein 1 (Keap1) pathway in astroglial-enriched cells. Thus, it appears that mild oxidative stress, occurring in an Nrf2-independent manner, is required for the downregulation of *Lcn2* expression. Taken together, our findings provide new insights into the regulatory mechanisms of LCN2 and suggest that mild oxidative stress may alter LCN2 homeostasis, even under neuroinflammatory conditions.

Keywords: astrocyte; lipocalin-2; oxidative stress; glial cells; sodium arsenite

1. Introduction

In the central nervous system, astrocytes and microglia perform diverse physiological functions, including the control of extracellular ion concentration, nutrient provision, and the regulation of blood flow by interacting with endothelial cells in the blood–brain barrier [1,2]. Moreover, they are involved in several neurological disorders through reactive astrogliosis [3]. The conventional classification of reactive astrocytes as A1 or A2 is based on their molecular and cellular characteristics [4]. According to a previous study, A1 astrocytes are induced by the activation of nuclear factor kappa B (NF-κB) signaling, while A2 astrocytes are induced by the activation of Janus kinase (JAK)/signal transducer and activator of transcription 3 (STAT3) signaling [5,6]. A1 and A2 astrocytes show high expression levels of pro-inflammatory and anti-inflammatory cytokines, respectively. A1 astrocytes are considered to be neurotoxic as they secrete neurotoxins, such as lipocalin-2 (LCN2), and A2 astrocytes are neuroprotective as they secrete neurotrophic factors, such as brain-derived neurotrophic factor (BDNF) or vascular endothelial growth factor (VEGF) [7–9]. In neuroinflammatory diseases, several cytokines derived from other cells, such as microglia, induce massive transcriptome alterations in quiescent astrocytes, resulting in a dominant population of A1 astrocytes and the secretion of neurotoxic molecules, including LCN2 [10].

However, it is widely accepted that the characteristics of reactive astrocytes should be considered in a context-dependent manner, rather than the traditional dichotomous concept that A1 astrocytes are neurotoxic and A2 astrocytes are neuroprotective [11,12].

LCN2, also known as neutrophil-associated gelatinase lipocalin or 24p3, has been identified as an acute-phase protein as its concentration dramatically increases in the blood under inflammatory stress conditions [13,14]. This secretory protein is conserved between mice and humans and it binds to several types of siderophores, which are strong iron utilizers found in some infectious bacteria and fungi, resulting in the sequestration of iron-utilizing molecules and inhibition of the growth of the bacteria and fungi. Therefore, increased LCN2 levels in plasma and several tissues, such as the kidney, lung, and brain, are often observed in pathological conditions [15–18]. *Lcn2* expression is regulated by several cell signaling pathways that are activated under inflammatory stress conditions, such as the NF-κB signaling and JAK/STAT3 signaling pathways [19]. Tumor necrosis factor alpha (TNF-α) and interferon gamma (IFN-γ) are required for the upregulation of *Lcn2* expression in various cell types [19,20]. Recently, several studies have found that LCN2 is produced and secreted by reactive astrocytes or activated microglia in the nervous system and is involved in the pathological progression of neuroinflammatory diseases or the restoration of nervous tissue after ischemic injury [21,22]. Of note, LCN2 secreted from reactive astrocytes can affect both damaged and healthy neurons [7].

Herein, we demonstrated that LCN2 production in primary astroglial-enriched cells can be significantly reduced under mild oxidative stress conditions induced by sodium arsenite treatment. Mild oxidative stress did not affect the viability of astroglial-enriched cells. Although activation of the nuclear factor erythroid-2-related factor 2 (Nrf2)–Kelch-like ECH-associated protein 1 (Keap1) pathway, the major antioxidant response pathway, has been suggested to suppress NF-κB signaling based on their molecular crosstalk, these two pathways may not interact or their interaction might be complex in astroglial-enriched cells [23]. Neither the activation of the Nrf2-Keap1 pathway nor the suppression of NF-κB signaling was the cause of reduced *Lcn2* expression under sodium arsenite treatment. Mild oxidative stress, induced by sodium arsenite treatment in an Nrf2-independent manner, seems to be responsible for the downregulation of *Lcn2* expression. Our current study suggests a novel strategy for reducing LCN2 levels to prevent its potentially neurotoxic effects without altering the NF-κB signaling pathway.

2. Results

2.1. Astroglial-Enriched Cells Are Resistant to Mild Oxidative Stress without Reduced Viability

When astroglial-enriched cells were exposed to sodium arsenite at concentrations of 100 μM or higher, their viability was markedly reduced (Figure 1A). According to previous studies, treatment with 10 μM sodium arsenite reduces the viability of primary neurons and mouse embryonic fibroblasts [24]. However, we found no reduction in the viability of astroglial-enriched cells when treated with the same concentration of sodium arsenite as used previously (10 μM), indicating that these cells may be more resistant to oxidative stress than other types of primary cells (Figure 1B). In fact, the LC50 of sodium arsenite was about 60 μM (Figure S1). Moreover, we observed that the MTT absorbance levels of astroglial-enriched cells did not markedly decrease until 50 μM sodium arsenite treatment, but rather increased with 10 μM sodium arsenite treatment (Figure 1B). Therefore, in this study, we defined the oxidative stress induced by sodium arsenite treatment of astroglial-enriched cells, without cytotoxicity, as "mild oxidative stress".

2.2. Mild Oxidative Stress Activates the Nrf2–Keap1 Pathway and Reduces Lcn2 Expression Levels in Astroglial-Enriched Cells

We cultured astroglial-enriched cells without mechanistically removing microglia and oligodendrocytes until passage number two (P2). We found that glial fibrillary acidic protein (Gfap)-positive cells (astrocytes) were the dominant population (Figure 2A), although Gfap-negative cells (e.g., microglia) were also present under these conditions. Of note, we observed that prolonged exposure of astroglial-enriched cells to fetal bovine serum during the culture period induced their activation and subsequent production of LCN2, even without lipopolysaccharide (LPS)-induced inflammatory stress (Figure 2B,D). Although consistent expression of *Lcn2* in astroglial-enriched cells may not fully recapitulate normal physiological conditions, it is highly relevant to pathophysiological conditions and is suitable for investigating strategies to reduce LCN2 levels.

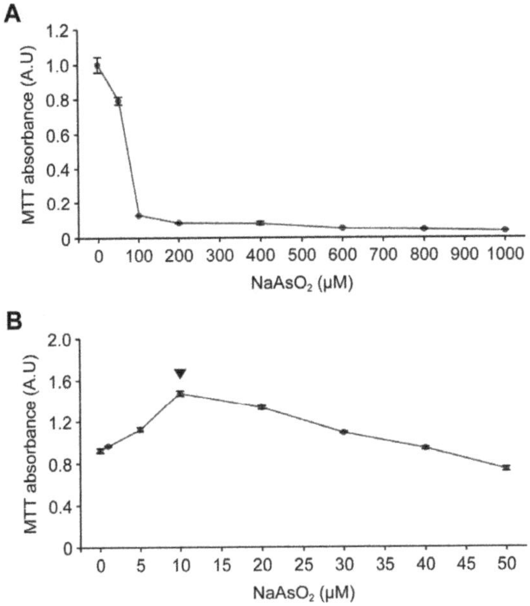

Figure 1. Astroglial-enriched cells survive under mild oxidative stress conditions without reduced viability. (**A**) astroglial-enriched cells were treated with sodium arsenite ($NaAsO_2$) at the indicated concentrations up to 1000 μM for 1 d. The cells were then subjected to MTT assays. After incubation and elution with DMSO, the converted MTT level in each well was measured using a microplate reader. Data are expressed as MTT absorbance levels (arbitrary units, A.U; $n = 3$). (**B**) MTT assays were performed using astroglial-enriched cells, which were treated with $NaAsO_2$ at the indicated concentrations up to 50 μM for 1 d. Data are expressed as described in (**A**) ($n = 3$). Arrowheads indicate the concentration of $NaAsO_2$ that showed the highest MTT absorbance levels. All data are expressed as the means ± standard error of the mean (SEM) from the indicated number of samples.

Next, we investigated whether mild oxidative stress in astroglial-enriched cells activated the antioxidant response pathway. The levels of heme oxygenase 1 (Hmox1) and NAD(P)H quinone dehydrogenase 1 (Nqo1) in these cells increased under mild oxidative stress, suggesting that the Nrf2–Keap1 pathway was activated (Figures 2B,C and S2A). Intriguingly, under the same conditions, intracellular LCN2 was reduced at the protein level, as confirmed by both immunoblot and immunofluorescence analyses (Figures 2A,B and S2B). Moreover, reduced LCN2 levels and increased Hmox1 levels were also observed in astroglial-enriched cells treated with LPS under mild oxidative stress conditions (Figures 2B and S2A,B). Thus, the effect of sodium arsenite seemed to override LPS-induced pro-inflammatory cytokine signaling, which can upregulate the expression of *Lcn2*.

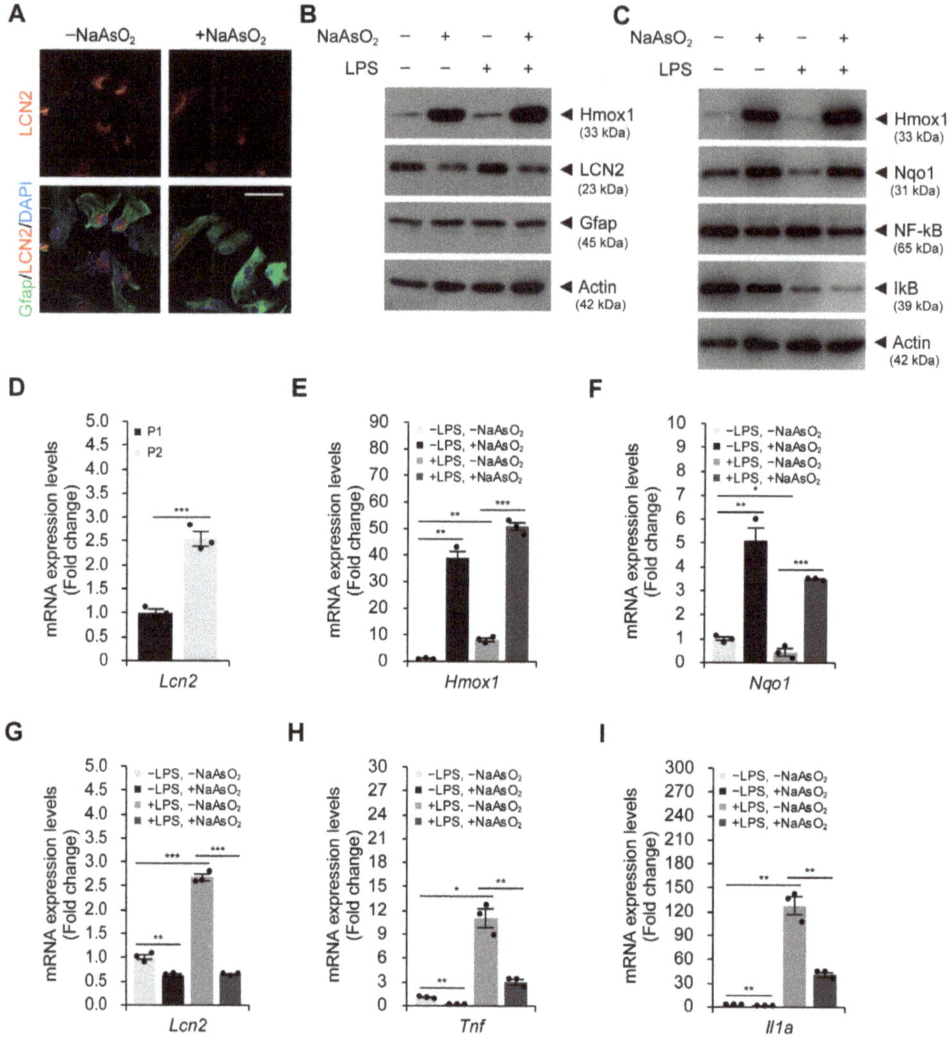

Figure 2. Mild oxidative stress activates the antioxidant response pathway and reduces LCN2 levels in astroglial-enriched cells. (**A**) immunofluorescence staining to detect intracellular LCN2 and Gfap in astroglial-enriched cells was performed with or without 10 μM $NaAsO_2$ treatment. DNA was visualized with DAPI. (**B**) immunoblot detection of Hmox1, LCN2, and Gfap was performed using astroglial-enriched cells treated with 10 μM sodium arsenite ($NaAsO_2$) or 100 ng/mL LPS for 1 d. Negative control cells were treated with an equal volume of PBS. (**C**) astroglial-enriched cells treated with 10 μM $NaAsO_2$ and/or 100 ng/mL LPS for 1 d were subjected to immunoblot detection of Hmox1, Nqo1, NF-κB, and IκB. (**D**) the expression levels of *Lcn2* in astroglial-enriched cells at passage number 1 (P1) and 2 (P2) were determined by qRT-PCR, normalized against *Gapdh* levels, and expressed as the fold change relative to the control (P1; $n = 3$). (**E–I**) astroglial-enriched cells treated as described in (**C**) were subjected to total RNA isolation and qRT-PCR analysis. The expression levels of *Hmox1*, *Nqo1*, *Lcn2*, *Tnf*, and *Il1a* were determined by qRT-PCR, normalized against *Gapdh* levels, and expressed as the fold change relative to the control (−LPS, −$NaAsO_2$; $n = 3$). For immunoblot analysis, β-actin was used as a loading control. Representative images of cells or immunoblots are

shown (n = 3). qRT-PCR data are expressed as the means ± SEM from the indicated number of samples. Scale bar, 50 μm. * $p < 0.05$, ** $p < 0.01$; *** $p < 0.001$ between two groups as indicated by the horizontal bars.

As expected, LPS treatment activated NF-κB signaling, as indicated by the reduced levels of inhibitory IκB (Figures 2C and S2D). However, activation of the Nrf2-Keap1 pathway via sodium arsenite treatment did not appear to affect NF-κB signaling in astroglial-enriched cells (Figures 2C and S2C). Furthermore, *Lcn2* expression levels increased significantly after passaging these cells to P2, even in the absence of NF-κB signaling activation due to LPS treatment (Figure 2D). Therefore, reduced LCN2 levels under mild oxidative stress do not seem to result from the suppression of NF-κB signaling (see Figure 2B). Consistent with the protein levels, *Hmox1* and *Nqo1* mRNA expression levels dramatically increased under mild oxidative stress conditions (Figure 2E,F). In addition, mild oxidative stress induced the downregulation of *Lcn2* expression in astroglial-enriched cells, regardless of LPS treatment conditions, which was consistent with the reduced LCN2 protein levels (Figure 2G). Interestingly, the increased expression levels of inflammatory markers due to LPS treatment were significantly decreased by sodium arsenite treatment (Figure 2H,I). Therefore, it is plausible that sodium arsenite may inhibit the transcription process of NF-κB target genes, either directly or indirectly, although it may not directly affect NF-κB signaling itself.

To further evaluate the effect of mild oxidative stress on astroglial-enriched cells, we treated the cells with sodium arsenite for different times or at different concentrations. We found that the expression levels of *Lcn2* decreased gradually with sodium arsenite treatment in a time-dependent manner for up to 24 h (Figure 3A). Furthermore, we observed that treatment with sodium arsenite at concentrations less than 10 μM did not sufficiently reduce the levels of LCN2 (Figures 3B and S3A,B). Reduced levels of LCN2 were observed when the cells were treated with 50 μM sodium arsenite, although the effect was less than the effect of treatment at 10 μM (Figures 3C and S4A,B). We also determined whether the reduction in LCN2 levels by sodium arsenite could be recapitulated with other oxidative stress inducers, such as hydrogen peroxide. Although the effect was relatively smaller than the effect of sodium arsenite, hydrogen peroxide also seemed to induce an antioxidant response pathway and slightly reduced LCN2 levels (Figures 3D and S4C,D). Furthermore, the reduced levels of LCN2 due to sodium arsenite treatment slightly increased in the presence of N-acetyl-L-cysteine (NAC), which is known as a reactive oxygen species scavenger, although co-treatment with NAC could not completely inhibit the sodium arsenite-induced reduction in LCN2 levels (Figures 3E and S4E). Therefore, the mild oxidative stress induced by sodium arsenite contributes to the reduction in LCN2 levels. However, we cannot exclude the possibility that other pathways or factors are involved in reducing LCN2 levels. Taken together, our results suggest that the induction of mild oxidative stress via sodium arsenite treatment induces the activation of the antioxidant response pathway and decreases LCN2 levels in astroglial-enriched cells at both the mRNA and protein level.

2.3. *The Nrf2-Mediated Antioxidant Response* per se *Cannot Reduce LCN2 Levels in Astroglial-Enriched Cells*

To determine if activation of the Nrf2-Keap1 pathway alone is sufficient to reduce LCN2 levels, we used tertiary-butylhydroquinone (tBHQ) as an Nrf2 stabilizer without inducing oxidative stress. We investigate whether the antioxidant response pathway is directly involved in the regulation of LCN2 levels. However, tBHQ treatment did not reduce LCN2 levels in astroglial-enriched cells, despite significantly increasing Hmox1 levels at both the mRNA and protein levels (Figures 4A–C and S5A,B). These results suggest that activation of the Nrf2–Keap1 pathway alone may not be sufficient to reduce LCN2 levels in astroglial-enriched cells. Thus, mild oxidative stress, induced in an Nrf2-independent manner, may be responsible for the reduced LCN2 levels in astroglial-enriched cells.

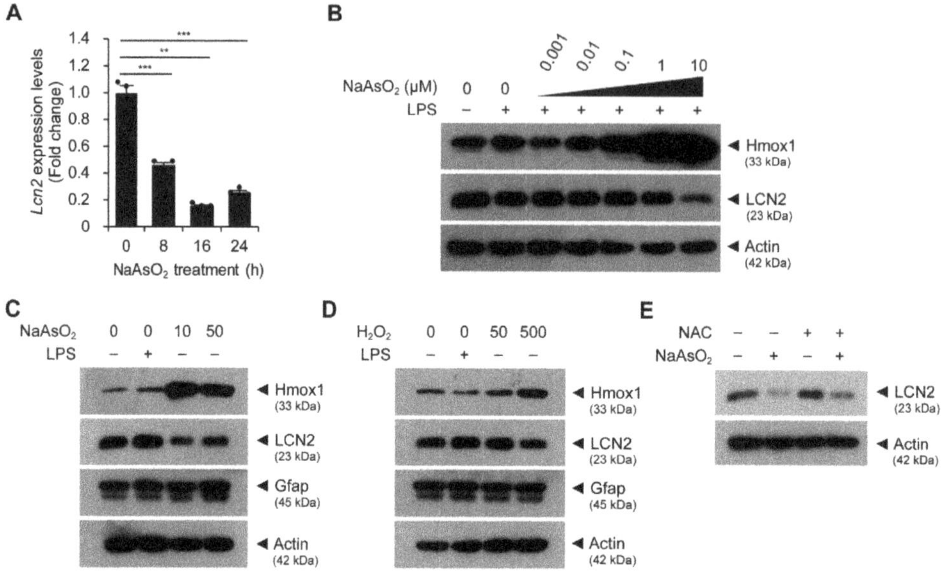

Figure 3. *Lcn2* expression levels are reduced by mild oxidative stress in a time-dependent manner. (**A**) astroglial-enriched cells were treated with 10 μM sodium arsenite ($NaAsO_2$) for 0 to 24 h. *Lcn2* expression levels were determined by qRT-PCR, normalized against *Gapdh* levels, and expressed as the fold change relative to the control (0 h; $n = 3$). (**B**) immunoblot detection of Hmox1 and LCN2 was performed using control astroglial-enriched cells (without LPS or $NaAsO_2$ treatment) and cells treated with 100 ng/mL LPS and various concentrations of $NaAsO_2$ up to 10 μM for 1 d. (**C**) astroglial-enriched cells were treated with 100 ng/mL LPS or with 10 or 50 μM $NaAsO_2$ for 1 d. The cells were then subjected to immunoblot detection of LCN2, Hmox1, and Gfap. (**D**) immunoblot analysis was performed, as described in (C), using astroglial-enriched cells treated with 100 ng/mL LPS or with 50 or 500 μM hydrogen peroxide (H_2O_2) for 1 d. (**E**) immunoblot detection of LCN2 in astroglial-enriched cells treated with 10 mM N-acetyl-L-cysteine (NAC) and/or 10 μM $NaAsO_2$ for 1 d. To investigate the effect of NAC, 10 mM NAC was added 2 h before $NaAsO_2$ treatment. For immunoblot analysis, β-actin was used as a loading control. Representative images of immunoblots are shown ($n = 2$). qRT-PCR data are expressed as the means ± SEM from the indicated number of samples. ** $p < 0.01$; *** $p < 0.001$ between two groups as indicated by the horizontal bars.

To investigate whether the reduction in *Lcn2* expression levels by sodium arsenite can be achieved without activation of the Nrf2–Keap1 pathway, we knocked down *Nrf2* in astroglial-enriched cells. We confirmed Nrf2 knockdown by the reduced expression levels of *Nrf2* and the reduced responses of the Nrf2 target genes, *Hmox1* and *Nqo1*, to sodium arsenite (Figure 4D–F). As *Hmox1* expression is also known to be regulated via an Nrf2-independent pathway, the effect of *Nrf2* knockdown on *Hmox1* expression was less significant than its effect on *Nqo1* expression (Figure 4E,F). When we knocked down *Nrf2* in astroglial-enriched cells, reduced expression levels of *Lcn2* due to mild oxidative stress were still observed (Figure 4G). Therefore, our data suggest that reduced LCN2 levels in response to sodium arsenite treatment in astroglial-enriched cells were not simply mediated by activation of the Nrf2–Keap1 pathway. Although further detailed investigation may be required, a method to reduce LCN2 levels in astroglial-enriched cells will provide insights into how to reduce reactive-astrocyte-induced neurotoxicity in various neurological disorders.

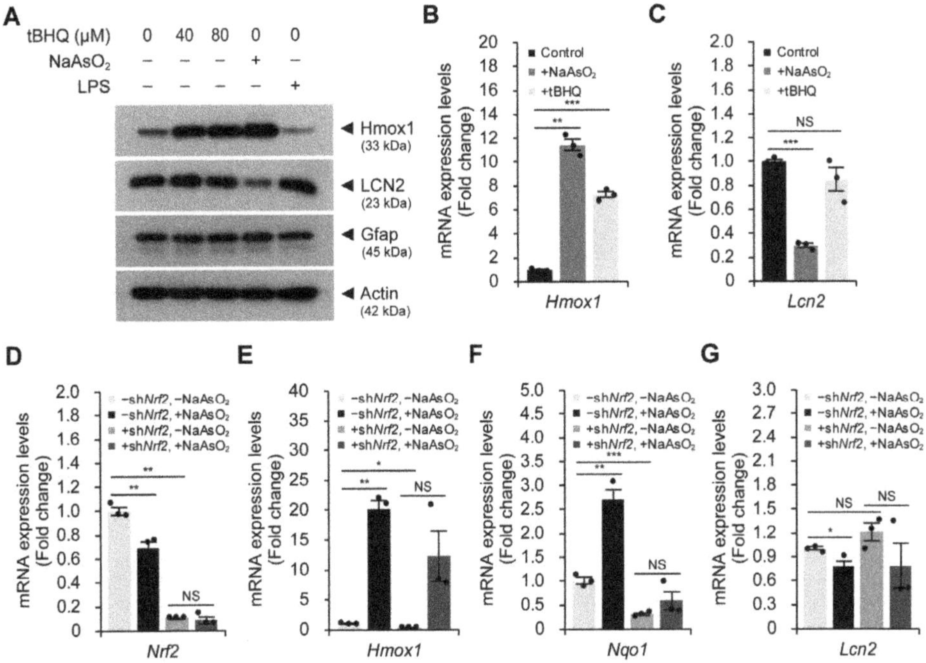

Figure 4. The Nrf2-mediated antioxidant response is not responsible for reduced *Lcn2* expression levels. (**A**) astroglial-enriched cells were treated with 40 or 80 μM tBHQ, 10 μM sodium arsenite ($NaAsO_2$) or 100 ng/mL LPS for 1 d. The cells were then subjected to immunoblot detection of LCN2, Hmox1, and Gfap. (**B**,**C**) astroglial-enriched cells were treated with 10 μM $NaAsO_2$ or 40 μM tBHQ for 1 d. The expression levels of *Lcn2* and *Hmox1* were determined by qRT-PCR, normalized against *Gapdh* levels, and expressed as the fold change relative to the control ($-NaAsO_2$, −tBHQ; $n = 3$). (**D**–**G**) astroglial-enriched cells were subjected to *Nrf2* knockdown (KD) via lentiviral transduction. Negative control cells were transduced with a lentivirus harboring scrambled shRNA sequences. One day after infection, control (−sh*Nrf2*) and *Nrf2* KD (+sh*Nrf2*) cells were treated with 10 μM $NaAsO_2$ or PBS for 1 d. Total RNA was isolated from the cells and subjected to qRT-PCR analysis. The expression levels of *Nrf2*, *Hmox1*, *Nqo1*, and *Lcn2* were determined by qRT-PCR, normalized against *Gapdh* levels, and expressed as the fold change relative to the control (−sh*Nrf2*, $-NaAsO_2$; $n = 3$). For immunoblot analysis, β-actin was used as a loading control. Representative images of immunoblots are shown ($n = 2$). qRT-PCR data are expressed as the means ± SEM from the indicated number of samples. * $p < 0.05$, ** $p < 0.01$; *** $p < 0.001$ between two groups as indicated by the horizontal bars. NS, not significant.

3. Discussion

To investigate the regulatory mechanisms of LCN2 in the central nervous system, we cultured primary glial cells isolated from the cortical regions of mouse brains. According to previous studies, cultured cortical cells are composed of mixed populations of glial cells, including astrocytes and microglia [25–27]. Based on the suggestion from a previous report, here, we refer to these cells as astroglial-enriched cells [25]. In this study, we treated these cells with sodium arsenite, which is generally regarded as a cytotoxic reagent that induces the production of reactive oxygen species, DNA damage, autophagosome formation, and even apoptosis [24,28,29]. However, sodium arsenite treatment at a concentration of 10 μM did not compromise the viability of astroglial-enriched cells.

Previous studies have shown that *Lcn2* expression is upregulated by the activation of NF-κB signaling, which is essential for the production of various inflammatory cytokines and for cell growth and survival [19,20]. Interestingly, intracellular crosstalk between NF-κB signaling and the Nrf2–Keap1 pathway has previously been reported [23]. Briefly, increased levels of Hmox1 induced by activation of the Nrf2–Keap1 pathway block the activation of inhibitory κB kinase (IKK), resulting in the stabilization of IκB and suppression of NF-κB signaling. Therefore, we first speculated that crosstalk between the Nrf2–Keap1 pathway and NF-κB signaling may be required for the mild oxidative stress-induced reduction in LCN2 levels. However, increased levels of Hmox1 did not seem to affect IκB levels in astroglial-enriched cells (see Figure 2C). Thus, reduced LCN2 levels under mild oxidative stress may not result from the suppression of NF-κB signaling. In fact, reduced *Lcn2* expression can also be observed without LPS stimulation or activation of the NF-κB signaling pathway.

We have recently reported that under neuroinflammatory conditions induced by LPS treatment, LCN2 production can be reduced by bortezomib treatment and autophagic degradation of LCN2 can be accelerated by torin 1 treatment [30]. Proteasome inhibition through bortezomib treatment suppresses NF-κB signaling due to the stabilization of IκB even under neuroinflammatory conditions, in contrast to sodium arsenite treatment. In reactive astrocytes, LCN2 is a short-lived protein with a half-life of ~30 min and is targeted to the autophagy–lysosome pathway for degradation. However, when secreted from the cells, it induces the loss of nearby neurons. We have clearly demonstrated that reduced secretion of LCN2 from reactive astrocytes has beneficial effects on neurons, leading to the restoration of neuronal viability under neuroinflammatory conditions.

In this study, we demonstrated that LCN2 levels can be reduced by mild oxidative stress in an Nrf2-independent manner, regardless of neuroinflammatory conditions. Mild oxidative stress may represent a novel strategy for reducing LCN2 levels in astroglial-enriched cells, thereby preventing its potentially neurotoxic effects without altering or impacting the NF-κB signaling pathway. Therefore, pharmaceutical approaches targeting LCN2 in neurological diseases have emerged and regulating LCN2 levels may become an important therapeutic approach for various neurological disorders.

4. Materials and Methods

4.1. Mouse Experiments

Wild-type CD-1 (ICR) mice were purchased from Raon Bio (Yongin, Republic of Korea) and were transferred to the mouse facility before the experiments. All mice were housed in plastic cages with ad libitum access to food and water under a 12 h light/dark cycle. All mouse experiments, including the isolation of primary cells from postnatal mouse brains, were approved by the University of Seoul Institutional Animal Care and Use Committee (approval no. UOS-IACUC-2020-03-A, UOS IACUC-2021-01-TA). All animal procedures were performed in accordance with relevant guidelines and regulations approved by the UOS IACUC.

4.2. Primary Astroglial-Enriched Cell Culture

Astrocytes were enriched and subsequent processes were performed as previously described [26]. Briefly, brains were isolated from pups on postnatal day 1 and placed in sterilized Petri dishes containing Hank's balanced salt solution (HBSS). The isolated brains were further processed by removal of the cerebellum, olfactory bulbs, and meninges to obtain the cortical regions. After cutting the cortex into small pieces (1–2 mm^3), the collected pieces (from 3 to 4 pups) were incubated in a 0.05% trypsin/EDTA solution (CellGro, Manassas, VA, USA) for 30 min at 37 °C. To prevent over-trypsinization, an equal volume of cell culture medium (Dulbecco's modified Eagle medium [DMEM] supplemented with 10% fetal bovine serum, 20 mM L-glutamine, and 1% antibiotics/antimycotics) was added and the samples were gently mixed. After centrifugation at 300× *g* for 5 min, the supernatant was removed and the pelleted cortical pieces were gently triturated in 5 mL of additional cell culture medium. After adjusting the volume of the cell suspension to 10 mL with additional cell culture medium, the cell suspension (from 3 to 4 pups) was transferred into a previously prepared 100 mm cell culture dish coated with poly-D-lysine (MW 30,000–70,000; Sigma-Aldrich, St. Louis, MO, USA). The medium was changed every 2 days. After 7–8 days, confluent astroglial-enriched cells were split in half and passaged. After 12–14 days, the cells were seeded on cell culture dishes or cover glasses, depending on the experiment, 48 h before beginning the experiment.

4.3. Immunoblot Analysis

Cell lysates were prepared from astroglial-enriched cells treated with sodium arsenite, tertiary-butylhydroquinone (tBHQ), lipopolysaccharide (LPS), or other chemicals, using radioimmunoprecipitation (RIPA) buffer. After lysis for 30 min on ice, the samples were centrifuged at 13,000× *g* for 15 min at 4 °C, and the resulting supernatant was collected. The protein concentration of each sample was determined using the Pierce™ BCA assay kit (Thermo Fisher Scientific, Waltham, MA, USA). Proteins (15–30 μg) were separated via sodium dodecyl sulfate (SDS) polyacrylamide gel (10%) electrophoresis at 25 mA (per gel) for 60–75 min. After electrophoresis, the separated proteins were transferred to polyvinylidene fluoride (PVDF) membranes at 100 V for 90 min. The transferred membranes were blocked with 5% skim milk in TBST (Tris-buffered saline [TBS] containing 0.05% Tween-20) for 1 h at room temperature and incubated with primary antibodies suspended in 5% skim milk/TBST at 4 °C overnight. After primary antibody incubation, the membranes were washed with TBST and subsequently incubated with secondary antibodies (horseradish-peroxidase [HRP]-conjugated goat anti-mouse IgG or rabbit IgG) suspended in 5% skim milk/TBST for 1 h at room temperature. Finally, chemiluminescent signals were detected and images were captured using an enhanced chemiluminescence solution (WSE-7120L; ATTO, Tokyo, Japan) with a ChemiDoc system (Bio-Rad, Hercules, CA, USA). The antibodies used in this study are as follows: anti-LCN2 (PA5-79590; rabbit polyclonal, 1:1000; Thermo Fisher Scientific); anti-Gfap (MAB360; mouse monoclonal, 1:1000; Millipore, Burlington, MA, USA); anti-Hmox1 (5061; rabbit polyclonal, 1:1000; Cell Signaling Technology, Danvers, MA, USA); anti-Nqo1 (ab34173; rabbit polyclonal, 1:2000; Abcam, Cambridge, UK); anti-NF-κB/p65 (51-0500; rabbit polyclonal, 1:1000; Thermo Fisher Scientific); anti-IκBα (MA5-15132; mouse monoclonal, 1:1000; Thermo Fisher Scientific); anti-β-actin (SC-47778; mouse monoclonal, 1:1000; Santa Cruz Biotechnology, Dallas, TX, USA); HRP-conjugated goat anti-mouse or anti-rabbit IgG (ADI-SAB-100-J or ADI-SAB-300-J; goat polyclonal, 1:10,000; Enzo Life Sciences, Farmingdale, NY, USA).

4.4. Quantitative Reverse Transcription Polymerase Chain Reaction (qRT-PCR) Analysis

After chemical treatment or lentivirus-mediated cell transduction, total RNA was isolated from cells using TRI Reagent (Molecular Research Center, Cincinnati, OH, USA) following the manufacturer's protocol and reconstituted in RNase-free ddH_2O. RNA concentration was measured using a NanoDrop One™ instrument (Thermo Fisher Scientific), and 1 μg of RNA was incubated with DNase I (Invitrogen, Carlsbad, CA, USA) for 15 min at room temperature. After inactivating DNase I, following the manufacturer's protocol, the prepared RNA samples were used as templates for reverse transcription using an oligo(dT) primer (18-mer) and SuperiorScript II reverse transcriptase (Enzynomics, Daejeon, Republic of Korea) following the manufacturer's protocol. cDNA samples generated from mRNA were used as templates for qRT-PCR using 2× SYBR Master Mix (Enzynomics), primer pairs, and an iCycler system (IQ5; Bio-Rad). The mRNA expression levels of *Lcn2*, *Nrf2*, *Hmox1*, and *Nqo1* were normalized to the levels of glyceraldehyde 3-phosphate dehydrogenase (*Gapdh*). The primers used in this study were as follows: *Lcn2*-F, 5′-CTG AAT GGG TGG TGA GTG TG-3′; *Lcn2*-R, 5′-GCT CTC TGG CAA CAG GAA AG-3′; *Nrf2*-F, 5′-ATC CAG ACA GAC ACC AGT GGA TC-3′; *Nrf2*-R, 5′-GGC AGT GAA GAC TGA ACT TTC A-3′; *Hmox1*-F, 5′-CCT GGT GCA AGA TAC TGC CC-3′; *Hmox1*-R, 5′-GAA GCT GAG AGT GAG GAC CCA-3′; *Nqo1*-F, 5′-GGT AGC GGC TCC ATG TAC TC-3′; *Nqo1*-R, 5′-CAT CCT TCC AGG ATC TGC AT-3′; *Tnf*-F, 5′-TCT CAT CAG TTC TAT GGC CC-3′; *Tnf*-R, 5′-GGG AGT AGA CAA GGT ACA AC-3′; *Il1a*-F, 5′-CAC CAA AGA ACA AAG TCG GG-3′; *Il1a*-R,

5′-GGA AGG TTC CTG TAC ATG GTA C-3′; *Gapdh*-F, 5′-GGC ATT GCT CTC AAT GAC AA-3′; and *Gapdh*-R, 5′-CTT GCT CAG TGT CCT TGC TG-3′.

4.5. MTT Assay

Astroglial-enriched cells were seeded into 24-well cell culture plates and treated with various concentrations of sodium arsenite. After 24 h of incubation, the medium was replaced with a thiazolyl blue tetrazolium bromide (MTT) solution (500 μg/mL of MTT in DMEM) and the cells were incubated for 3 h at 37 °C. The converted MTT was solubilized with dimethyl sulfoxide (DMSO) and measured using a microplate reader at a wavelength of 570 nm with background subtraction at 650 nm.

4.6. Immunofluorescence Analysis

Astroglial-enriched cells were seeded on poly-D-lysine-coated coverslips 2 days before beginning the experiments and treated with sodium arsenite overnight. Cells were washed briefly with phosphate-buffered saline (PBS), fixed with 4% paraformaldehyde/PBS for 10 min at room temperature, permeabilized with 0.3% Triton X-100/PBS for 10 min at room temperature, and blocked with 3% bovine serum albumin/PBS for 1 h at room temperature. The cells were then incubated with primary antibodies (1:500, anti-LCN2 and anti-Gfap) at 4 °C overnight. After primary antibody incubation, the cells were washed with PBS and incubated with fluorophore-conjugated secondary antibodies (1:1000, Alexa Fluor 488-conjugated goat anti-mouse IgG and 555-conjugated goat anti-rabbit IgG) and 4, 6-diamidino-2-phenylindole (DAPI) for 1 h at room temperature. The cells were washed with PBS and mounted using the Prolong Gold Antifade Reagent (Thermo Fischer Scientific).

4.7. Lentivirus Production

Lentivirus production in Lenti-X HEK293T cells (Takara Bio, Kusatsu, Japan) was performed using the general calcium phosphate transfection method, as previously described [31]. pLKO.1-scrambled was purchased from Addgene and pLKO.1-sh*Nrf2* was generated using the oligonucleotide targeting mouse *Nrf2* [23]. Briefly, one day before transfection, HEK293T cells were split and seeded on poly-D-lysine-coated 100 mm cell culture dishes at 2.0×10^6 cells/dish. Three hours before transfection, the medium was changed to achieve optimal transfection conditions. The concentrations of all plasmids (pLKO.1-scrambled shRNA, pLKO.1-sh*Nrf2*, psPAX2, and pMD2.G) were measured using a NanoDrop One™ instrument. Ten μg of pLKO.1-sh*Nrf2* (or pLKO.1-scrambled shRNA as a negative control), 8 μg of psPAX2, and 3 μg of pMD2.G were mixed with 500 μL of HEPES-buffered saline (HEBS) and 250 μL of $CaCl_2$, and the volume of the mixture was adjusted with ddH_2O to 1 mL. After incubation for 30 min at room temperature, the mixture was sprinkled evenly onto the prepared cells and mixed with the medium by gentle shaking. After an overnight incubation, the medium was replaced with fresh medium, incubated for an additional 24 h, and then collected. The collected medium (containing lentivirus) was filtered through a 0.45 μm low-protein-binding filter (Pall Corporation, Port Washington, NY, USA), mixed with 3 mL of Lenti-X™ concentrator (Takara Bio), and incubated at 4 °C overnight. The mixture was centrifuged at 1500× *g* for 45 min at 4 °C, and the pelleted lentivirus was resuspended in PBS and stored at −80 °C before use. Before infection, the lentiviral titer was measured using a qPCR Lentivirus Titer Kit (Applied Biological Materials, Richmond, BC, Canada), and the multiplicity of infection (MOI) was calculated based on the titer. Astroglial-enriched cells were transduced with lentiviral vectors at an MOI of 10 in the presence of 4 μg/μL polybrene. The shRNA sequences used in this study were as follows: scrambled-sense, 5′-CCT AAG GTT AAG TCG CCC TCG CTC GAG CGA GGG CGA CTT AAC CTT AGG-3′ and sh*Nrf2*-sense, 5′-CCA AAG CTA GTA TAG CAA TAA CTC GAG TTA TTG CTA TAC TAG CTT TGG-3′.

4.8. Statistical Analysis

Two-tailed unpaired Student's *t*-tests were used to compare data between the two groups. Differences were considered statistically significant at $p < 0.05$.

Supplementary Materials: The supporting information can be downloaded at:https://www.mdpi.com/article/10.3390/ijms242115864/s1.

Author Contributions: Conceptualization, Y.-J.C., S.-H.P. and K.-Y.R.; validation, Y.-J.C. and S.-H.P.; formal analysis, Y.-J.C. and S.-H.P.; investigation, Y.-J.C. and S.-H.P.; writing—original draft preparation, Y.-J.C., S.-H.P. and K.-Y.R.; writing—review and editing, K.-Y.R.; visualization, Y.-J.C. and S.-H.P.; supervision, K.-Y.R.; project administration, K.-Y.R.; funding acquisition, K.-Y.R. All authors have read and agreed to the published version of the manuscript.

Funding: This work was supported by the 2021 Research Fund of the University of Seoul (Grant No. 202104241083) to KYR.

Institutional Review Board Statement: ICR mice were maintained in plastic cages with ad libitum access to food and water. All experimental procedures were approved by the University of Seoul Institutional Animal Care and Use Committee (UOS IACUC; approval no. UOS-IACUC-2020-03-A, UOS IACUC-2021-01-TA) and were performed in accordance with the relevant guidelines and regulations approved by the UOS IACUC.

Informed Consent Statement: Not applicable.

Data Availability Statement: Not applicable.

Conflicts of Interest: The authors declare no conflict of interest. The funders had no role in the design of the study; in the collection, analyses, or interpretation of data; in the writing of the manuscript; or in the decision to publish the results.

References

1. Freeman, M.R. Specification and morphogenesis of astrocytes. *Science* **2010**, *330*, 774–778. [CrossRef]
2. Aguzzi, A.; Barres, B.A.; Bennett, M.L. Microglia: Scapegoat, saboteur, or something else? *Science* **2013**, *339*, 156–161. [CrossRef]
3. Lee, H.G.; Wheeler, M.A.; Quintana, F.J. Function and therapeutic value of astrocytes in neurological diseases. *Nat. Rev. Drug Discov.* **2022**, *21*, 339–358. [CrossRef]
4. Li, K.; Li, J.; Zheng, J.; Qin, S. Reactive Astrocytes in Neurodegenerative Diseases. *Aging Dis.* **2019**, *10*, 664–675. [CrossRef]
5. Zamanian, J.L.; Xu, L.; Foo, L.C.; Nouri, N.; Zhou, L.; Giffard, R.G.; Barres, B.A. Genomic analysis of reactive astrogliosis. *J. Neurosci.* **2012**, *32*, 6391–6410. [CrossRef]
6. Lian, H.; Yang, L.; Cole, A.; Sun, L.; Chiang, A.C.; Fowler, S.W.; Shim, D.J.; Rodriguez-Rivera, J.; Taglialatela, G.; Jankowsky, J.L.; et al. NFkappaB-activated astroglial release of complement C3 compromises neuronal morphology and function associated with Alzheimer's disease. *Neuron* **2015**, *85*, 101–115. [CrossRef]
7. Bi, F.; Huang, C.; Tong, J.; Qiu, G.; Huang, B.; Wu, Q.; Li, F.; Xu, Z.; Bowser, R.; Xia, X.G.; et al. Reactive astrocytes secrete lcn2 to promote neuron death. *Proc. Natl. Acad. Sci. USA* **2013**, *110*, 4069–4074. [CrossRef]
8. Anderson, M.A.; Burda, J.E.; Ren, Y.; Ao, Y.; O'Shea, T.M.; Kawaguchi, R.; Coppola, G.; Khakh, B.S.; Deming, T.J.; Sofroniew, M.V. Astrocyte scar formation aids central nervous system axon regeneration. *Nature* **2016**, *532*, 195–200. [CrossRef]
9. Liddelow, S.A.; Barres, B.A. Reactive Astrocytes: Production, Function, and Therapeutic Potential. *Immunity* **2017**, *46*, 957–967. [CrossRef]
10. Liddelow, S.A.; Guttenplan, K.A.; Clarke, L.E.; Bennett, F.C.; Bohlen, C.J.; Schirmer, L.; Bennett, M.L.; Munch, A.E.; Chung, W.S.; Peterson, T.C.; et al. Neurotoxic reactive astrocytes are induced by activated microglia. *Nature* **2017**, *541*, 481–487. [CrossRef]
11. Sofroniew, M.V. Astrocyte Reactivity: Subtypes, States, and Functions in CNS Innate Immunity. *Trends Immunol.* **2020**, *41*, 758–770. [CrossRef]
12. Escartin, C.; Galea, E.; Lakatos, A.; O'Callaghan, J.P.; Petzold, G.C.; Serrano-Pozo, A.; Steinhauser, C.; Volterra, A.; Carmignoto, G.; Agarwal, A.; et al. Reactive astrocyte nomenclature, definitions, and future directions. *Nat. Neurosci.* **2021**, *24*, 312–325. [CrossRef]
13. Yang, J.; Goetz, D.; Li, J.Y.; Wang, W.; Mori, K.; Setlik, D.; Du, T.; Erdjument-Bromage, H.; Tempst, P.; Strong, R.; et al. An iron delivery pathway mediated by a lipocalin. *Mol. Cell* **2002**, *10*, 1045–1056. [CrossRef]
14. Flo, T.H.; Smith, K.D.; Sato, S.; Rodriguez, D.J.; Holmes, M.A.; Strong, R.K.; Akira, S.; Aderem, A. Lipocalin 2 mediates an innate immune response to bacterial infection by sequestrating iron. *Nature* **2004**, *432*, 917–921. [CrossRef]
15. Cowland, J.B.; Borregaard, N. Molecular characterization and pattern of tissue expression of the gene for neutrophil gelatinase-associated lipocalin from humans. *Genomics* **1997**, *45*, 17–23. [CrossRef]
16. Liu, Q.; Ryon, J.; Nilsen-Hamilton, M. Uterocalin: A mouse acute phase protein expressed in the uterus around birth. *Mol. Reprod. Dev.* **1997**, *46*, 507–514. [CrossRef]

17. Bennett, M.; Dent, C.L.; Ma, Q.; Dastrala, S.; Grenier, F.; Workman, R.; Syed, H.; Ali, S.; Barasch, J.; Devarajan, P. Urine NGAL predicts severity of acute kidney injury after cardiac surgery: A prospective study. *Clin. J. Am. Soc. Nephrol.* **2008**, *3*, 665–673. [CrossRef]
18. Ip, J.P.; Nocon, A.L.; Hofer, M.J.; Lim, S.L.; Muller, M.; Campbell, I.L. Lipocalin 2 in the central nervous system host response to systemic lipopolysaccharide administration. *J. Neuroinflamm.* **2011**, *8*, 124. [CrossRef]
19. Zhao, P.; Stephens, J.M. STAT1, NF-kappaB and ERKs play a role in the induction of lipocalin-2 expression in adipocytes. *Mol. Metab.* **2013**, *2*, 161–170. [CrossRef]
20. Fujino, R.S.; Tanaka, K.; Morimatsu, M.; Tamura, K.; Kogo, H.; Hara, T. Spermatogonial cell-mediated activation of an IkappaBzeta-independent nuclear factor-kappaB pathway in Sertoli cells induces transcription of the lipocalin-2 gene. *Mol. Endocrinol.* **2006**, *20*, 904–915. [CrossRef]
21. Lee, S.; Lee, J.; Kim, S.; Park, J.Y.; Lee, W.H.; Mori, K.; Kim, S.H.; Kim, I.K.; Suk, K. A dual role of lipocalin 2 in the apoptosis and deramification of activated microglia. *J. Immunol.* **2007**, *179*, 3231–3241. [CrossRef] [PubMed]
22. Lee, S.; Park, J.Y.; Lee, W.H.; Kim, H.; Park, H.C.; Mori, K.; Suk, K. Lipocalin-2 is an autocrine mediator of reactive astrocytosis. *J. Neurosci.* **2009**, *29*, 234–249. [CrossRef] [PubMed]
23. Wardyn, J.D.; Ponsford, A.H.; Sanderson, C.M. Dissecting molecular cross-talk between Nrf2 and NF-kappaB response pathways. *Biochem. Soc. Trans.* **2015**, *43*, 621–626. [CrossRef] [PubMed]
24. Kim, M.N.; Choi, J.; Ryu, H.W.; Ryu, K.Y. Disruption of polyubiquitin gene Ubc leads to attenuated resistance against arsenite-induced toxicity in mouse embryonic fibroblasts. *Biochim. Biophys. Acta* **2015**, *1853*, 996–1009. [CrossRef]
25. Saura, J. Microglial cells in astroglial cultures: A cautionary note. *J. Neuroinflamm.* **2007**, *4*, 26. [CrossRef]
26. Foo, L.C.; Allen, N.J.; Bushong, E.A.; Ventura, P.B.; Chung, W.S.; Zhou, L.; Cahoy, J.D.; Daneman, R.; Zong, H.; Ellisman, M.H.; et al. Development of a method for the purification and culture of rodent astrocytes. *Neuron* **2011**, *71*, 799–811. [CrossRef]
27. Schildge, S.; Bohrer, C.; Beck, K.; Schachtrup, C. Isolation and culture of mouse cortical astrocytes. *J. Vis. Exp.* **2013**, *71*, e50079. [CrossRef]
28. Ruiz-Ramos, R.; Lopez-Carrillo, L.; Rios-Perez, A.D.; De Vizcaya-Ruiz, A.; Cebrian, M.E. Sodium arsenite induces ROS generation, DNA oxidative damage, HO-1 and c-Myc proteins, NF-kappaB activation and cell proliferation in human breast cancer MCF-7 cells. *Mutat. Res.* **2009**, *674*, 109–115. [CrossRef]
29. Zhu, X.X.; Yao, X.F.; Jiang, L.P.; Geng, C.Y.; Zhong, L.F.; Yang, G.; Zheng, B.L.; Sun, X.C. Sodium arsenite induces ROS-dependent autophagic cell death in pancreatic beta-cells. *Food Chem. Toxicol.* **2014**, *70*, 144–150. [CrossRef]
30. Jung, B.K.; Park, Y.; Yoon, B.; Bae, J.S.; Han, S.W.; Heo, J.E.; Kim, D.E.; Ryu, K.Y. Reduced secretion of LCN2 (lipocalin 2) from reactive astrocytes through autophagic and proteasomal regulation alleviates inflammatory stress and neuronal damage. *Autophagy* **2023**, *19*, 2296–2317. [CrossRef]
31. Jung, B.K.; Park, C.W.; Ryu, K.Y. Temporal downregulation of the polyubiquitin gene Ubb affects neuronal differentiation, but not maturation, in cells cultured in vitro. *Sci. Rep.* **2018**, *8*, 2629. [CrossRef] [PubMed]

International Journal of
Molecular Sciences

MDPI

Article

Probing the Effect of Acidosis on Tether-Mode Mechanotransduction of Proprioceptors

Yuan-Ren Cheng [1,2], Chih-Hung Chi [2], Cheng-Han Lee [2], Shing-Hong Lin [2], Ming-Yuan Min [1,*] and Chih-Cheng Chen [2,3,*]

1 Department of Life Science, National Taiwan University, Taipei 10090, Taiwan; infinityster@gmail.com
2 Institute of Biomedical Sciences, Academia Sinica, Taipei 11529, Taiwan
3 Neuroscience Program of Academia Sinica, Academia Sinica, Taipei 11529, Taiwan
* Correspondence: mymin@ntu.edu.tw (M.-Y.M.); chih@ibms.sinica.edu.tw (C.-C.C.)

Abstract: Proprioceptors are low-threshold mechanoreceptors involved in perceiving body position and strain bearing. However, the physiological response of proprioceptors to fatigue- and muscle-acidosis-related disturbances remains unknown. Here, we employed whole-cell patch-clamp recordings to probe the effect of mild acidosis on the mechanosensitivity of the proprioceptive neurons of dorsal root ganglia (DRG) in mice. We cultured neurite-bearing parvalbumin-positive (Pv+) DRG neurons on a laminin-coated elastic substrate and examined mechanically activated currents induced through substrate deformation-driven neurite stretch (SDNS). The SDNS-induced inward currents (I_{SDNS}) were indentation depth-dependent and significantly inhibited by mild acidification (pH 7.2~6.8). The acid-inhibiting effect occurred in neurons with an I_{SDNS} sensitive to APETx2 (an ASIC3-selective antagonist) inhibition, but not in those with an I_{SNDS} resistant to APETx2. Detailed subgroup analyses revealed I_{SDNS} was expressed in 59% (25/42) of Parvalbumin-positive (Pv+) DRG neurons, 90% of which were inhibited by APETx2. In contrast, an acid (pH 6.8)-induced current (I_{Acid}) was expressed in 76% (32/42) of Pv+ DRG neurons, 59% (21/32) of which were inhibited by APETx2. Together, ASIC3-containing channels are highly heterogenous and differentially contribute to the I_{SNDS} and I_{Acid} among Pv+ proprioceptors. In conclusion, our findings highlight the importance of ASIC3-containing ion channels in the physiological response of proprioceptors to acidic environments.

Keywords: acidosis; APETx2; ASIC3; mechanotransduction; proprioceptors; SDNS

1. Introduction

Proprioception is an awareness that underlies body–limb coordination and determines both unconscious reflexes and conscious motor tasks [1,2]. Specialized mechanoreceptors innervate muscle spindles (MSs) and Golgi tendon organs (GTOs), which are collectively called "proprioceptors" [3]. Analysis of molecular identity in proprioceptors has revealed that calcium-binding proteins parvalbumin (Pv), Runt-domain transcription factor (RUNX3), and vesicular glutamate transporter (VGULT2) are expressed predominantly in MS afferent proprioceptors [3–8]. These MS afferents consist of mechanically activated channels (Macs) that mediate the mechanotransduction of the peripheral nerves into the central nervous system. Based on the channel gating mechanism, Macs can be gated by either membrane tension change (bilayer model) or tethering proteins that link to the extracellular matrix and/or cytoskeletons (tether model) [9]. Although plenty of explanations about how Macs transduce the mechanical strain into electrical signals have been proposed, there are a limited number of methods for describing neurosensory mechanotransduction at the cellular level [10]. A common approach to assessing Mac activity is to directly apply force on the cell soma via glass pipette indentation, through which Piezo1 and Piezo2 have been identified as non-selective cation channels in whole-cell patch-clamp recordings [11].

Piezo2 is expressed in parvalbumin-positive (Pv+) dorsal root ganglion (DRG) proprioceptive neurons, and the conditional knockout of Piezo2 in Pv+ DRG neurons shows that Piezo2 is the major contributor to rapid adaptive mechanically activated responses to the direct soma indentation in proprioceptors [12]. Meanwhile, the intermediary adaptive mechanically activated responses are slightly upregulated by Piezo2 ablation, which suggests that Piezo2-independent Macs may also be involved in the mechanotransduction of proprioceptors [13].

Contrarily, to identify the tether-mode Macs of proprioceptors, we developed a method in which neurites on an extracellular matrix (ECM)-coated elastomeric substrate are stretched by imposing localized substrate deformation, so that the electrical responses of tether-mode Macs can be recorded via whole-cell patch clamping [14]. This method can provide a deeper understanding of mechanotransduction at the subcellular level of proprioceptors [14]. The targeted knockout of acid-sensing ion channel 3 (Asic3) on Pv+ DRG neurons disrupts MS afferent sensitivity and impairs substrate deformation-driven neurite stretch (SDNS) mechanotransduction, but not the soma indentation-induced mechanotransduction [15]. Behaviorally, Asic3 knockout mice maintain normal locomotion function as shown in open-field activity and rotarod performance, but the animals exhibit a proprioceptive deficit in the grid walking test in the dark and a balance task using a beam with a size of less than 6 fields [15]. Together, these findings suggest that ASIC3 is involved in mechanosensing associated with the tether-mode mechanotransduction of proprioceptors [15–17]. Since ASIC3 has a dual function of acid sensing and mechanosensing, how ASIC3-dependent proprioception is influenced by tissue acidosis should be a topic of scrutiny.

The fine adjustment of body coordination relies on precise proprioceptive input, which is crucial to achieving balance. Balance can be impaired by fatigue, intermittent training, and intensive exercise [1,18–22]; however, the molecular and cellular mechanisms underlying fatigue-associated balance disturbances remain under-explored. Previous microdialysis studies on humans and rodents have demonstrated that the levels of muscle interstitial lactate and protons substantially increase during exercise as compared with rest, and that they tend to increase further with exercise intensity [23]. Moreover, when the rat hindlimb is subjected to a supramaximal stimulation, H^+ increases earlier than lactate, which leads a minor acidic level (pH = 7.2) in the muscle [24]. When human subjects undergo intermittent training, the H^+ released at exhaustion is significantly higher in the legs of the trained subjects than in those of the untrained subjects at the 8th week of training, and at the 30th week, the blood pH and muscle pH decrease to 7.13 and 6.82, respectively, in the trained legs [25]. It can be concluded from these results that exercise and fatigue affect the homogenous blood flow and muscle metabolism. However, it is still unknown how this micro-environment affects proprioception and leads to balance impairment.

This study sought to explore how the micro-environment of fatigued muscles influences proprioception. Specifically, we used whole-cell patch-clamp recordings of Pv+ proprioceptors to probe how mild acidosis modulates the tether-mode mechanotransduction of proprioceptors.

2. Results

2.1. Pv-Expressing DRG Nerve Terminals Are Proprioceptive Fibers Innervating MSs and GTOs

To investigate the effect of acidosis on the neurosensory mechanotransduction of proprioceptors, we used a genetic model to label the proprioceptors of DRG. We also used the Cre::LoxP reporting system to characterize the muscle afferents in the soleus muscle of Pv-Cre::CAG-cat-EGFP mice. In the immunofluorescence study, we examined six Pv-expressing muscle afferents and found that they can be identified as one of three types of proprioceptive nerve terminals, namely groups Ia and II muscle spindle afferents and group Ib Golgi tendon organ afferents in the hind limb soleus muscle (Figure 1). On the muscle edge, Pv+ eGFP immunoreactivity was observed in free nerve terminals in the soleus muscle, which are the GTO afferents or group Ib proprioceptors (Figure 1a), and in the annunospiral fibers and the free nerve endings in the intrafusal bag, which are group Ia and group II proprioceptive nerve terminals, respectively (Figure 1b,c). All three of these types of DRG nerve terminals were neurofilament heavy chain-positive, which indicates myelinated nerve fibers.

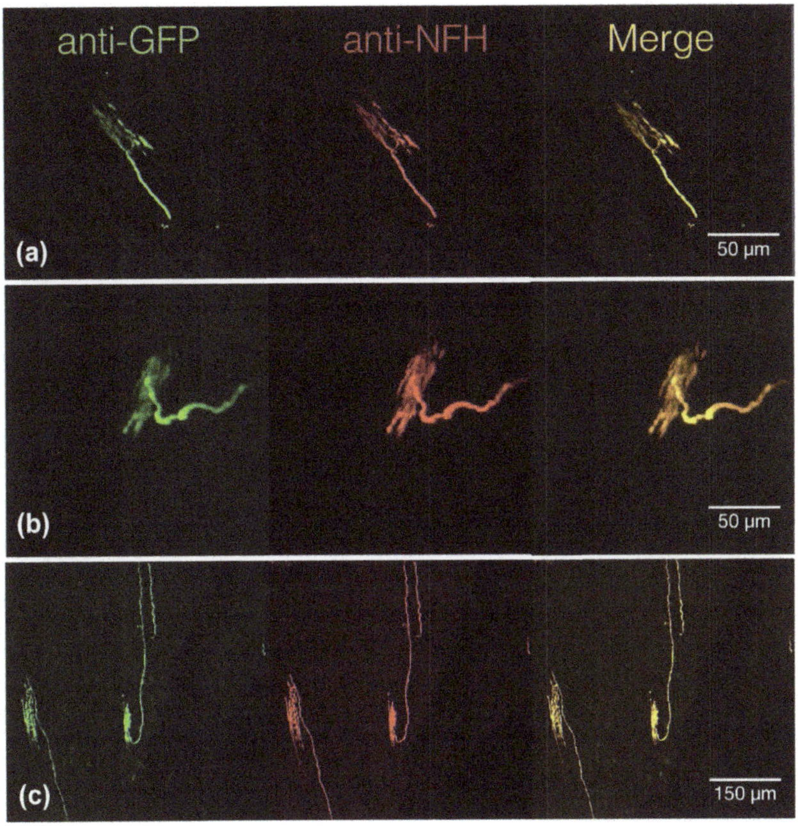

Figure 1. Parvalbumin+ proprioceptive muscle afferents are neurofilament heavy chain immunoreactive muscle afferents with specialized mechanosensing structures. (**a**) Extrafusal muscle afferent showed the co-localized expression of green fluorescence protein (GFP) and neurofilament heavy chain (NFH). (**b**) Intrafusal muscle spindle afferents, which can be classified as type II muscle spindle afferents, displayed co-expression of parvalbumin-triggered GFP and NFH. (**c**) The annunospiral type I muscle spindle afferents in the intrafusal bag also presented strong parvalbumin-triggered GFP and NFH.

2.2. Acidic Environment Attenuates the Mechanically Activated Currents Induced by SDNS

We used whole-cell patch clamping to record the mechanically activated currents of SDNS in tdTomato+ neurons isolated from Pv-Cre::tdTomato mice (Figure 2a). Laminin-coated polydimethylsiloxane (PDMS) served as an elastic basement for Pv+ neurite growth. For each recording, action potentials (APs) and rheobase were first analyzed in the normal artificial cerebral spinal fluid (ACSF). Through this procedure, we identified at least three proprioceptor subtypes, based on their firing patterns, as having single, burst, and tonic APs (Table 1). Then, the normal ACSF was replenished with tetrodotoxin (TTX, 300 nM)-containing ACSF for mechano-clamping. To deliver a series of mechanical stimuli to the PDMS surface, a 40 μm-wide glass electrode with an interior opening of less than 1 MOhm was used as an indentation pipette and connected to a 3-axis mechanical actuator

(Figure 2b). After mechano-clamping, we used a six-channel valve controller and three-barrel pipette to apply acidic perfusion (Figure 2c). The perfusion area covered the soma and neurites of interest. The tip of the perfusion pipette was 1 cm away from the recording soma (Figure 2d). One of the three-barrel pipettes constantly delivered the exact pH of ACSF as a bathing solution. These procedures were meant to reduce the effect of mechanical stress from perfusion. The six-channel valve controller regulated the inlet of acidic ACSF (Figure 2e).

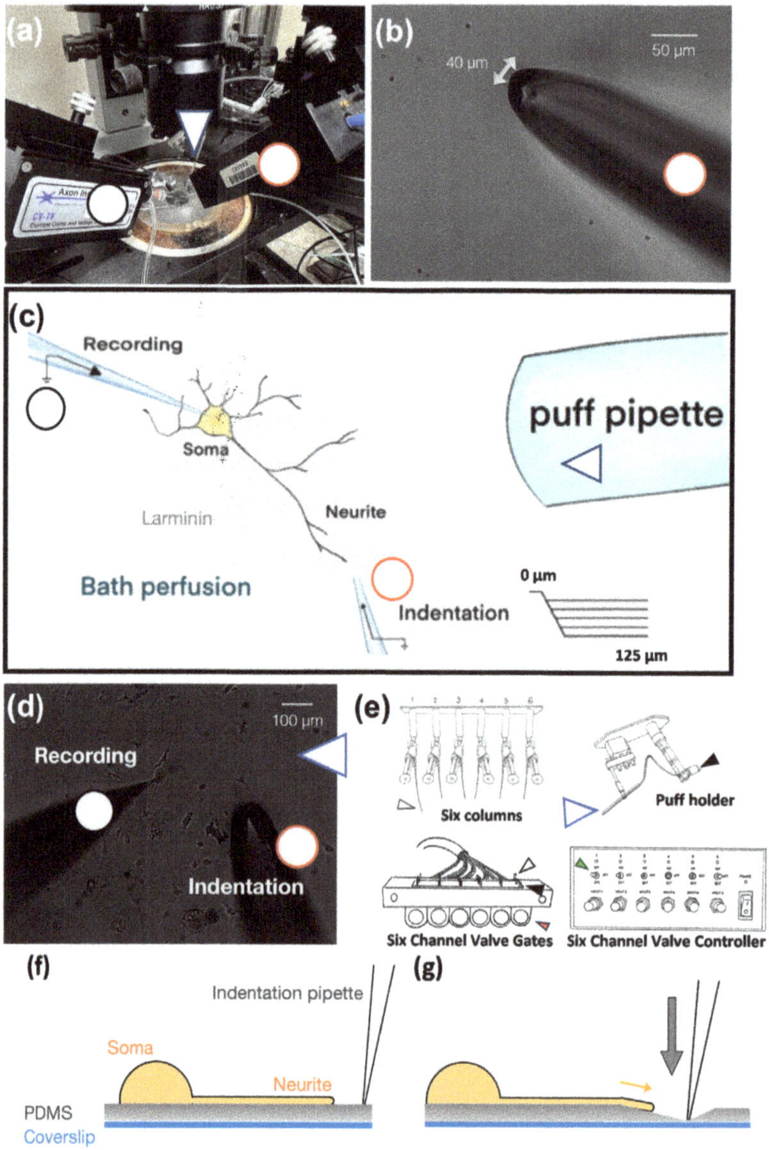

Figure 2. The assembly of the substrate deformation-induced neurite stretch (SDNS) approach. (**a**) Top view of the SDNS mechano-clamp system. The black circle points to the whole-cell patch-clamp recording pipette. The red circle denotes the mechanical actuator-driven indentation pipette.

The blue arrow indicates the acidic solution perfusion pipette. (**b**) The mechanical actuator-driven indentation pipette was a hollow borosilicate glass pipette whose tip was forged into a 40 µm blunt opening ending. (**c**) Schematic view of the whole SDNS setup. The neurite-bearing DRG neuron was cultivated on laminin-coated polydimethylsiloxane (PDMS). The neurite process extended to a distal region, and the indentation pipette was positioned 10 µm away to avoid direct contact with the neurite. The recording pipette was sealed with the soma. The perfusion (puff) pipette was far away from the soma to prevent the sheer force against the cell. (**d**) An actual recording and the puff pipette are shown on the right side of the image. (**e**) The six columns contained artificial cerebral spinal fluid (ACSF) with different pH levels to provide an acidic challenge to the cell, controlled by six-channel valve gates. The controller received an electrical signal, opened the gate (green arrow, switch; red arrow, gates), and then moved the puff holder (blue arrows, puff pipette) to deliver acidic challenges (the plastic pipes were linked between the black arrow and white arrow respectively to conduct ACSF). (**f**) The schematic diagram describes the SDNS recording setup. At the beginning, an indentation pipette was close to a target neurite and the surface of the PDMS substrate. (**g**) The schematic diagram depicts the SDNS procedure. After the indentation began, the indentation pipette was inserted into the PDMS substrate (gray arrow indicates the direction of the force), which caused the deformation of the substrate and thus stretched the neurite (yellow arrow indicates the direction of the stretching).

Table 1. Action potential profile of parvalbumin-positive dorsal root ganglion neurons.

		Total			Single			Burst				Tonic			
Parameter	Unit	Mean	SEM	N	Mean	SEM	N	Mean	SEM	N		Mean	SEM	N	
Cell size	µm	34.48	1.76	100	38.19	2.44	63	29.54	2.52	23		25.93	3.08	14	*
Resting potential	mV	−53.77	0.82	105	−56.03	0.71	66	−48.30	2.41	24		−52.56	1.99	15	
Cm	pF	55.81	4.22	106	64.31	6.17	67	43.21	4.17	24		38.00	5.76	15	
Rm	MOhm	38.30	17.72	106	49.07	28.01	67	21.04	2.06	24		17.80	1.82	15	
Rheobase	pA	827.83	72.03	106	1064.18	101.30	67	441.67	56.28	24		390.00	44.77	15	
Rise slope	ms	22.68	3.21	104	25.16	4.23	67	19.34	6.56	24		16.06	6.28	13	
Decay slope	ms	−60.08	2.82	84	−61.22	3.86	51	−55.53	4.10	21		−63.19	8.55	12	
Peak amplitude	mV	80.24	1.99	105	86.51	2.41	66	70.17	3.36	24	**	68.80	4.82	15	**
Half-width	ms	5.48	2.71	103	8.01	4.34	64	1.33	0.06	24		1.33	0.11	15	

* $p < 0.05$, ** $p < 0.01$ vs. single.

The Pv+ neurites, which were tightly bound to the PDMS surface, were stretched by substrate deformation via a vertical pipette indentation (Figure 2f,g). We performed whole-cell patch clamping to measure the substrate deformation-driven neurite stretch-induced currents (I_{SDNS}) of neurite-bearing Pv+ DRG neurons and examined the effect of acidosis (pH 6.8) on the I_{SDNS} (Figure 3a). We first identified seven Pv+ neurons that expressed I_{SDNS} in response to mechanical stimuli with an indentation greater than 100 µm in the neutral pH 7.4 ACSF, and the I_{SDNS} was significantly attenuated when the local pH was dropped to 6.8 (Figure 3b,c). The I_{SDNS} was force-dependently increased as the indentation depth increased in both neutral and acidic pH (Figure 4a–c). In the pH 7.4 ACSF, the I_{SDNS} peak amplitudes were (expressed in pA) 0.52 ± 1.63, -2.02 ± 1.80, -6.01 ± 2.31, -13.98 ± 2.27, and -18.96 ± 4.67 in response to indentation depths of 25, 50, 75, 100, and 125 µm, respectively (Figure 4d). In the pH6.8 ACSF, the SDNS currents were only detectable with amplitudes of -3.42 ± 1.58 pA and -7.08 ± 2.66 pA with indentation depths of 100 and 125 µm, respectively.

Although there were notable differences among the AP firing patterns, an analysis of the I_{SDNS} current amplitudes at pH 7.4 in relation to cell size, rheobase, and AP firing patterns revealed no correlation between these factors (Supplementary Figure S2). This finding indicates that the current amplitude is not associated with the properties of the AP.

To investigate the effect of acidosis on Pv+ proprioceptors in depth, we examined the I_{SDNS} in different pH environments ranging from 6.8 to 7.4 (Figure 5a) using 100 µm indentations. The bath solution was pre-conditioned to pH 8.0 ACSF with TTX, and the perfusion solution was applied to the recording cell two minutes before indentation. The I_{SDNS} current amplitudes were pH-dependently attenuated as the pH dropped from 7.2 to 6.8 (Figure 5b). Quantitative analyses showed that the average current amplitude of the I_{SDNS} was -13.79 ± 1.85 pA at pH 7.4 ($n = 32$), which was significantly reduced to -9.32 ± 1.24 pA at pH 7.2, -7.16 ± 1.36 pA at pH 7.0, and -4.97 ± 1.07 pA at pH 6.8 (Figure 5c). Thus, the inhibitory effects of acidosis on the I_{SDNS} were more prominent in the Pv+ proprioceptors with single APs or tonic APs than in the proprioceptor subtype with burst APs (Figure 5d–f).

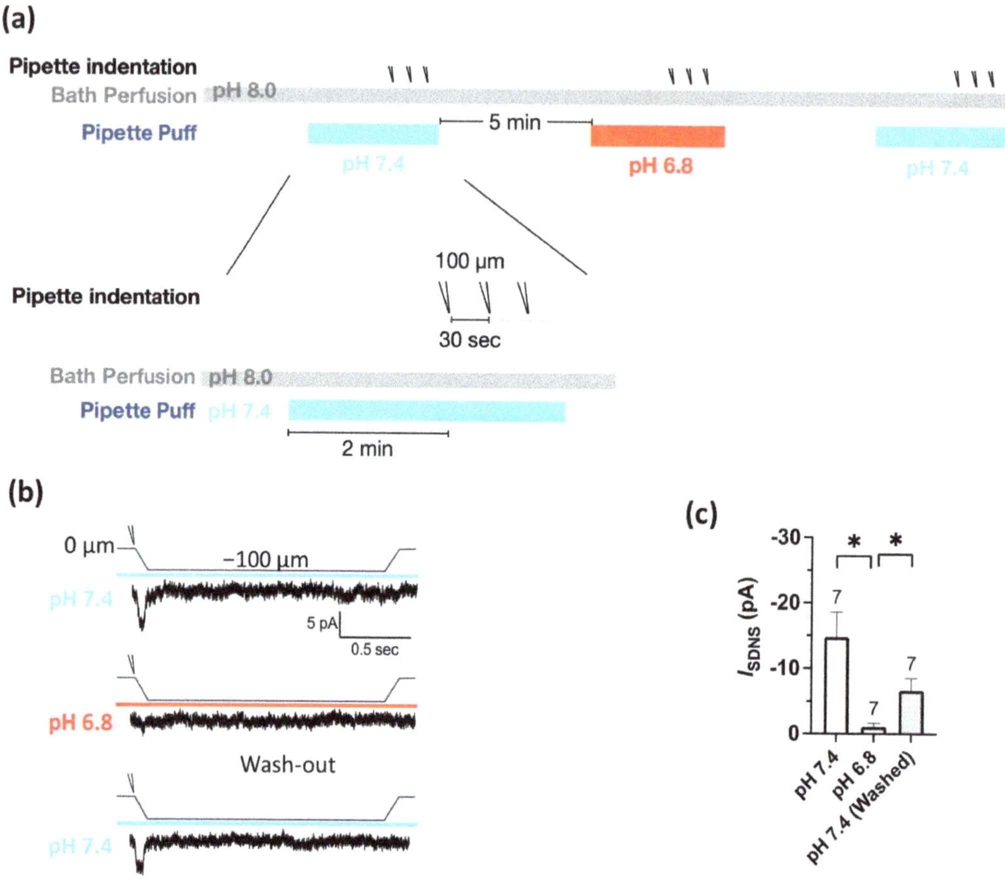

Figure 3. Mild acidosis at pH 6.8 attenuated the substrate deformation-driven neurite stretch (SDNS)-induced currents with a partial reversibility. (**a**) Experimental protocol for testing the effect of acidosis (pH 6.8) on SDNS-induced currents in Pv+ DRG proprioceptors. The bath was a pH 8.0 ACSF, and the perfusion pipette contained either a pH 7.4 or pH 6.8 ACSF during the recordings of SNDS-induced currents. Three repetitive indentations were applied to stretch the neurite to generate SDNS currents; pH 7.4 (colored in light blue) and pH 6.8 (colored in red). (**b**) Three normalized representative SDNS current traces with a 100 µm depth at pH 7.4 or pH 6.8 (with three repetitive indentation traces merged for normalization). (The scale bar indicates 5 pA and 500 ms.) (**c**) The amplitude of I_{SDNS} (expressed in pA) during pH 7.4, pH 6.8, and pH 7.4 washings is shown. One-way ANOVA revealed statistically significant differences in the mean test score between the above three groups (F (1.245, 7.470) = 9.044, p = 0.0153). Multiple comparisons showed that the mean value of I_{SDNS} was significantly different between the pH 7.4 and pH 6.8 washings (p = 0.0234) and between the pH 6.8 and pH 7.4 washings (p = 0.0255). The cell numbers were labeled above the column bars. * $p < 0.05$ vs. pH 6.8.

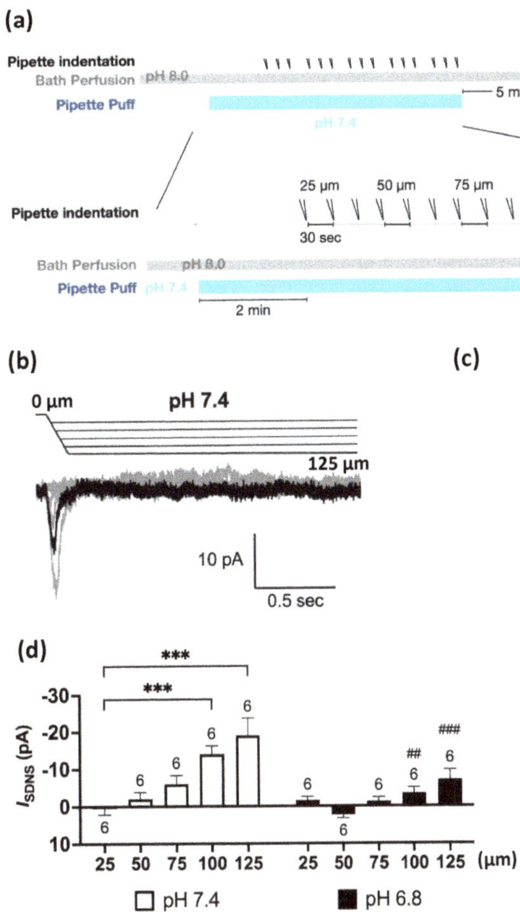

Figure 4. The force dependency of substrate deformation-driven neurite stretch (SDNS) currents was significantly attenuated by acidic perfusion. (**a**) Experimental protocol for testing the effect of acidosis (pH 6.8) on the force dependency of SDNS currents (I_{SDNS}) in Pv+ DRG proprioceptors. The bath was a pH 8.0 ACSF, and the perfusion pipette contained either a pH 7.4 or pH 6.8 ACSF during the recordings of I_{SDNS}. A series of three repeated force indentations with a depth of 25 to 125 µm was applied with the indentation pipette during the recording. The recording started at 0, and the acidic challenge began at the 5th second. The repetitive indentations stretched the neurite during the acidic challenge to generate an SDNS current; pH 7.4 (colored in light blue) and pH 6.8 (colored in red). (**b**) Force-dependent I_{SDNS} traces of Pv+ DRG neurons in the neutral (pH 7.4) environment. (The scale bar indicates 10 pA and 0.5 s in the trace graph.) (**c**) Force-dependent I_{SDNS} traces of Pv+ DRG neurons in the acidic (pH 6.8) environment. (**d**) Bar graphs of I_{SDNS} with different indentation depths (ranging from 25 to 125 µm) in pH 7.4 and pH 6.8 conditions. Two-way ANOVA revealed statistically significant differences in the mean test score between different depths ($F(4, 20) = (12.53)$, $p < 0.001$), pH ($F(1, 5) = (10.28)$, $p = 0.02$), and depth × pH ($F(4, 20) = (7.911)$, $p < 0.001$). The Holm–Sidak test for multiple comparisons found that the mean value of I_{SDNS} was significantly different between 25 and 100 µm at pH 7.4 ($p < 0.001$); between 25 and 125 µm ($p < 0.001$); between pH 7.4 and pH 6.8 with 100 µm ($p < 0.001$); and between pH 7.4 and pH 6.8 with 125 µm ($p < 0.001$). The cell numbers were labeled above the column bars. *** $p < 0.001$, differences within the same pH group; ## $p < 0.01$ and ### $p < 0.001$, differences between the pH 7.4 and pH 6.8 groups.

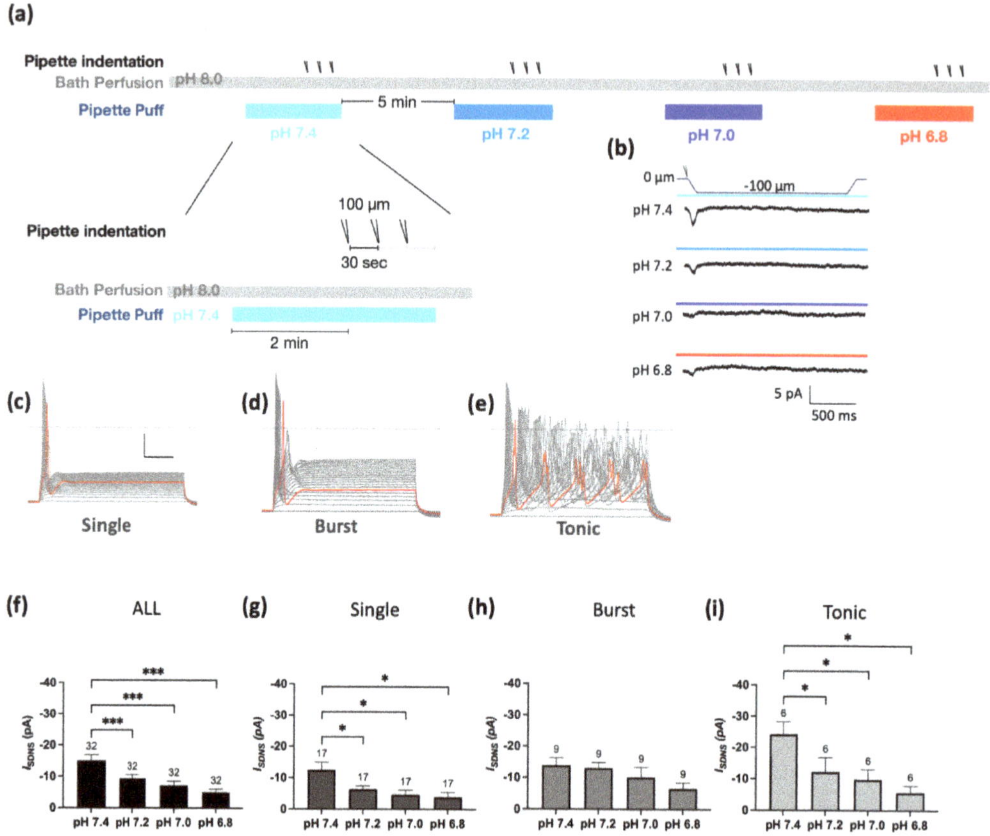

Figure 5. Mild acidosis pH-dependently attenuates SDNS-induced currents in Pv+ DRG proprioceptors. (**a**) Experimental protocol for testing the effect of a series of pH gradients (from 7.4 to 6.8) on SDNS-induced current (I_{SDNS}). The bath was a pH 8.0 ACSF, and a series of pH gradients was applied using the perfusion pipette during the recording. The recording started at 0, and the acidic challenge began at the 5th second. Three repetitive indentations stretched the neurite during the acidic challenge to generate an SDNS current; pH 7.4 (colored in light blue), pH 7.2 (colored in gray-blue), pH 7.0 (colored in blue), and pH 6.8 (representing mild acidosis, colored in red). (**b**) Four normalized representative I_{SDNS} traces with a 100 μm depth, with three repetitive indentation traces merged for normalization. (The scale bar indicates 5 pA and 500 ms). (**c**) Representative single-firing AP traces were held at −70 mV and depolarized via 40 steps with a 10 pA increment. The gray dotted line represents zero potential level, and the red line represents a trace at the rheobase. The x- and y-axes of the scale denote 20 ms and 20 pA, respectively. (**d**) Representative burst-firing AP traces were held at −70 mV and depolarized via 40 steps with a 10 pA increment. The gray dotted line represents zero potential level, and the red line represents a trace at the rheobase. (**e**) Representative tonic-firing AP traces were held at −70 mV and depolarized via 40 steps with a 10 pA increment. The gray dotted line represents zero potential level, and the red line represents a trace at the rheobase. (**f**) pH-dependent effect of acidosis on I_{SDNS} of all firing types of neurons ranged from pH 7.4 to 6.8. One-way ANOVA analysis revealed statistically significant differences in the mean test score between different pH levels (F (2.195, 68.06) = 13.59 $p < 0.0001$). The Holm–Sidak test for multiple comparisons found that the mean value of I_{SDNS} was significantly different between pH 7.4 and pH 6.8 ($p < 0.001$); between pH 7.4 and pH 7.0 ($p = 0.001$); and between pH 7.4 and pH 7.2 ($p < 0.001$). (**g**) pH-dependent effect of acidosis on I_{SDNS} of Pv+ neurons with single-firing APs. One-way ANOVA analysis revealed

statistically significant differences in the mean test score between different pH levels (F (1.586, 25.38) = 6.063, p = 0.0107). The Holm–Sidak test for multiple comparisons found that the mean values of I_{SDNS} were significantly different between pH 7.4 and pH 6.8 (p = 0.0015); between pH 7.4 and pH 7.0 (p = 0.0239); and between pH 7.4 and pH 7.2 (p = 0.0239). (**h**) pH-dependent effect of acidosis on I_{SDNS} of Pv+ DRG neurons with burst-firing APs. One-way ANOVA analysis revealed no statistically significant differences in the mean test score between different pH levels (F (1.467, 11.73) = 3.560, p = 0.0725). (**i**) pH-dependent effect of acidosis on I_{SDNS} of Pv+ DRG neurons with tonic-firing APs. One-way ANOVA analysis revealed statistically significant differences in the mean test score between different pH levels (F (1.974, 9.872) = 5.649, p = 0.02). The Holm–Sidak test for multiple comparisons found that the mean values of I_{SDNS} were significantly different between pH 7.4 and pH 6.8 (p = 0.05); between pH 7.4 and pH 7.0 (p = 0.05); and between pH 7.4 and pH 7.2 (p = 0.05). The cell numbers were labeled above the column bars in each subfigure. * $p < 0.05$ *** $p < 0.001$.

2.3. Role of ASIC3 in Tether-Mode Mechanotransduction during Acidosis

Previous studies have shown that ASIC3 is not only a sensitive acid sensor for mild acidosis but also a mechanic sensor involved in the tether-mode mechanotransduction of Pv+ DRG proprioceptors [15,26]. To explore how acidosis modulates the ASIC3-dependent I_{SDNS}, we designed a protocol to determine the relationship of ASIC3-dependent currents (I_{ASIC3}) with the I_{SDNS} and acid-induced current (I_{Acid}) in different Pv+ DRG proprioceptors (Figure 6a). We tested the effects of APETx2 (2 μM), a selective antagonist for ASIC3 [27], on the I_{SDNS} in both pH 7.4 and pH 6.8 conditions, and on the I_{Acid} in pH 6.8 conditions. In pH 7.4 conditions, the I_{SDNS} was expressed in 60% (25/42) of Pv+ DRG neurons, and the current was inhibited by APETx2 in 92% (23/25) of I_{SDNS}-expressing neurons (Figure 6b, Table S1). Quantitative analyses revealed that APETx2 significantly reduced the I_{SDNS} in Pv+ DRG neurons with single or burst APs, but not in those with tonic APs (Figure 7a–d). In addition, among 25 I_{SDNS}-expressing Pv+ DRG neurons, pH 6.8 acidosis attenuated the APETx2-sensitive I_{SDNS} (n = 23) but had no effect on the APETx2-resistant I_{SDNS} (n = 2) (Figure 7e). Intriguingly, among these 23 APETx2-sensitive I_{SDNS}-expressing Pv+ neurons, acid perfusion of pH 6.8 only induced I_{Acid} in 19 neurons, where the I_{Acid} was inhibited by APETx2 in 11 of these neurons and defined as I_{ASIC3} (Figure 7f,g). Mild acidosis of pH 6.8 significantly attenuated the I_{SDNS} in both groups of neurons, respectively expressing I_{SDNS}/I_{Acid} and I_{SDNS} only (Figure 7f). For those 19 neurons with I_{SDNS}/I_{Acid}, pH 6.8 acidosis significantly attenuated the I_{SDNS} in neurons expressing I_{ASIC3} (n = 11), but not in neurons without I_{ASIC3} (n = 9) (Figure 7g). All these results indicate a strong association between the inhibitory effect of acidosis on the I_{SDNS} and on whether the I_{SDNS} or I_{Acid} can be inhibited by APETx2.

Regarding the acid sensitivity, 76% (32/42) of Pv+ DRG neurons expressed the I_{Acid} in response to the pH 6.8 acid stimulation (Figure 6b). Interestingly, the I_{Acid} was expressed in 84% (21/25) of I_{SDNS}-expressing neurons and 65% (11/17) of I_{SDNS}-negative neurons. The average peak amplitudes of the I_{Acid} showed no difference between Pv+ DRG neurons with and without the I_{SDNS} (Figure 7h) or between I_{ASIC3}-positive and I_{ASIC3}-negative groups (Figure 7i). Based on the expression of the I_{SDNS}, I_{Acid}, and I_{ASIC3}, we organized these 42 Pv+ DRG neurons into 7 subgroups to highlight the heterogeneous functionality among Pv+ DRG proprioceptors (Figure 8).

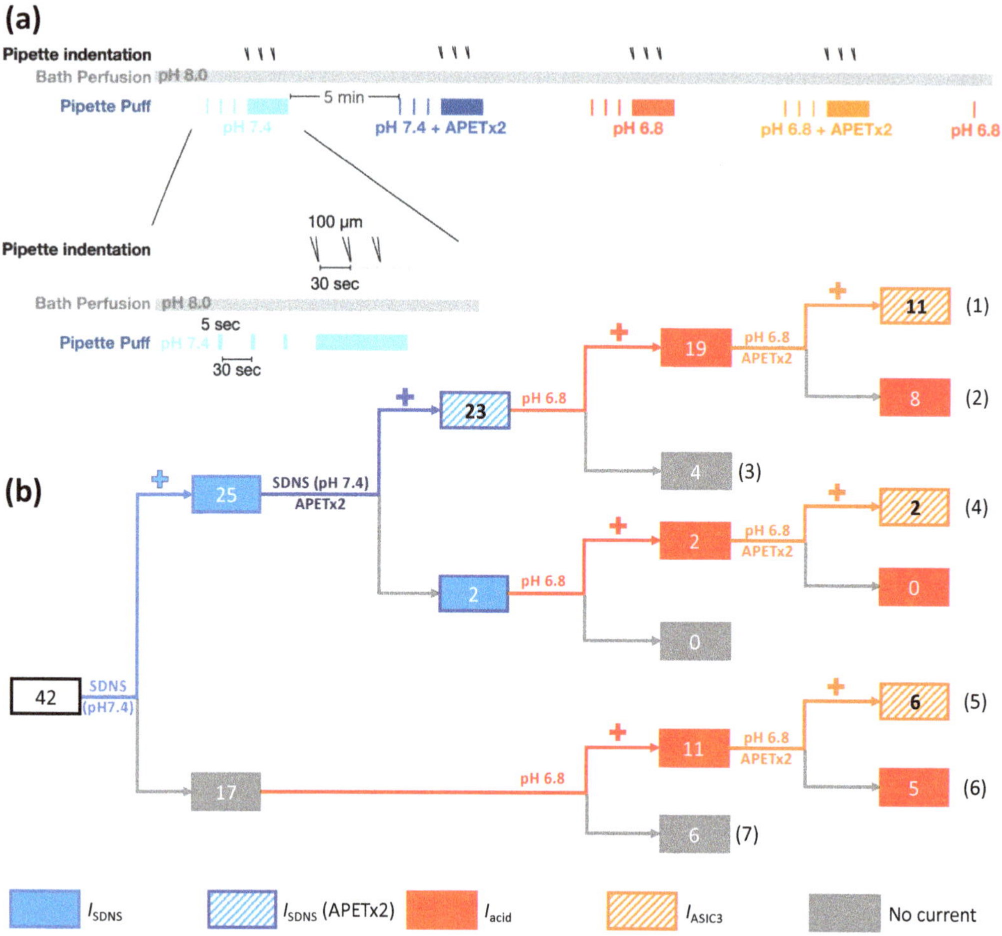

Figure 6. The involvement of ASIC3-containing channels in SDNS-induced currents in different Pv+ DRG proprioceptor subtypes. (**a**) Experimental design for testing the role of ASIC3 in SDNS-induced currents (I_{SDNS}) and acid-induced currents (I_{Acid}) during acidosis in Pv+ DRG proprioceptors. The neurons were bathed in a pH 8.0 ACSF, and different environmental conditions were established using a puff pipette with ACSF pH 7.4 (colored in light blue), ACSF pH 7.4 + APETx2 (colored in dark blue), ACSF pH 6.8 (colored in red), and ACSF pH 6.8 + APETx2 (colored in orange). In each environmental condition, the neuron first received a 5 sec puff 3 times (with a 30 sec interval) and was then subjected to three pipette indentations (with a 30 sec interval) during continuous puffing with the defined pH condition. With each pipette buffer change, the bath condition was changed back to pH 8.0 ACSF for 5 min. After all treatments, pH 6.8 ACSF was applied to confirm that APETx2 had been washed out. (**b**) Illustration of the subgrouping of Pv+ DRG neurons in response to SDNS and mild acidosis. The cell numbers were labeled in the bricks and the numbers of subgroups were labeled in the brackets.

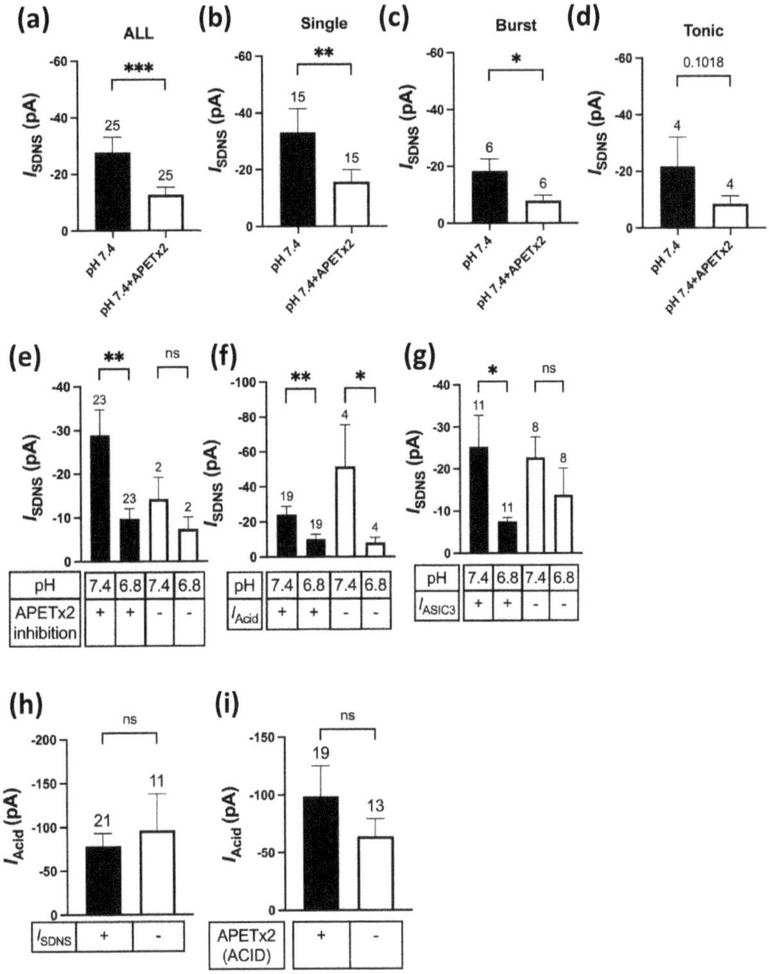

Figure 7. Effects of APETX2 and mild acidosis at pH 6.8 on SDNS-induced currents in different Pv+ DRG neuron subgroups. (**a**) The average peak current amplitude of the SDNS-induced currents (I_{SDNS}) of Pv+ DRG neurons are shown. A paired t-test found that the mean value of I_{SDNS} was significantly different ($p = 0.0002$). (**b**) The average peak current amplitudes of I_{SDNS} (expressed in pA) of Pv+ DRG neurons with single AP are shown. A paired t-test found that the mean value of I_{SDNS} was significantly different ($p = 0.0074$). (**c**) The average peak amplitude amplitudes of I_{SDNS} (expressed in pA) of Pv+ DRG neurons with burst AP are shown. A paired t-test found that the mean value of I_{SDNS} was significantly different ($p = 0.0386$). (**d**) The average peak amplitude amplitudes of I_{SDNS} (in pA) of Pv+ DRG neurons with tonic AP are shown. A paired t-test found no significant difference in the mean value of I_{SDNS} ($p = 0.1018$). (**e**) The effect of pH 6.8 acidosis on I_{SDNS} between Pv+ DRG neurons with APETx2-sensitive and APETx2-resistant I_{SDNS}. This graph indicates which SDNS current (I_{SDNS}) was APETx2-sensitive or -insensitive. A paired two-tailed t-test found that the mean value of I_{SDNS} was significantly different between the pH levels of 6.8 and 7.4 in the APETx2 inhibition group ($p = 0.0066$). (**f**) The effect of pH 6.8 acidosis on I_{SDNS} between I_{SDNS}-expressing Pv+ DRG neurons with and without acid-induced currents (I_{Acid}). A paired two-tailed t-test found that the mean value of I_{SDNS} was significantly different between the pH levels of 6.8 and 7.4, whether in the I_{acid}-conducted ($p = 0.0097$) or I_{acid}-non-conducted ($p = 0.0286$) group. (**g**) The effect of pH 6.8 acidosis

on I_{SDNS} between I_{SDNS}-expressing Pv+ DRG neurons with APETx2-sensitive I_{Acid} (I_{ASIC3}) and APETx2-resistant I_{Acid}. A paired two-tailed t-test found that the mean value of I_{SDNS} was significantly different between pH 6.8 and pH 7.4 in the I_{ASIC3}-containing group ($p = 0.02$). (**h**) The I_{acid} amplitudes of Pv+ DRG neurons with and without I_{SDNS} are shown. An unpaired two-tailed t-test found no significant difference in the mean value of I_{SDNS} ($p = 0.6237$). (**i**) The I_{acid} amplitudes of Pv+ DRG neurons with and without I_{ASIC3} are shown. An unpaired two-tailed t-test found that the mean value of I_{SDNS} was not significantly different, with $p = 0.3223$. The cell numbers were labeled above the column bars in each subfigure. * $p < 0.05$ ** $p < 0.01$ *** $p < 0.001$.

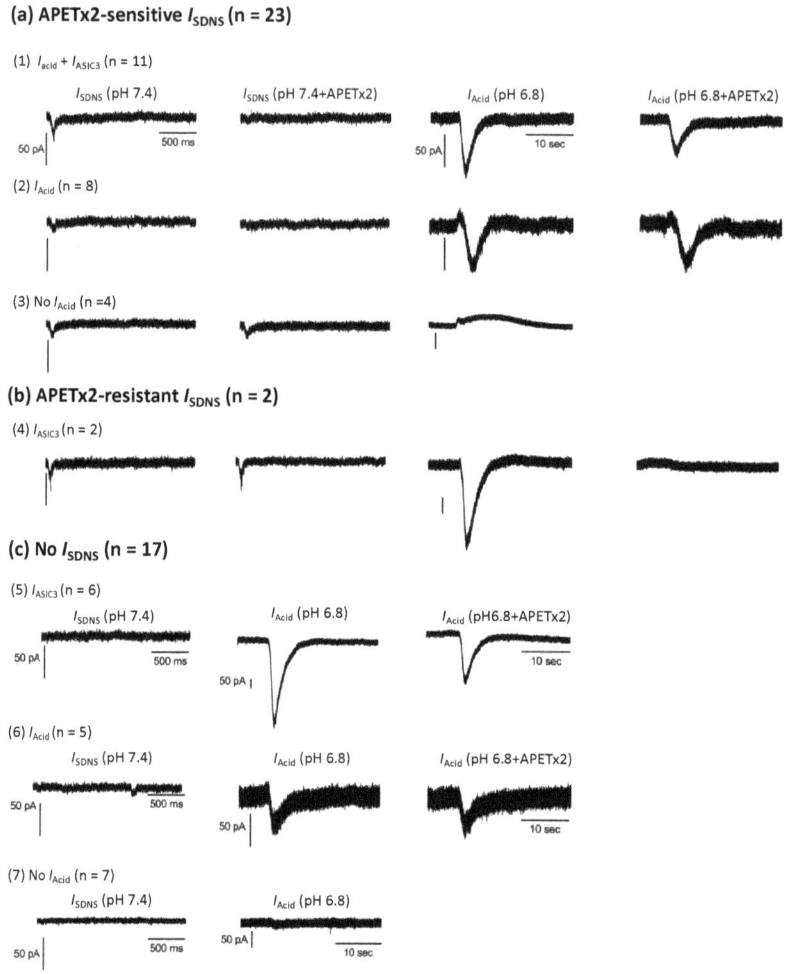

Figure 8. Representative traces of SDNS-induced currents (I_{SDNS}) and acid-induced currents (I_{Acid}) in seven groups of Pv+ DRG neurons. For the SDNS traces, the x- and y-axes of the scale bar denote 500 ms and 50 pA, respectively. For the acidic current traces, the x- and y-axes of the scale denote 10 s and 50 pA, respectively. (**a**) Pv+ DRG neurons with APETx2-sensitive I_{SDNS} were organized into Group (1), which expressed I_{Acid} + I_{ASIC3}; Group (2), which expressed I_{Acid} only; and Group (3), which expressed no I_{Acid}. (**b**) Pv+ DRG neurons with APETx2-resistant I_{SDNS} were organized into Group (4), which expressed I_{ASIC3}. (**c**) Pv+ DRG neurons that did not express I_{SDNS} were organized into Group (5) neurons, which expressed I_{Acid} + I_{ASIC3}; Group (6), which expressed I_{Acid}; and Group (7), which expressed no I_{Acid}.

3. Discussion

In this study, the SDNS approach was employed to probe the effect of mild acidosis on the tether-mode mechanotransduction of proprioceptive neurites. One advantage of this approach is that it delivers a specific mechanical mode emulating physiologically relevant conditions, in which the proprioceptive terminals of MSs or GTOs are stretched during muscle contraction [10]. It was found that 59% (25/42) of Pv+ DRG proprioceptors expressed the I_{SDNS}, and that ASIC3 was the major molecular determinant contributing to the I_{SDNS}. This finding is consistent with previous studies that showed amiloride-sensitive sodium channels were the main mechanically activated ion channels in response to MS stretching in ex vivo models [28]. Moreover, using a series of experimental designs, this study showed that mild acidosis significantly attenuated the I_{SDNS} in most ASIC3-positive proprioceptors. The I_{SDNS}-expressing Pv+ DRG proprioceptors were functionally heterogeneous in terms of their AP profiles, sensitivity to neurite stretching, and acidosis. In particular, not all I_{SDNS}-expressing Pv+ DRG proprioceptors expressed the I_{Acid} in response to the pH 6.8 ACSF. Of the 23 I_{SDNS}-expressing Pv+ proprioceptors, 19 expressed the I_{Acid}, but the I_{Acid} was only inhibited by the ASIC3-selective antagonist APETx2 in 61% (11/18) of Pv+ proprioceptors, suggesting that the ASIC3-containing ion channels are highly heterogenous among I_{SDNS}-expressing proprioceptors.

The fact that mild acidosis attenuated the I_{SDNS} in proprioceptive neurons provides valuable insights into the impairment of balance caused by fatigue or intermittent exercise [18]. Proprioceptive neurons play a crucial role in providing sensory information about body position and movement for the central nervous system. The attenuation of the I_{SDNS} due to mild acidosis suggests that acidosis could disrupt the mechanotransduction of proprioceptors, potentially leading to a diminished ability to accurately sense and respond to mechanical stimuli. This impairment in proprioceptive function could have significant consequences for maintaining balance and coordinating movements during exercise or in situations involving metabolic acidosis.

The inhibitory effect of mild acidosis on the I_{SDNS} appears to have been highly associated with the expression of ASIC3, as the acid-inhibiting effect occurred in the I_{SDNS} sensitive to APETx2 inhibition but not in that resistant to APETx2 (Figure 7e). APETx2 is a 42-amino-acid peptide isolated from the sea anemone *Anthopleura elegantissima*, known to inhibit ASIC3 homomeric channels and ASIC3-containing heteromeric channels [27]. Although APETx2 reversibly inhibits rat ASIC3 without affecting ASIC1a, ASIC1b, or ASIC2a, it exerts stronger inhibitory effects on the I_{Acid} of homomeric ASIC3 or heteromeric ASIC2b+3 than on that of ASIC1a+3 and ASIC1b+3 channels [27]. In addition, APETx2 was more effective in inhibiting the I_{SDNS} than inhibiting the I_{Acid}, as APETx2 failed to inhibit the I_{Acid} in 35% (8/23) of I_{SDNS}-expressing Pv+ proprioceptors (Figure 6b). Interestingly, mild acidosis did not significantly attenuate the I_{SDNS} in neurons expressing an APETx2-resistant I_{Acid} (Figure 7g). This suggests that mild acidosis might differentially influence the I_{SDNS} between Pv+ proprioceptors expressing an APETx2-sensitive I_{Acid} (possibly mediated by ASIC3 or ASIC2b+3) and those expressing an APETx2-resistant I_{Acid} (possibly mediated by ASIC1a+3 or ASIC1b+3). In brief, the composition of heteromeric ASIC3-containing channels may determine the effects of mild acidosis on proprioceptive nerve activity during muscle contraction. Further studies should be conducted to understand the functional properties and specific roles of different heteromeric ASIC3 channels in proprioceptive neurons, as well as their modulation of proprioception through acidosis.

Taking into account the pH dependency of ASIC subtypes and heteromeric compositions in heterologous expression systems, the presence of ASIC2a in the channel composition consistently leads to a half-maximum pH activation that is lower than 6.0 [29]. This indicates that ASIC2a-containing channels require a more concentrated acidosis to activate the proton-gated current compared with other subtype combinations. Accordingly, the ASIC2a-containing channels (ASIC2a+3) may account for the Pv+ proprioceptor subtype that expresses an I_{SDNS} but not an I_{Acid} at pH 6.8 (Figure 6b). Still, ASIC2a-containing channels might also be expressed in some Pv+ proprioceptors without an I_{SDNS}.

Mild-acidosis-induced structural changes in ASIC channels may underlie the observed attenuation of SDNS currents. Acidosis alters the pH environment surrounding the ASIC channels, potentially impacting their conformation and functional properties. ASIC3, as a homologue to chicken ASIC1a, shares similarities in its activation and desensitization mechanisms in response to low-pH environments [29,30]. Studies on chicken ASIC1a crystal structures have elucidated the structural changes associated with channel activation and desensitization [31]. In the resting state, the thumb domain of ASIC1a moves outward relative to its position in the open and desensitized state, resulting in the expansion of the acidic pocket. Activation involves the closure of the thumb domain into the acidic pocket, the expansion of the lower palm domain, and an iris-like opening of the channel gate. The beta11-12 linker, which separates the upper and lower palm domains, acts as a molecular clutch and undergoes rearrangement to facilitate rapid desensitization, effectively stopping the proton sensing current [32]. Although ASIC3 and ASIC1a may have divergent characteristics, it is reasonable to assume that ASIC3 might exhibit similar structural rearrangements and conformational changes in response to low pH and acidosis. These alterations in the thumb and palm domains, along with the rearrangement of the beta11-12 linker, could potentially influence the mechanical gating properties of ASIC3 channels [31]. As a result, the attenuation of the I_{SDNS} observed in mild acidosis may be attributed, at least in part, to the effect of acidosis on the mechanical gating of ASIC3 channels.

We proposed three possible mechanisms to determine how the SDNS response of ASIC3-containing channels in proprioceptive neurons is modulated by acidosis through different mechanisms, depending on the channel composition. In the case of ASIC3 homomeric channels, mild acidosis promoted channel opening, allowing the influx of ions and the initiation of a proton-gated response. However, due to the inherent desensitization properties of ASIC3, the channels were transitioned into a rapid desensitized state, resulting in the attenuation of the SDNS current (Figure 9a). In the case of ASIC3+1a and ASIC3+1b heteromeric channels, mild acidosis also facilitated channel opening, leading to the initiation of a proton-gated response and thus attenuating the SDNS current. However, the presence of ASIC1a or ASIC1b altered the kinetic and pharmacological responses, potentially influencing the magnitude of the SDNS current (Figure 9b). The combination of ASIC3 and ASIC2a in a heteromeric channel resulted in a direct transition into the desensitized state upon exposure to mild acidosis, bypassing the open state of the channel. This indicates that the SDNS response in ASIC3+2a heteromeric channels is primarily mediated by the channels being in a desensitized state without being transitioned into the open state (Figure 9c).

In summary, this study explored the complex interplay between acidosis, channel composition, and the mechanosensitivity of ASIC3-containing channels in proprioceptive neurons. Understanding the specific responses of ASIC3-containing channels to acidosis and their effect on the mechanosensitivity of proprioceptive neurons may provide insight into the regulation of sensory processing and motor control during exercise. Still, this study had some limitations: (1) the SDNS approach only induced an I_{SDNS} in 59% of Pv+ proprioceptors, which may have only represented a subgroup of proprioceptors sensitive to the SDNS stimuli; (2) some SDNS-insensitive neurons expressed an I_{ASIC3}, suggesting they are involved in tether-mode mechanotransduction but require either stronger SDNS stimuli or a different mechanical modality; and (3) we did not know the exact compositions of the ASIC3-containing channels among the I_{SDNS}-expressing and I_{SDNS}-insensitive Pv+ proprioceptors. Further research should be undertaken to unravel the precise molecular and biophysical mechanisms underlying the acidosis-induced modulation of ASIC3 channel gating, as well as its implications for proprioceptive function and balance control. Such research may pave the way for the development of targeted therapeutic interventions aimed at improving proprioception in individuals with acidosis-induced balance impairments.

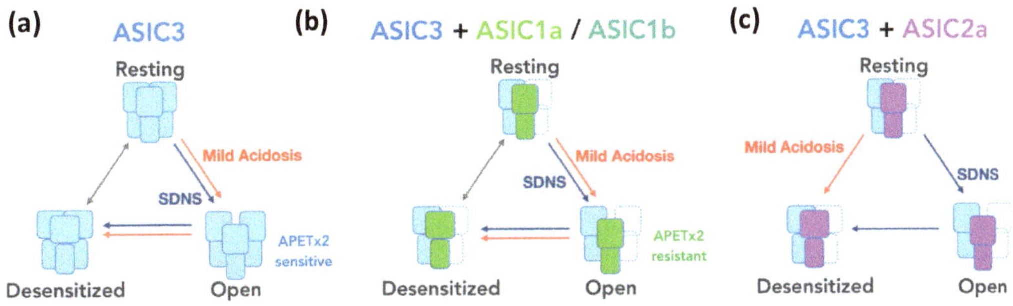

Figure 9. Hypothetical models illustrating the involvement of ASIC3 in the attenuation of SDNS-induced currents through mild acidosis. (**a**) A tentative model depicting the mechanistic process underlying the attenuation of SDNS-induced currents (I_{SDNS}) in response to mild acidosis, with a particular focus on the role of acid-sensing ion channel 3 (ASIC3). The model assumes that, through mild acidosis, ASIC3-containing channels are activated by SDNS, leading to the initiation of I_{SDNS} in proprioceptive neurons. Mild acidosis also triggers ASCI3 homomeric channels and thus the activation of proton-gated currents (which present APETx2-sensitive acid-induced currents (I_{Acid})). However, due to the rapid desensitization property of ASIC3, the channel is transitioned into a desensitized state, leading to the attenuation of the I_{SDNS}. (**b**) According to the expression of APETx2-resistant I_{Acid} in some Pv+ DRG proprioceptor neurons, the subunits ASIC1a and ASIC1b may contribute to neurons with I_{SDNS} that are less affected by mild acidosis. Pharmacological studies using APETx2, a selective ASIC3 antagonist, have shown that APETx2 effectively inhibits ASIC3 (IC_{50} = 63 nM) without affecting ASIC1a, ASIC1b, or ASIC2a, whereas ASIC1a+3 and ASIC1b+3 heteromeric channels showed low affinity to APETx2 as compared with ASIC3 or ASIC2b+3 channels. Therefore, the inclusion of the subunit ASIC1a or ASIC1b in the heteromeric channels may modify their response to the pH gradient of mild acidosis and kinetics to the attenuation of I_{SDNS}. (**c**) In I_{SDNS}-expressing Pv+ DRG neurons without I_{Acid}, ASIC2a may be incorporated in the composition of the heteromeric ASIC3 channels. Previous studies have indicated that tASIC2a-containing channels have a lower half-maximum pH activation (<pH 6.0) than other combinations lacking ASIC2a [29]. This suggests that mild acidosis (with a pH level of 6.8) cannot directly activate the proton-gated current in these neurons but rather attenuates I_{SDNS} by transforming the resting-state channels into a desensitized state, thus reducing their responsiveness to mechanical stimuli.

4. Materials and Methods

4.1. Animals

All the animal protocols followed in this study complied with the Guide for the Use of Laboratory Animals (National Academy of Sciences Press) and were approved by the Institutional Animal Care and Use Committee of the Academia Sinica, Taiwan. Both male and female mice at ages of 12–16 weeks were used. The Pv-Cre::Td (parvalbumin-Cre and tdTomato reporter) and Pv-Cre::CAG-cat-EGFP (parvalbumin-Cre and EGFP reporter) transgenic lines were backcrossed to the C57BL6 background and kept as heterozygotes for all experiments.

4.2. Immunostaining of MSs

Procedures for immunostaining and muscle tissue preparation were modified from a picric acid fixative method. The mice were anesthetized with a combination of Zoletil and dexmedetomidine hydrochloride; they were then perfused transcardially with 25 mL of 0.02 M phosphate-buffered saline (PBS, Omics Bio, Taipei, Taiwan) (pH = 7.4, at 4 °C) and cold fixative (4% (*w*/*v*) formaldehyde, 14% (*v*/*v*) saturated picric acid, and 0.1 M PBS (pH = 7.4, at 4 °C)). The animals' soleus muscles were dissected and post-fixed in the cold fixative solution (4 °C) for 1 h. The fixative was washed out with PBS for a while and cryoprotected in 20% sucrose PBS (pH = 7.3, at 4 °C) for 24 h. After being embedded in the optimal cutting temperature (OCT) compound (Leica Biosystem, Deer Park, IL, USA), the muscle tissues were frozen and sectioned into 12–16 μm thick sections, placed on a gelatin-coated slide glass, and stored at −80 °C before usage. Before immunostaining, the muscle sections were washed three times with PBST (PBS + 0.1% Triton X-100), blocked with PBST containing 3% bovine serum albumin (Sigma-Aldrich, St. Louis, MO, USA) and 5% goat serum (Sigma-Aldrich, St. Louis, MO, USA), and incubated with primary antibodies (rabbit anti-GFP 1:5000 in blocking solution, Abcam, Cambridge, UK; mouse anti-NF-H, 1:2000 in blocking solution, Chemicon, St. Louis, MO, USA) overnight at 4 °C. After PBST washing was performed 3 times, the muscle sections were incubated in the secondary antibodies (Goat anti-rabbit 1:500, Invitrogen, Waltham, MA, USA) for 1 h (at room temperature).

4.3. Primary Culture of DRG Neurons

Mice at the age of 12–16 weeks were sacrificed with CO_2, and their DRG neurons were collected and cultured in accordance with the procedures described in Cheng et al., (2010) [14] and Lin et al., (2016) [15]. The dissected DRGs were digested with 0.125% collagenase (type I, Sigma-Aldrich, St. Louis, MO, USA) and 2 units/mL dispase II (Sigma-Aldrich, St. Louis, MO, USA) for 30 min at 37 °C. The segregated neurons were triturated using a flame-polished Pasteur pipette and seeded on laminin (Sigma-Aldrich, St. Louis, MO, USA)-coated polydimethylsiloxane (PDMS, UNI WARD, New Taipei City, Taiwan) substrate, which was prepared on a 12 mm coverslip with a base-to-curing-agent ratio of 35:1. Before the DRG neurons were seeded, the PDMS-covered coverslips were exposed to ultraviolet light for 15 min and coated with poly-L-lysine (0.01%, Sigma-Aldrich, St. Louis, MO, USA) for 10 min; then, they were coated with laminin (10 μg/mL, Sigma-Aldrich, St. Louis, MO, USA) for 2 h. The neurons were cultured in a 3.5 cm Petri dish with DMEM plus 10% fetal bovine serum and maintained in an incubator with 5% CO_2 at 37 °C for 2–3 days. The cultured DRG neurons were then subjected to electrophysiological recordings of acid-induced currents or mechanically activated currents.

4.4. Electrophysiology

4.4.1. Whole-Cell Patch-Clamp Recordings

The acid-induced currents (I_{Acid}) in the Pv-Cre::Td-positive DRG neurons were measured. Six-channel valve gates (white arrow) were attached to the perfusion columns, the other side was conducted to a three-barrel pipette using a six-in one-out device (black arrow), and the acidic artificial cerebral spinal fluid (ACSF) controlled by the six-channel valve controller (Warner, VC-6, Science Products, Hesse, Germany) (red arrow) was applied (Figure 2). Whole-cell patch clamping was conducted on neurite-bearing DRG neurons cultured for 48–72 h. The recording pipette (with a resistance of 6–8 MOhm) was filled with internal solution (in mM, 100 KCl, 2 Na-ATP, 3 Na-GTP, 10 EGTA, 5 $MgCl_2$, and 40 HEPES, with the pH adjusted to 7.4 using KOH; osmolality 290–310 mOsm) (Sigma-Aldrich, St. Louis, MO, USA), and the neurons were externally bathed in the ACSF solution (in mM, 130 NaCl, 5 KCl, 1 $MgCl_2$, 2 $CaCl_2$, 10 glucose, and 20 HEPES, with the pH adjusted to 7.4 or 8.0) (Sigma-Aldrich, St. Louis, MO, USA). The puff pipette was filled with acidic ACSF, whose pH was adjusted to 6.8, 7.0, 7.2, and 7.4. Mechanical stimuli were delivered using a mechanical pipette on an electronically controlled micromanipulator (Scientifica PatchStar Micromanipulator pc8200c, Scientifica, East Sussex, UK). The pipette was microforge-polished (MF-900, Narishige, Tokyo, Japan) to 4–5 MOhm resistance. Loaded with ACSF, the pipette was held at −5 to −10 mV (the leak current slowed to less than −1500 pA) using a Multiclamp 700B (Axon instrument, San Jose, CA, USA) so that we could monitor the contact with the PDMS surface. Once the mechanical pipette was lowered down and touched the PDMS surface, a leak current drop was observed, indicating increased resistance. Thus, we synchronized the timing of the whole-cell recording with the starting point of the PDMS indentation.

4.4.2. Rheobase Analysis of AP Threshold

The neurons were bathed in an ACSF solution (pH 7.4) and formed by the whole-cell patch. The membrane potential was then held at −70 mV by injecting current in the current clamp mode, and a series of stimuli (from 100 to 2000 pA for 20 sweeps, with a 100 pA increase per step, over a 500 ms duration) was provided to trigger an AP. The rheobase current was determined using the minimal injected current to trigger an AP (Supplementary Figure S1).

4.4.3. Mechanically Activated Current Recording

After the rheobase current was recorded, the neurons were transferred to be externally bathed in an ACSF solution (pH 8.0) containing TTX (300 nM, Tocris Bioscience, Bristol, UK) to inhibit the voltage-gated sodium channel activity, and the mechanical pipette was placed around the distal neurite terminal in a region on the PDMS surface that satisfied the following criteria: (i) the region did not contain any other cells such as glia or fibroblasts; (ii) the region was at least 100 μm away from the Pv-Cre::Td-selected neuron cell soma; and (iii) the region was 15–25 μm away from the distal terminal (Figure 2b,c). To indent the PDMS substrate, the mechanical pipette was first anchored on the surface as a starting position (p0) and then hoisted on the z-axis for 100 μm (p+100). The indentation of the mechanical pipette was programmed so that the pipette was (i) dropped down on the z-axis for 200 μm (that is, 100 μm below the initial surface plane (p-100)) at a velocity of 1.6 $\mu m\ ms^{-1}$; (ii) kept at (p-100) for 1000 ms; and (iii) elevated at the same velocity to (p+100). After pipette indentation was performed on PDMS, the deformation of the substrate stretched local neurites. A series of indentation depths (25, 50, 75, 100, and 125 μm) were tested within the same Pv+ DRG neuron to analyze the force dependency of the SDNS-induced currents (I_{SDNS}) (Figure 2d).

4.4.4. Mechanically Activated Currents during Acidosis

To study the effect of acidosis on mechanically activated currents, we first recorded the I_{SDNS} with indentation depths of 25, 50, 75, 100, and 125 μm in a pH 7.4 bath. A pH 6.8 acidic bath was then applied to the same neurons, and we waited at least 5 min for the acidic ACSF to fully immerse the neurons before measuring the I_{SDNS} under the same indentation protocol. This allowed us to evaluate the effect of acidosis on the I_{SDNS} (Figure 3).

To further investigate the effect of proton concentrations on different proprioceptor groups, we triggered APs and tested the rheobase of Pv+ neurons in neutral ACSF; then, we used a pH 8.0 TTX-containing ACSF solution for the I_{SDNS} recordings. The neurons were perfused with a puff pipette preloaded with TTX-containing ACSF at the pH levels of 6.8, 7.0, 7.2, and 7.4. They were sequentially exposed to acidic challenges from the ACSF with pH levels of 7.2, 7.0, and 6.8 to assess the effect of proton concentrations on the I_{SDNS} (with a 100 μm indentation) (Figure 4).

4.4.5. ASIC3 Dependency

To determine whether ASIC3 contributed to the expression of the I_{Acid} and I_{SDNS} in Pv+ DRG neurons, the ASIC3-selective antagonist APETx2 (2 μM, Alomone Labs, Jerusalem, Israel) was included in the puff pipette containing ACSF with a pH of either 7.4 or 6.8 [15]. After the AP profile measurement, the Pv+ DRG neurons were subjected to a series of recordings of the I_{Acid} and I_{SDNS} and tested for APETx2 inhibition (Figure 6).

4.5. Statistical Analysis

Data in all figures are presented in the form of mean ± sem. Statistical comparison was performed using Prism (ver. 10.0.2) (ANOVA and *t*-test). A non-parametric two-tailed Mann–Whitney test was used in some electrophysiology experiments where the raw data did not pass the normal distribution or equal variance test. The Holm–Sidak test was employed for post hoc tests with two-way ANOVA and paired one-way ANOVA, because the Holm–Sidak test is more powerful than the Tukey or Bonferroni test. The criterion for a significant difference was set to $p < 0.05$.

5. Conclusions

This study focused on the significant role of ASIC3, a proton-gated ion channel with the dual gating properties of acid sensing and mechanosensing, in modulating proprioception under acidic physiological conditions. Understanding the role of ASIC3 in these processes may help to clarify the underlying mechanisms of exercise-induced balance impairments, to develop strategies for mitigating these effects, and to optimize athletic performance. Furthermore, investigating the heterogeneity of Pv+ DRG neurons in the context of ASIC3 and its potential interactions with other ASIC subtypes may improve the understanding of the mechanosensing capabilities of proprioceptive neurons and inform the development of targeted interventions for conditions involving impaired proprioception.

Supplementary Materials: The following supporting information can be downloaded at: https://www.mdpi.com/article/10.3390/ijms241612783/s1.

Author Contributions: Conceptualization, Y.-R.C. and C.-C.C.; Formal analysis, C.-H.C.; Investigation, C.-H.C.; Resources, S.-H.L.; Data curation, C.-H.C.; Writing – original draft, Y.-R.C. and C.-H.C.; Writing—review & editing, C.-C.C.; Supervision, C.-H.L., M.-Y.M. and C.-C.C.; Project administration, C.-C.C. All authors have read and agreed to the published version of the manuscript.

Funding: This study was undertaken with the financial support of Academia Sinica (IBMS-CRC111-P02, AS-IA111-L06) and the National Science and Technology Council of Taiwan (-, NSTC112-2321-B-001-009).

Institutional Review Board Statement: Not applicable.

Informed Consent Statement: Not applicable.

Data Availability Statement: Not applicable.

Conflicts of Interest: The authors declare no conflict of interest.

References

1. Proske, U. Exercise, fatigue and proprioception: A retrospective. *Exp. Brain Res.* **2019**, *237*, 2447–2459. [CrossRef] [PubMed]
2. Proske, U.; Gandevia, S.C. The proprioceptive senses: Their roles in signaling body shape, body position and movement, and muscle force. *Physiol. Rev.* **2012**, *92*, 1651–1697. [CrossRef] [PubMed]
3. Zampieri, N.; de Nooij, J.C. Regulating muscle spindle and Golgi tendon organ proprioceptor phenotypes. *Curr. Opin. Physiol.* **2021**, *19*, 204–210. [CrossRef] [PubMed]
4. Oliver, K.M.; Florez-Paz, D.M.; Badea, T.C.; Mentis, G.Z.; Menon, V.; de Nooij, J.C. Molecular correlates of muscle spindle and Golgi tendon organ afferents. *Nat. Commun.* **2021**, *12*, 1451. [CrossRef]
5. Santuz, A.; Akay, T. Muscle spindles and their role in maintaining robust locomotion. *J. Physiol.* **2023**, *601*, 275–285. [CrossRef]
6. Wu, D.; Schieren, I.; Qian, Y.; Zhang, C.; Jessell, T.M.; de Nooij, J.C. A Role for Sensory end Organ-Derived Signals in Regulating Muscle Spindle Proprioceptor Phenotype. *J. Neurosci. Off. J. Soc. Neurosci.* **2019**, *39*, 4252–4267. [CrossRef] [PubMed]
7. Macefield, V.G.; Knellwolf, T.P. Functional properties of human muscle spindles. *J. Neurophysiol.* **2018**, *120*, 452–467. [CrossRef]
8. Dietrich, S.; Company, C.; Song, K.; Lowenstein, E.D.; Riedel, L.; Birchmeier, C.; Gargiulo, G.; Zampieri, N. Molecular identity of proprioceptor subtypes innervating different muscle groups in mice. *Nat. Commun.* **2022**, *13*, 6867. [CrossRef]
9. Nilius, B.; Honore, E. Sensing pressure with ion channels. *Trends Neurosci.* **2012**, *35*, 477–486. [CrossRef]
10. Bewick, G.S.; Banks, R.W. Mechanotransduction in the muscle spindle. *Pflug. Arch. Eur. J. Physiol.* **2015**, *467*, 175–190. [CrossRef]
11. Coste, B.; Mathur, J.; Schmidt, M.; Earley, T.J.; Ranade, S.; Petrus, M.J.; Dubin, A.E.; Patapoutian, A. Piezo1 and Piezo2 are essential components of distinct mechanically activated cation channels. *Science* **2010**, *330*, 55–60. [CrossRef] [PubMed]
12. Assaraf, E.; Blecher, R.; Heinemann-Yerushalmi, L.; Krief, S.; Carmel Vinestock, R.; Biton, I.E.; Brumfeld, V.; Rotkopf, R.; Avisar, E.; Agar, G.; et al. Piezo2 expressed in proprioceptive neurons is essential for skeletal integrity. *Nat. Commun.* **2020**, *11*, 3168. [CrossRef] [PubMed]
13. Woo, S.H.; Lukacs, V.; de Nooij, J.C.; Zaytseva, D.; Criddle, C.R.; Francisco, A.; Jessell, T.M.; Wilkinson, K.A.; Patapoutian, A. Piezo2 is the principal mechanotransduction channel for proprioception. *Nat. Neurosci.* **2015**, *18*, 1756–1762. [CrossRef]
14. Cheng, C.M.; Lin, Y.W.; Bellin, R.M.; Steward, R.L., Jr.; Cheng, Y.R.; LeDuc, P.R.; Chen, C.C. Probing localized neural mechanotransduction through surface-modified elastomeric matrices and electrophysiology. *Nat. Protoc.* **2010**, *5*, 714–724. [CrossRef] [PubMed]
15. Lin, S.H.; Cheng, Y.R.; Banks, R.W.; Min, M.Y.; Bewick, G.S.; Chen, C.C. Evidence for the involvement of ASIC3 in sensory mechanotransduction in proprioceptors. *Nat. Commun.* **2016**, *7*, 11460. [CrossRef]

16. Lin, J.H.; Hung, C.H.; Han, D.S.; Chen, S.T.; Lee, C.H.; Sun, W.Z.; Chen, C.C. Sensing acidosis: Nociception or sngception? *J. Biomed. Sci.* **2018**, *25*, 85. [CrossRef] [PubMed]
17. Cheng, Y.R.; Jiang, B.Y.; Chen, C.C. Acid-sensing ion channels: Dual function proteins for chemo-sensing and mechano-sensing. *J. Biomed. Sci.* **2018**, *25*, 46. [CrossRef] [PubMed]
18. Douris, P.C.; Handrakis, J.P.; Gendy, J.; Salama, M.; Kwon, D.; Brooks, R.; Salama, N.; Southard, V. Fatiguing upper body aerobic exercise impairs balance. *J. Strength Cond. Res.* **2011**, *25*, 3299–3305. [CrossRef] [PubMed]
19. Kroger, S. Proprioception 2.0: Novel functions for muscle spindles. *Curr. Opin. Neurol.* **2018**, *31*, 592–598. [CrossRef]
20. Beenhakker, M.P.; Kirby, M.S.; Nusbaum, M.P. Mechanosensory gating of proprioceptor input to modulatory projection neurons. *J. Neurosci. Off. J. Soc. Neurosci.* **2007**, *27*, 14308–14316. [CrossRef]
21. Moon, K.M.; Kim, J.; Seong, Y.; Suh, B.C.; Kang, K.; Choe, H.K.; Kim, K. Proprioception, the regulator of motor function. *BMB Rep.* **2021**, *54*, 393–402. [CrossRef] [PubMed]
22. Lyle, M.A.; Nichols, T.R. Evaluating intermuscular Golgi tendon organ feedback with twitch contractions. *J. Physiol.* **2019**, *597*, 4627–4642. [CrossRef] [PubMed]
23. Bangsbo, J.; Johansen, L.; Graham, T.; Saltin, B. Lactate and H^+ effluxes from human skeletal muscles during intense, dynamic exercise. *J. Physiol.* **1993**, *462*, 115–133. [CrossRef] [PubMed]
24. Pilegaard, H.; Bangsbo, J.; Henningsen, P.; Juel, C.; Richter, E.A. Effect of blood flow on muscle lactate release studied in perfused rat hindlimb. *Am. J. Physiol.* **1995**, *269 Pt 1*, E1044–E1051. [CrossRef]
25. Juel, C.; Klarskov, C.; Nielsen, J.J.; Krustrup, P.; Mohr, M.; Bangsbo, J. Effect of high-intensity intermittent training on lactate and H^+ release from human skeletal muscle. *Am. J. Physiol. Endocrinol. Metab.* **2004**, *286*, E245–E251. [CrossRef]
26. Deval, E.; Noel, J.; Lay, N.; Alloui, A.; Diochot, S.; Friend, V.; Jodar, M.; Lazdunski, M.; Lingueglia, E. ASIC3, a sensor of acidic and primary inflammatory pain. *EMBO J.* **2008**, *27*, 3047–3055. [CrossRef]
27. Diochot, S.; Baron, A.; Rash, L.D.; Deval, E.; Escoubas, P.; Scarzello, S.; Salinas, M.; Lazdunski, M. A new sea anemone peptide, APETx2, inhibits ASIC3, a major acid-sensitive channel in sensory neurons. *EMBO J.* **2004**, *23*, 1516–1525. [CrossRef]
28. Simon, A.; Shenton, F.; Hunter, I.; Banks, R.W.; Bewick, G.S. Amiloride-sensitive channels are a major contributor to mechanotransduction in mammalian muscle spindles. *J. Physiol.* **2010**, *588 Pt 1*, 171–185. [CrossRef]
29. Hesselager, M.; Timmermann, D.B.; Ahring, P.K. pH Dependency and desensitization kinetics of heterologously expressed combinations of acid-sensing ion channel subunits. *J. Biol. Chem.* **2004**, *279*, 11006–11015. [CrossRef]
30. Boscardin, E.; Alijevic, O.; Hummler, E.; Frateschi, S.; Kellenberger, S. The function and regulation of acid-sensing ion channels (ASICs) and the epithelial Na(+) channel (ENaC): IUPHAR Review 19. *Br. J. Pharmacol.* **2016**, *173*, 2671–2701. [CrossRef]
31. Yoder, N.; Yoshioka, C.; Gouaux, E. Gating mechanisms of acid-sensing ion channels. *Nature* **2018**, *555*, 397–401. [CrossRef] [PubMed]
32. Rook, M.L.; Williamson, A.; Lueck, J.D.; Musgaard, M.; Maclean, D.M. beta11-12 linker isomerization governs acid-sensing ion channel desensitization and recovery. *Elife* **2020**, *9*, e51111. [CrossRef] [PubMed]

International Journal of
Molecular Sciences

MDPI

Article

Sodium Leak Channel in Glutamatergic Neurons of the Lateral Parabrachial Nucleus Modulates Inflammatory Pain in Mice

Lin Wu [1,2], Yujie Wu [1,2], Jin Liu [1,2], Jingyao Jiang [1,2], Cheng Zhou [1,2,*] and Donghang Zhang [1,2,*]

1 Department of Anesthesiology, West China Hospital, Sichuan University, Chengdu 610041, China; wulin_1998@163.com (L.W.); yujie-wu@outlook.com (Y.W.); scujinliu@gmail.com (J.L.); 18983476564@163.com (J.J.)

2 Laboratory of Anaesthesia and Critical Care Medicine, National-Local Joint Engineering Research Centre of Translational Medicine of Anesthesiology, West China Hospital, Sichuan University, Chengdu 610041, China

* Correspondence: zhouc@scu.edu.cn (C.Z.); zhangdh911@scu.edu.cn (D.Z.)

Abstract: Elevated excitability of glutamatergic neurons in the lateral parabrachial nucleus (PBL) is associated with the pathogenesis of inflammatory pain, but the underlying molecular mechanisms are not fully understood. Sodium leak channel (NALCN) is widely expressed in the central nervous system and regulates neuronal excitability. In this study, chemogenetic manipulation was used to explore the association between the activity of PBL glutamatergic neurons and pain thresholds. Complete Freund's adjuvant (CFA) was used to construct an inflammatory pain model in mice. Pain behaviour was tested using von Frey filaments and Hargreaves tests. Local field potential (LFP) was used to record the activity of PBL glutamatergic neurons. Gene knockdown techniques were used to investigate the role of NALCN in inflammatory pain. We further explored the downstream projections of PBL using cis-trans-synaptic tracer virus. The results showed that chemogenetic inhibition of PBL glutamatergic neurons increased pain thresholds in mice, whereas chemogenetic activation produced the opposite results. CFA plantar modelling increased the number of C-Fos protein and NALCN expression in PBL glutamatergic neurons. Knockdown of NALCN in PBL glutamatergic neurons alleviated CFA-induced pain. CFA injection induced C-Fos protein expression in central nucleus amygdala (CeA) neurons, which was suppressed by NALCN knockdown in PBL glutamatergic neurons. Therefore, elevated expression of NALCN in PBL glutamatergic neurons contributes to the development of inflammatory pain via PBL-CeA projections.

Keywords: chemogenetics; glutaminergic neurons; lateral parabrachial nucleus; NALCN; inflammatory pain

1. Introduction

Inflammatory pain is a complex condition characterised by multiple mechanisms, such as hyperexcitability and/or sensitisation of primary nociceptive neurons or nociceptors and neural-immune-endocrine interactions [1–4]. Inflammatory pain is one of the major contributors to the health care burden and is associated with many chronic diseases [5,6]. Despite the high prevalence and severity of inflammatory pain, the medications currently available to treat inflammatory pain are unsatisfactory, such as nonsteroidal anti-inflammatory drugs (NSAIDs), opioids and some adjuvant medications [7]. The combination of different drugs with different mechanisms of action improves the therapeutic effect, but their use is limited due to their increased risk of side effects [8]. Clinically, the "pain ladder" therapy published by the WHO is suitable for the treatment of acute and chronic pain but has little relevance to inflammatory pain [9]. The underlying cause of the problem is that the exact molecular mechanisms associated with inflammatory pain are not fully understood and involve complex peripheral and central mechanisms [2]. Therefore, it is clinically important to investigate the molecular mechanisms mediating inflammatory pain and to develop novel therapeutic agents.

The parabrachial nucleus (PBN) is a heterogeneous nucleus located on the dorsal side of the pons around the superior cerebellar peduncle (scp) [10]. PBN is involved in regulating many physiological functions of the body, such as respiration, circulation and pain [11,12]. PBN consists of three distinct subnuclei, including PBL, PBM (parabrachial medial nucleus) and KFN (Kölliker-Fuse nucleus) [11]. PBL mainly receives nociceptive input from dorsal projection neurons in the spinal cord and is most relevant to pain regulation [13–16]. Yang et al. [17] reported that PBL is essential for the transmission of pain signals from the spinal cord to the substantia nigra of the reticular formation (SNR). SNR-projecting PBL neurons can be activated by noxious stimuli [17]. In fact, most neurons in the PBL are glutamatergic [11,18]. A previous study [15] found that spinal cord projections to PBL are strictly glutamatergic. Taken together, these studies suggest an important role for PBL glutamatergic neurons in pain regulation. However, the molecular mechanism by which PBL glutaminergic neurons are involved in inflammatory pain is not fully understood.

The sodium leak channel (NALCN) is an unselective cation background channel that regulates resting membrane potential and neuronal activities [4,19]. NALCN is associated with the regulation of many important biological functions, such as respiration, pain, and anaesthesia [4,19,20]. Ford et al. [21] demonstrated that NALCN in spino-PBN neurons is involved in the upstream transmission of nociception and therefore may affect pain perception. Moreover, Zhang and colleagues [19] found that elevated expression and function of NALCN in the peripheral dorsal root ganglion and spinal cord contributes to the development of both chronic constriction injury (CCI)-induced neuropathic pain and complete Freund's adjuvant (CFA)-induced inflammatory pain in rodents [4], highlighting NALCN as a potential molecular target for the treatment of pain. However, whether NALCN in brain nuclei is implicated in the modulation of pain remains unclear.

Based on the above evidence, this study hypothesised that NALCN in PBL glutamatergic neurons is involved in the modulation of CFA-induced inflammatory pain.

2. Results

- Chemogenetic inhibition of PBL glutamatergic neurons increases the pain threshold in mice.

Chemogenetic manipulations were used to clarify whether the activities of PBL glutamatergic neurons are involved in pain regulation (Figure 1A). First, we injected viruses encoding hM4D(Gi) receptors (AAV2/9-mCaMKIIa-hM4D(Gi)-mCherry-ER2-WPRE-pA) or its control (AAV2/9-mCaMKIIa- mCherry-WPRE-pA) into the bilateral PBL of mice (Figure 1B, left). Three weeks later, mCherry expression was observed in PBL (Figure 1B, right). Immunofluorescence staining of C-Fos was performed after 2.5 mg·kg^{-1} clozapine N-oxide (CNO) injection (Figure 1C,D). Compared with the control group, there was a significant decrease in the number of C-Fos-positive cells in the PBL of the hM4D(Gi) group (Figure 1E, 30.3 $\pm$ 6.1 vs. 15.4 $\pm$ 4.4; n = 10 in both groups, $p < 0.001$), suggesting that PBL glutamatergic neuronal activity was significantly inhibited after CNO injection.

Next, we measured the von Frey threshold and thermal withdrawal latency in mice under chemogenetic manipulations. We intraperitoneally injected 2.5 mg·kg^{-1} CNO or an equal volume of saline into the mice 1 h before behavioural testing, and then the mice were placed on a metal elevated net for acclimatisation. In the control group, there was no significant difference in the von Frey threshold (Figure 1F, 0.4 $\pm$ 0.2 vs. 0.4 $\pm$ 0.2 g; n = 10, $p > 0.05$) or thermal withdrawal latency (Figure 1G, 10.2 $\pm$ 1.2 vs. 11.2 $\pm$ 1.5 s; n = 10, $p > 0.05$) after intraperitoneal injection of saline or CNO into mice. However, after intraperitoneal injection of CNO, the von Frey threshold (Figure 1H, 0.3 $\pm$ 0.2 vs. 0.6 $\pm$ 0.2 g; n = 11, $p < 0.001$) and thermal withdrawal latency (Figure 1I, 10.1 $\pm$ 1.4 vs. 12.8 $\pm$ 1.3 s; n = 11, $p < 0.001$) were significantly higher in the hM4D(Gi) group compared with intraperitoneal saline injection. The von Frey threshold and thermal withdrawal latency in control and hM4D(Gi) mice did not change significantly after saline injection compared with the baseline (Supplementary Figure S1A–D).

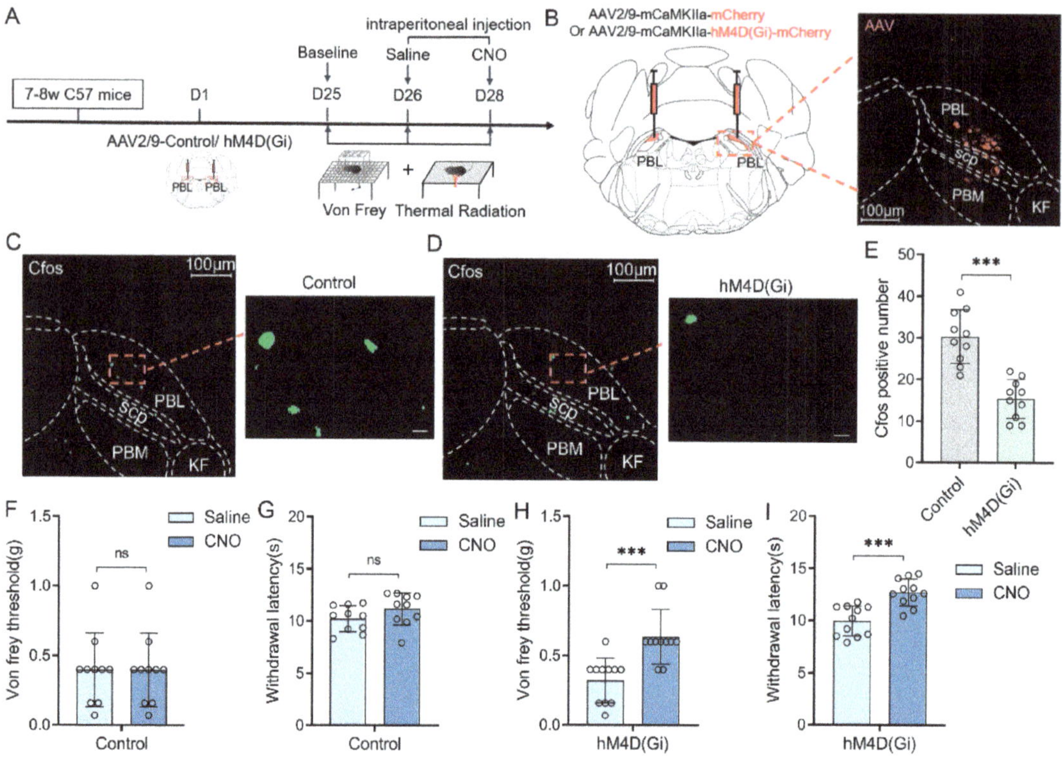

Figure 1. Chemogenetic inhibition of PBL glutamatergic neurons increases the pain threshold in mice. (**A**) Flow chart of the behavioural measurements in control and genetically manipulated mice. (**B**) Schematic of virus injection and hM4D(Gi) and mCherry expression in PBL glutamatergic neurons. Scale = 100 µm. (**C**,**D**) C-Fos immunofluorescence (green) in PBL after CNO injection in control (**C**) and genetically manipulated mice (**D**). Scale = 100 µm (**left**) or 10 µm (**right**). (**E**) C-Fos count comparison after CNO injection in control and hM4D(Gi) mice after immunofluorescence staining (n = 10 in both groups). (**F**,**G**) Von Frey threshold (**F**) and withdrawal latency (**G**) in control mice after saline or CNO injection (n = 10). (**H**,**I**) Von Frey threshold (**H**) and withdrawal latency (**I**) in hM4D(Gi) mice after saline or CNO injection (n = 11). Data are presented as the mean ± SD. *** $p < 0.001$, ns: no significance by a two-tailed independent samples t test (**E**–**I**).

- Chemogenetic activation of PBL glutamatergic neurons decreases the pain threshold in mice.

Next, we chemogenetically activated PBL glutamatergic neurons (Figure 2A) by injecting AAV2/9-mCaMKIIa-hM3D(Gq)-mCherry-ER2-WPRE-pA or AAV2/9-mCaMKIIa-mCherry-WPRE-pA into the bilateral PBL of C57BL/6J mice (Figure 2B, left). The expression of mCherry was observed in PBL after 3 weeks (Figure 2B, right). Immunofluorescence staining of C-Fos (Figure 2C,D) showed a significant increase in the hM3D(Gq) group compared with the control group after intraperitoneal injection of CNO into mice (Figure 2E, 29.5 ± 8.2 vs. 61.4 ± 12.2; n = 10 in both groups, $p < 0.001$), suggesting that PBL glutamatergic neuronal activity was significantly activated by chemogenetic activation.

Then, we measured the von Frey threshold and thermal withdrawal latency of the mice. In the control group, the von Frey threshold (Figure 2F, 0.4 ± 0.2 vs. 0.4 ± 0.2 g; n = 10, $p > 0.05$) and thermal withdrawal latency (Figure 2G, 11.2 ± 1.6 vs. 11.5 ± 1.6 g; n = 10, $p > 0.05$) of mice did not change significantly after intraperitoneal injection of CNO compared with saline. In the hM3D(Gq) group, the von Frey threshold (Figure 2H,

0.4 ± 0.3 vs. 0.1 ± 0.1 g; n = 11, $p < 0.001$) and thermal withdrawal latency (Figure 2I, 10.3 ± 1.1 vs. 6.7 ± 1.5 s; n = 11, $p < 0.001$) were significantly decreased after intraperitoneal injection of CNO compared with saline. There was no significant difference in the von Frey threshold and thermal withdrawal latency in control and hM4D(Gi) mice after saline injection compared with the baseline (Supplementary Figure S1E–H).

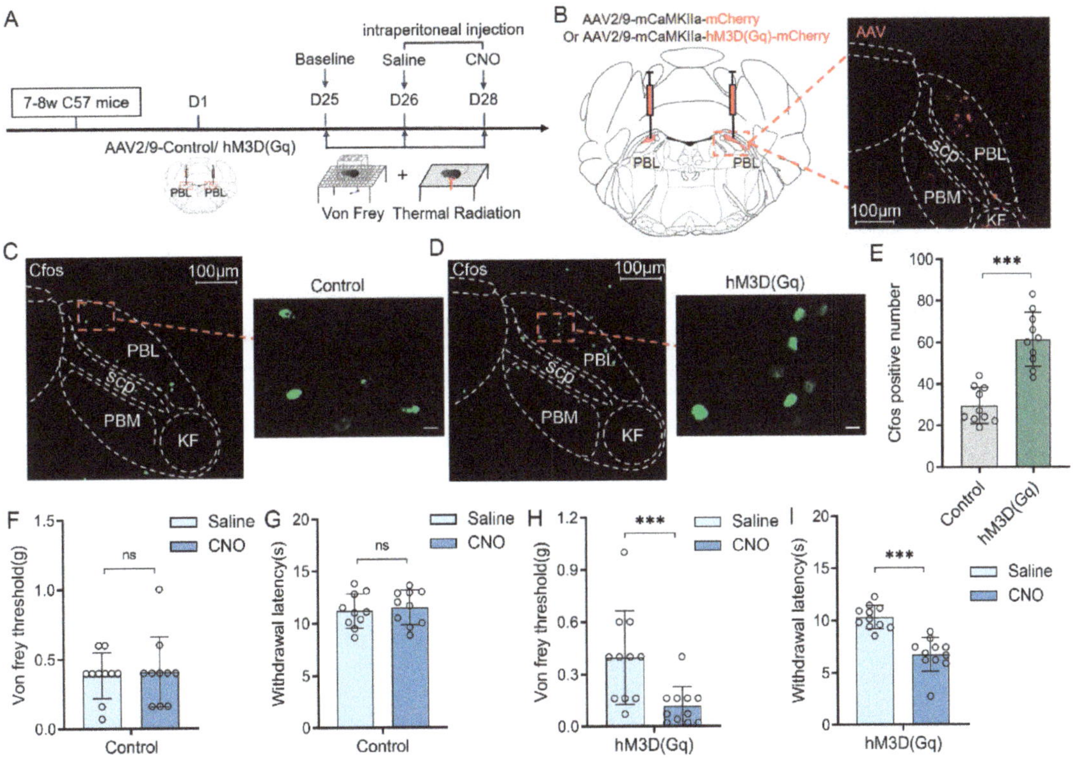

Figure 2. Chemogenetic activation of PBL glutamatergic neurons decreases the pain threshold in mice. (**A**) Flow chart of the behavioural measurements in control and genetically manipulated mice. (**B**) Schematic of virus injection and hM3D(Gq) and mCherry expression in PBL glutamatergic neurons. Scale = 100 μm. (**C**,**D**) C-Fos immunofluorescence (green) in PBL after CNO injection in control (**C**) and hM3D(Gq) (**D**) mice. Scale = 100 μm (**left**) or 10 μm (**right**). (**E**) C-Fos count comparison after CNO injection in control and hM3D(Gq) mice after immunofluorescence staining (n = 10 in both groups). (**F**,**G**) Von Frey threshold and thermal withdrawal latency in control mice after saline or CNO injection (n = 10). (**H**,**I**) Von Frey threshold and thermal withdrawal latency in hM3D (Gq) mice after saline or CNO injection (n = 11). Data are presented as the mean ± SD. *** $p < 0.001$, ns: no significance by a two-tailed independent samples *t* test (**E**–**I**).

- Inflammatory pain excites PBL glutaminergic neurons.

To further explore whether inflammatory pain was associated with PBL glutamatergic neuronal activity, we used complete Freund's adjuvant (CFA) to construct an inflammatory pain model in mice (Figure 3A). After CFA pain modelling on the left plantar of C57 BL/6J mice, we found that the von Frey threshold and thermal withdrawal latency were significantly lower compared with equivalent saline modelling (Figure 3B, n = 8–9, $p < 0.001$), indicating successful model establishment. Immunofluorescence staining of C-Fos in the PBL was then performed in saline- and CFA-modelled mice (Figure 3C,D). The results showed that C-Fos significantly increased in PBL after CFA modelling compared with saline (Figure 3E, 9.4 ± 2.2 vs. 19.8 ± 5.2; n = 9–10, $p < 0.001$).

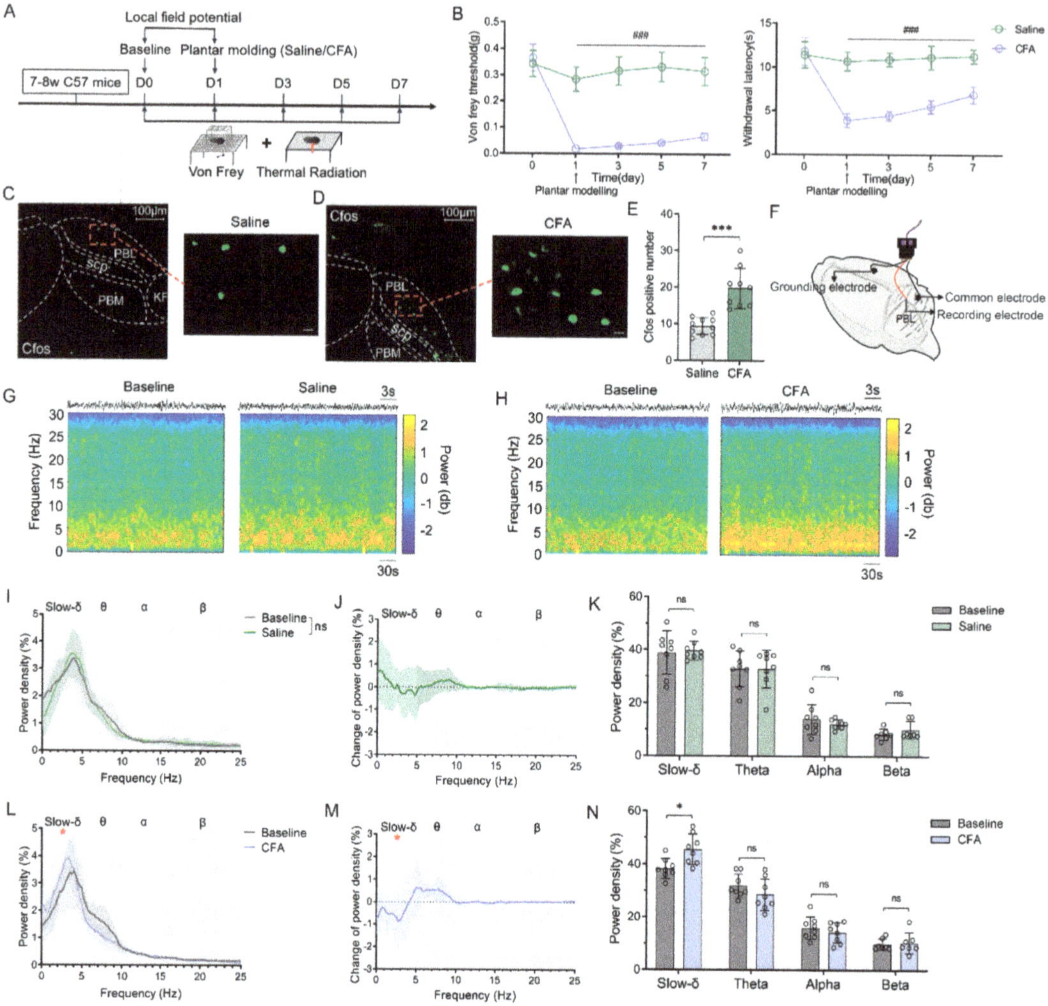

Figure 3. Inflammatory pain excites PBL glutaminergic neurons. (**A**) Flow chart of the behavioural measurements in wild-type mice. (**B**) Von Frey threshold and thermal withdrawal latency in wild-type mice after saline or complete Freund's adjuvant (CFA) plantar modelling (n = 8–9). (**C**,**D**) C-Fos immunofluorescence (green) in PBL after saline (**C**) and CFA plantar modelling (**D**) in wild-type mice. Scale = 100 μm (**left**) or 10 μm (**right**). (**E**) C-Fos count comparison after saline and CFA plantar modelling in wild-type mice after immunofluorescence staining (n = 9–10). (**F**) Schematic of PBL local field potential electrode implantation. (**G**,**H**) A representative spectrogram of PBL in wild-type mice before and after saline (**G**) or CFA modelling (**H**). (**I**–**K**) Power density of LFP (**I**,**K**) and its group difference (**J**) in slow-delta, theta, alpha, and beta oscillations before and after saline plantar modelling in wild-type mice (n = 8). (**L**–**N**) Power density of LFP (**L**,**N**) and its group difference (**M**) in slow-delta, theta, alpha, and beta oscillations before and after CFA plantar modelling in wild-type mice (* $p < 0.05$ for slow-delta between baseline and CFA group; n = 8). Data are presented as the mean ± SD or mean ± SEM ((**B**), **left**). ### $p < 0.001$ by two-way repeated-measures ANOVA (**B**). * $p < 0.05$, *** $p < 0.001$, ns: no significance by a two-tailed independent samples t test (**E**,**I**–**N**).

Next, we explored the changes in neuronal activity in response to pain stimulation by measuring local field potentials (LFPs) in PBL (Figure 3F). The results showed that slow-delta oscillations did not change significantly before and after saline modelling (Figure 3G,I–K, n = 8, $p > 0.05$) but were significantly enhanced after CFA modelling (Figure 3H,L–N, n = 8, $p < 0.05$). However, theta, alpha and beta oscillations did not change notably in the saline and CFA groups (Figure 3G–N, n = 8, $p > 0.05$).

- Knockdown of NALCN in PBL glutamatergic neurons alleviates CFA-induced pain.

To identify whether NALCN on PBL glutamatergic neurons is involved in pain regulation in mice, we specifically knocked down the expression of NALCN on PBL glutamatergic neurons (Figure 4A). First, we performed immunofluorescence staining of NALCN in the PBL after saline and CFA plantar pain modelling in wild-type mice (Figure 4B,C). The results showed that NALCN was significantly increased in the PBL of mice after CFA modelling compared with saline (Figure 4D, 20.6 ± 3.3 vs. 39.6 ± 5.4; $p < 0.\ 001$; n = 9 in both groups).

Then, to knock down NALCN in PBL glutamatergic neurons, we microinjected NALCN gene silencing virus (AAV2/9-mCaMKIIa-mCherry-mIRNAI (NALCN)) and its control virus (AAV2/9-mCaMKIIa-mCherry-mIRNAI (NC)) (Figure 4E, left). We confirmed whether the virus was correctly expressed in PBL after 3 weeks (Figure 4E, right). We also performed fluorescence quantification of NALCN in the control and NALCN knockdown groups of PBL using immunofluorescence staining to ensure that NALCN had been successfully knocked down in PBL in the knockdown group (Figure 4F–H, 23.4 ± 4.5 vs. 13.4 ± 3.2; n = 10 in both groups, $p < 0.001$). Meanwhile, we found that CFA-induced C-Fos expression was significantly reduced after CFA modelling in the NALCN knockdown group (Figure 4I–K, 35.3 ± 6.7 vs. 17.8 ± 5.2; n = 10 in both groups, $p < 0.001$).

Next, we conducted behavioural tests, including the von Frey threshold and thermal withdrawal latency, in the control and NALCN knockdown groups. The von Frey threshold (Figure 4L, n = 9–10, $p > 0.05$) and thermal withdrawal latency (Figure 4M, n = 9–10, $p > 0.05$) were not significantly different between the two groups of mice after saline plantar modelling. There was no difference in the basal values of the von Frey threshold (Figure 4L, 0.4 ± 0.2 vs. 0.3 ± 0.2 g; n = 8–11, $p > 0.05$) and thermal withdrawal latency (Figure 4M, 12.4 ± 1.6 vs. 12.1 ± 1.4 s; n = 8–11, $p > 0.05$) between control and NALCN knockdown group mice. However, after CFA modelling, the von Frey threshold (Figure 4L, n = 8–11, $p < 0.05$) and thermal withdrawal latency (Figure 4M, n = 8–11, $p < 0.001$) were significantly increased in the NALCN knockdown group compared with the control group during recovery. It returned to baseline levels faster than the control group.

Similarly, to further understand whether there were differences in PBL glutamatergic neuronal activity in control and NALCN knockdown mice before and after CFA pain modelling, we conducted local field potential analysis of PBL glutamatergic neurons. The results showed that slow-delta oscillations were significantly increased in control mice after CFA modelling (Figure 5A,C–E, n = 10, $p < 0.05$), but the NALCN knockdown group did not show such an increase (Figure 5B,F–H, n = 9, $p > 0.05$). Theta, alpha and beta oscillations were not significantly different in control and NALCN knockdown mice before and after CFA modelling (Figure 5E,H, n = 8–11, $p > 0.05$).

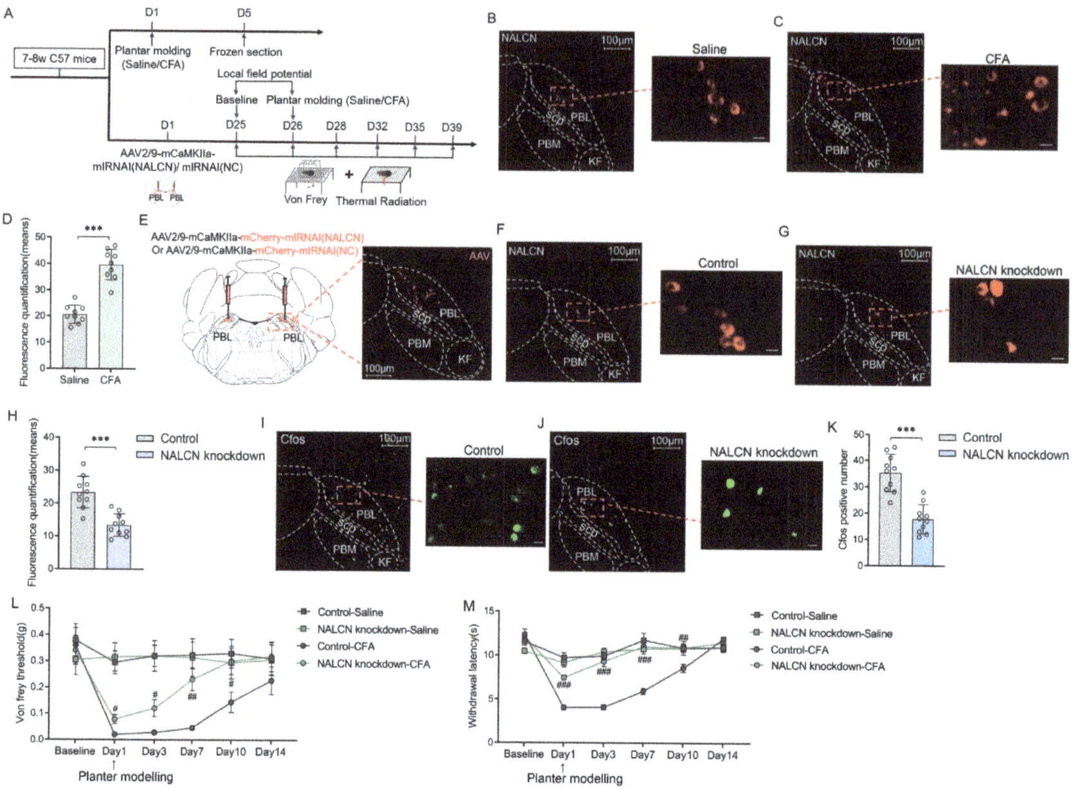

Figure 4. Knockdown of NALCN in PBL glutamatergic neurons alleviates CFA-induced pain. (**A**) Flow chart of virus injection and behavioural measurements in mice. (**B**,**C**) NALCN immunofluorescence (red) after saline (**B**) and CFA plantar modelling (**C**) in wild-type mice after immunofluorescence staining. Scale = 100 μm (**left**) or 25 μm (**right**). (**D**) NALCN fluorescence quantification comparison after saline and CFA plantar modelling in wild-type mice (n = 9 in both groups). (**E**) Injection of control and NALCN knockdown viruses into PBL of C57 6J mice. Scale = 100 μm. (**F**,**G**) NALCN fluorescence (mCherry) in PBL in control (**F**) and NALCN knockdown mice (**G**). Scale = 100 μm (**left**) or 25 μm (**right**). (**H**) NALCN fluorescence quantification comparison in PBL of control and NALCN knockdown mice (n = 10 in both groups). (**I**,**J**) C-Fos fluorescence (green) in PBL in control (**I**) and NALCN knockdown mice (**J**) after CFA plantar modelling. Scale = 100 μm (**left**) or 10 μm (**right**). (**K**) C-Fos count comparison in PBL of control and NALCN knockdown mice before and after CFA plantar modelling (n = 10 in both groups). (**L**) The von Frey threshold in control and NALCN knockdown mice before and after saline or CFA plantar modelling at 14 days (n = 8–11). (**M**) Withdrawal latency in control and NALCN knockdown mice before and after saline or CFA plantar modelling at 14 days (n = 8–11). Data are presented as the mean ± SD or mean ± SEM (**L**). *** $p < 0.001$ by a two-tailed independent samples *t* test (**D**,**H**,**K**). $^{###}$ $p < 0.001$, $^{##}$ $p < 0.01$, $^{#}$ $p < 0.05$ by two-way repeated-measures ANOVA (**L**,**M**).

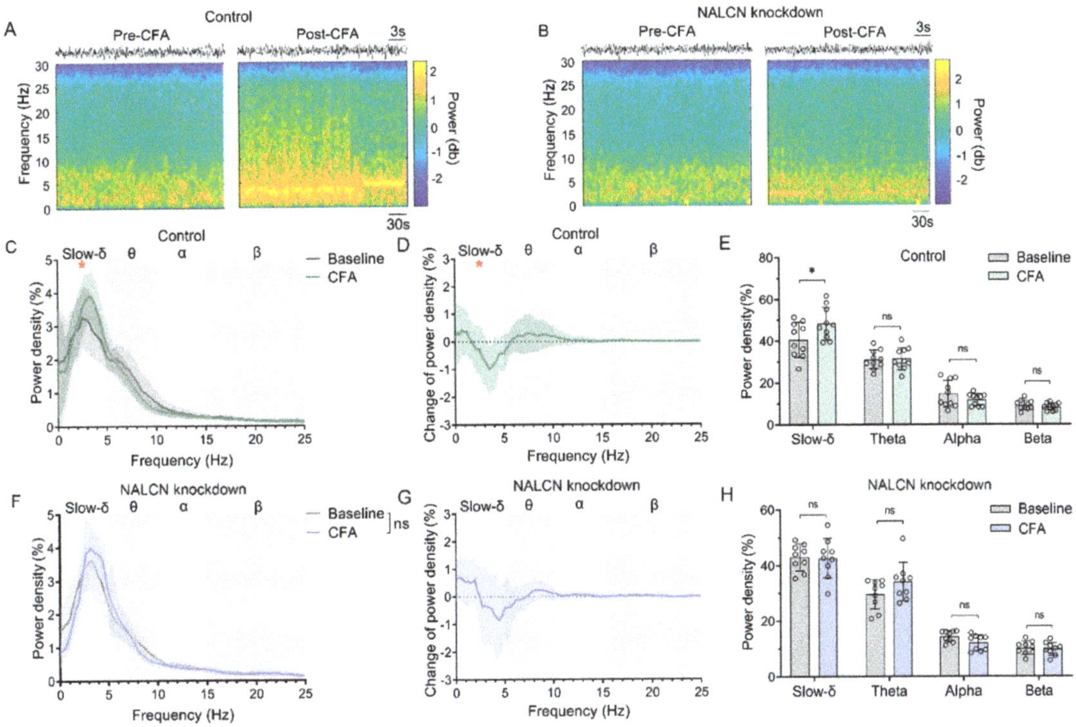

Figure 5. Knockdown of NALCN in PBL glutamatergic neurons inhibits CFA-induced slow-delta oscillations. (**A**,**B**) A representative spectrogram of PBL in control (**A**) and NALCN knockdown mice (**B**) before and after CFA plantar modelling. (**C**–**E**) Power density of LFP (**C**,**E**) and its group difference (**D**) in slow-delta, theta, alpha, and beta oscillations before and after CFA plantar modelling in control mice (* $p < 0.05$ for slow-delta between baseline and CFA group). (**F**–**H**) Power density of LFP (**F**,**H**) and its group difference (**G**) in slow-delta, theta, alpha, and beta oscillations before and after CFA plantar modelling in NALCN knockdown mice. Data are presented as the mean ± SD. * $p < 0.05$, ns: no significance by a two-tailed independent samples t test (**C**–**H**).

- NALCN modulates CFA-induced pain via PBL glutamatergic neuron-central nucleus amygdala (CeA) projections.

Next, we used anterograde viral tracers and immunofluorescence staining to explore the downstream projections of PBL neurons involved in CFA-induced inflammatory pain (Figure 6A). First, we injected self-EGFP-labelled scAAV2/1-hSyn-EGFP-WPRE-pA anterograde tracer virus into the bilateral PBL (Figure 6B) and then observed each brain region 2 weeks later, after confirming the correct expression of the tracer virus in the PBL region (Figure 6C). We detected the projection fluorescence of PBL neurons in several brain regions, including the bed nucleus of the stria terminalis (BNST) (Figure 6D), central nucleus amygdala (CeA) (Figure 6E), central nucleus of the amygdala (PAG) (Figure 6F), and central nucleus of the inferior colliculus (CNIC) (Figure 6G).

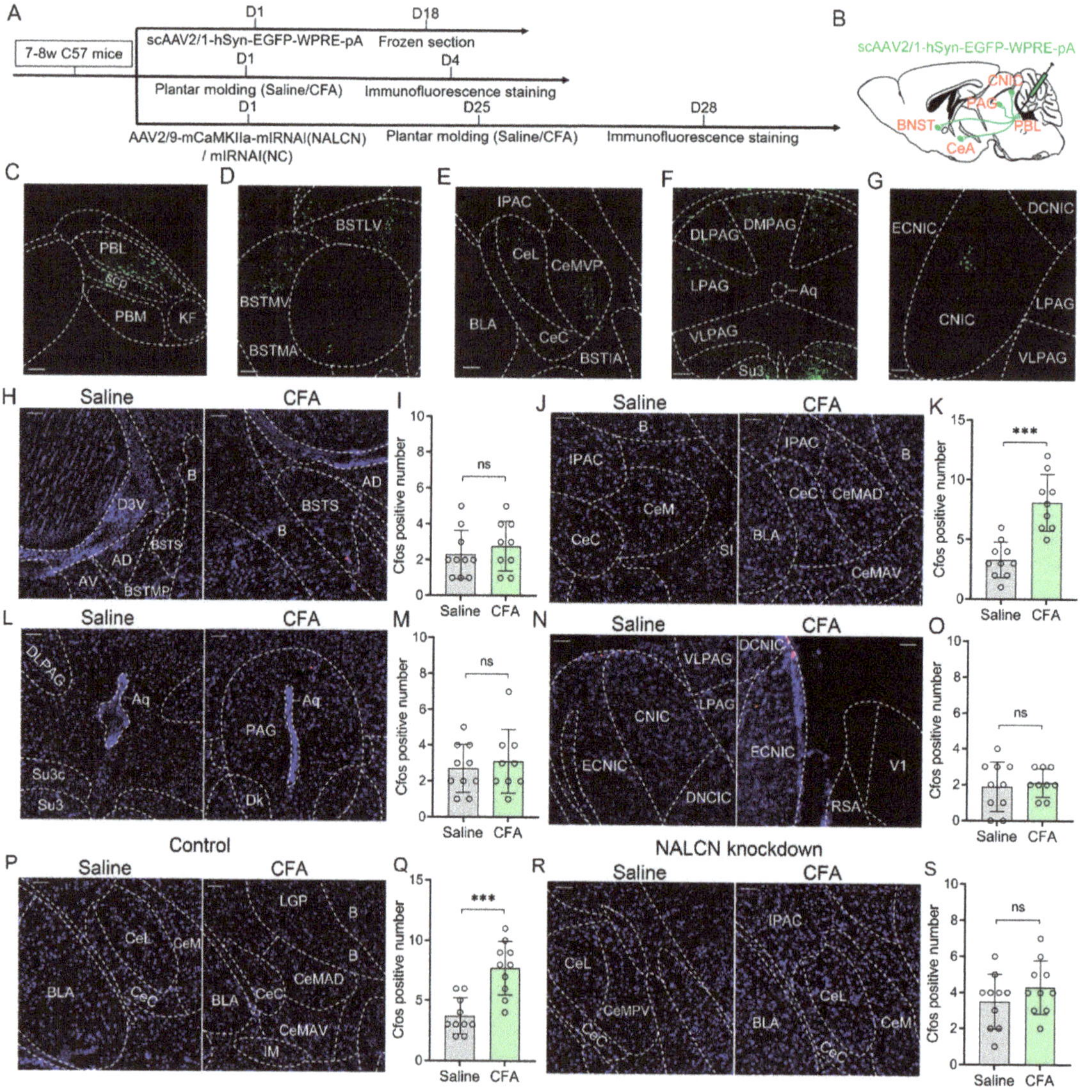

Figure 6. NALCN modulates CFA-induced pain via PBL glutamatergic neuron-central nucleus amygdala (CeA) projections. (**A**) Flow chart of virus injection and related experimental operations. (**B**) Schematic of anterograde tracer virus injection and downstream projection of PBL. (**C–G**) After injection of scAAV2/1-hSyn-EGFP-WPRE-pA anterograde tracer virus into the bilateral PBL (**C**), viral fluorescence (EGFP) was distributed in the BNST (**D**), CeA (**E**), PAG (**F**) and CNIC (**G**). Scale = 50 μm. PBL, lateral parabrachial nucleus; scp, superior cerebellar peduncle; PBM, medial parabrachial nucleus; KF, Kölliker-Fuse nucleus; BSTLV, bed nucleus of the stria terminalis, lateral division, ventral part; BSTMV, bed nucleus of the stria terminalis, medial division, ventral part; BSTMA, bed nucleus of the stria terminalis, medial division, anterior part; IPAC, interstitial nucleus of the posterior limb of the anterior commissure; CeL, central amygdaloid nucleus, lateral division; CeMVP, central amygdaloid nucleus, medial posteroventral part; CeC, central amygdaloid nucleus, capsular part; BSTIA, bed nucleus of the stria terminalis, intra-amygdaloid division; BLA, basolateral amygdaloid nucleus, anterior part; DMPAG, dorsomedial periaqueductal gray; DLPAG, dorsolateral periaqueductal gray;

LPAG, lateral periaqueductal gray; VLPAG, ventrolateral periaqueductal gray; Su3, supraoculomotor periaqueductal gray; Aq, aqueduct; ECNIC, external central nucleus of the inferior colliculus; CNIC, central nucleus of the inferior colliculus; DCNIC, dorsal central nucleus of the inferior colliculus. (**H,J,L,N**) C-Fos fluorescence (red) and DAPI fluorescence (blue) in the BNST (**H**), CeA (**J**), PAG (**L**) and CNIC (**N**) after saline and CFA plantar modelling in wild-type mice after immunofluorescence staining. Scale = 50 μm. D3V, dorsal 3rd ventricle; B, basal nucleus; BSTS, bed nucleus of stria terminalis, supracapsular part; AD, anterodorsal thalamic nucleus; AV, anteroventral thalamic nucleus; BSTMP, bed nucleus of the stria terminalis, medial division, posterior part; CeM, central amygdaloid nucleus, medial division; SI, substantia innominata; CeMAD, central amygdaloid nucleus, medial division, anterodorsal part; CeMAV, central amygdaloid nucleus, medial division, anteroventral part; Su3, supraoculomotor periaqueductal gray; Dk, nucleus of Darkschewitsch; V1, primary visual cortex; RSA, retrosplenial agranular cortex. (**I,K,M,O**) C-Fos count comparison in the BNST (**I**), CeA (**K**), PAG (**M**) and CNIC (**O**) after saline and CFA plantar modelling in wild-type mice (n = 9–10). (**P,R**) C-Fos fluorescence (red) and DAPI fluorescence (blue) in the CeA after saline and CFA plantar modelling in control (**P**) and NALCN knockdown mice (**R**) after immunofluorescence staining. Scale = 50 μm. LGP, lateral globus pallidus; IM, intercalated amygdaloid nucleus, main part. (**Q,S**) C-Fos count comparison in the CeA after saline and CFA plantar modelling in control (**Q**) and NALCN knockdown mice (**S**) (n = 10 in both groups). Data are presented as the mean ± SD. *** $p < 0.001$, ns: no significance by a two-tailed independent samples t test (**I,K,M,O,Q,S**).

To further identify the PBL downstream nuclei linked to pain transmission, we performed CFA plantar inflammatory pain modelling in wild-type mice. Immunofluorescence staining of C-Fos protein in the BNST (Figure 6H), CeA (Figure 6J), PAG (Figure 6L) and CNIC (Figure 6N) was performed after saline and CFA modelling. The results showed that C-Fos protein in the CeA was significantly increased after CFA modelling compared with saline (Figure 6K, 3.3 ± 1.4 vs. 8.1 ± 2.2; n = 9–10, $p < 0.001$). However, no significant difference in C-Fos protein was observed in the BNST (Figure 6I, 2.3 ± 1.3 vs. 2.8 ± 1.3; n = 9–10, $p > 0.05$), PAG (Figure 6M, 2.7 ± 1.3 vs. 3.1 ± 1.7; n = 9–10, $p > 0.05$) and CNIC (Figure 6O, 1.9 ± 1.3 vs. 2.1 ± 0.7; n = 9–10, $p > 0.05$) in CFA modelling compared with saline.

Subsequently, we conducted saline and CFA plantar pain modelling in PBL NALCN knockdown and control mice, followed by immunofluorescence staining for C-Fos protein in the CeA region of both groups (Figure 6P,R). The results showed that C-Fos was significantly increased in the CeA of the control group after CFA modelling compared with saline (Figure 6Q, 3.7 ± 1.4 vs. 7.7 ± 2.1; n = 10 in both groups, $p < 0.001$), whereas it did not change significantly in the NALCN knockdown group (Figure 6S, 3.5 ± 1.4 vs. 4.3 ± 1.4; n = 10 in both groups, $p > 0.05$). This suggests that NALCN knockdown in the PBL decreased neuronal activity in the CeA, which might be a downstream nucleus for PBL pain transmission.

3. Discussion

In this study, we showed that upregulated expression of NALCN in PBL glutamatergic neurons is involved in the development of inflammatory pain in mice. First, using chemogenetic inhibition or activation of PBL glutamatergic neurons, we demonstrated that PBL glutamatergic neurons were associated with sensory transduction in mice under physiological conditions. Then, we showed that PBL glutamatergic neuronal activity was enhanced when mice experienced inflammatory pain. To investigate the underlying molecular mechanism, we specifically knocked down NALCN in PBL glutamatergic neurons and found that NALCN knockdown alleviated CFA-induced pain. We further explored the downstream pathways of the PBL using antegrade virus tracing, which included the BNST, CeA, CNIC, and PAG, and finally showed that NALCN may modulate inflammatory pain via PBL glutamatergic neuron-CeA projections.

Roeder et al. [22] reported that activities of the rostral ventromedial medulla (RVM) evoked by nociceptive inputs were significantly attenuated by inactivation of PBL neurons. One study [18] showed that PBL sends excitatory glutamate to norepinephrine neurons in the A7 region of rats to participate in central analgesia and mediate pain inhibition at the spinal dorsal horn level. In the present study, chemogenetic activation or inhibition of PBL glutamatergic neurons reduced or increased basal pain thresholds, respectively, in mice, suggesting a close link between PBL glutamatergic neurons and pain sensation. However, one study [16] showed that chemogenetic inhibition of PBL^{Oprm1} neurons did not change the thermal sensitivity of mice. Two factors may contribute to the inconsistent findings between their study and ours. First, a hot plate was used in their study [16] to measure the thermal sensitivity of mice, whereas we used an infrared heat pain tester. The two types of thermal pain measurement are not identical in manner or specific nature. Second, PBL^{Oprm1} neurons represent only a portion of PBL neurons that express mu-opioid receptors. However, we manipulated all glutamatergic neurons in the PBL, which may generate a much larger effect. Therefore, PBL glutamatergic neurons other than $Oprm1^{+}$ neurons are suggested to affect sensory sensation.

Zhang et al. [19] reported that elevated NALCN expression and function in the DRG and spinal cord were associated with chronic constriction injury (CCI)-induced neuropathic pain. Li et al. [4] showed that NALCN in the DRG and spinal cord was also implicated in CFA-induced inflammatory pain. This evidence highlights the role of NALCN in the peripheral DRG and spinal cord in the regulation of chronic pain conditions. However, it is currently not known whether NALCN in brain nuclei and/or circuits regulates pain. The present study provided evidence that NALCN in PBL glutamatergic neurons modulated CFA-induced inflammatory pain.

Previous studies have shown that mutations in the NALCN gene are associated with many neurological diseases, such as infantile hypotonia [23,24], severe mental retardation [23,24], psychomotor retardation [24–26], and epilepsy [27]. Recently, one study showed that NALCN modulates inflammation-induced depression by maintaining the activity of glutamatergic neurons in the ventral dentate gyrus (DG) [28]. Future studies will investigate whether NALCN is associated with other inflammation-related neurological diseases.

The mechanisms by which NALCN modulates inflammatory pain are not fully understood. Evidence has shown that pain induces the release of substance P from injurious afferent nerve endings [29,30], which can activate NALCN currents in neurons by binding to tachykinin receptor 1 (TACR1) in a G protein-independent manner that requires the Src family of kinases (SFKs) and UNC80 [31–33]. A previous study also revealed that the cAMP/PKA signalling pathway may participate in the regulation of NALCN in neuropathic pain [19]. It will be interesting to explore the underlying pathways by which NALCN modulates inflammatory pain.

Previous studies have shown that PBN is involved in the regulation of various functions via different projection circuits [11,34]. For example, in the regulation of respiratory activity, the PBN projects to the nucleus tractus solitarius (NST), preBöt complex, rostral ventral respiratory group (rVRG), etc. [7,35]. In pain regulation, the PBN mainly projects to the CeA, BNST, PAG, intralaminar thalamic nucleus (ILN), ventral tegmental area (VTA), etc. [14,17,34,36]. Deng et al. [14] showed that PBN directly transmits injurious signals from the spinal cord to the intralaminar thalamic nuclei. One study [34] reported that different efferent neurons involved in pain regulation in PBL are associated with different components of pain regulation. For example, activation of efferent signals from the ventral medial hypothalamus (VMH) or the lateral periaqueductal grey matter (lPAG) drives avoidance behaviour, whereas activation of efferent signals from the ventral medial hypothalamus (VMH) to the BNST or CeA produces aversive memory [34]. Another study [17] also showed that a subpopulation of PBL neurons is essential for transmitting nociceptive signals from the spinal cord to the substantia nigra pars reticulata (SNR). In this study, we used cis-trans-synaptic tracer virus to bilaterally label the mouse PBL, and green fluorescence of the tracer virus was found at the CeA, BNST, PAG, and CNIC after 2 weeks. CeA, BNST and PAG are well-known downstream nuclei of PBL that have been reported in previous studies [14,34,36]. We further confirmed that NALCN knockdown in

PBL glutamatergic neurons suppressed CFA-induced C-Fos expression in the CeA, suggesting that NALCN may modulate CFA-induced inflammatory pain via PBL glutamatergic neuron-CeA projections. Interestingly, we also found a small amount of tracer fluorescence in the central nucleus of the inferior colliculus (CNIC), which was not previously reported. The CNIC is a critical component of the auditory pathway that receives input from all inferior auditory nuclei located in the brainstem and projects upwards to the medial geniculate nucleus (MGN) of the thalamus [37]. However, C-Fos protein in the CNIC did not change before and after CFA pain modelling, suggesting that PBL-CNIC projections might not be involved in the regulation of CFA-induced pain. It will be interesting to determine their function in future studies.

Although only male mice were used in this study, we speculate that there might be no sex difference in the mechanisms of inflammatory pain associated with NALCN. In our previous studies, we showed no sex differences in either neuropathic [19] or inflammatory pain [4] associated with NALCN in the dorsal root ganglia (DRG) and spinal cord. In these two studies, we also showed that neuropathic or inflammatory pain could increase NALCN currents and neuronal excitability in both neonatal and adult rats. However, it is not clear whether this mechanism regulates inflammatory pain in aging rodents. Nevertheless, it will be interesting to confirm whether NALCN modulates inflammatory pain in a sex- or age-dependent manner in future studies.

There are several limitations in this study. First, NALCN is related to the regulation of several physiological functions, and it is not clear whether NALCN knockdown in PBL glutamatergic neurons causes other phenotypic abnormalities. Second, the present study did not examine the differentially expressed genes in NALCN knockdown mice. Examining differentially expressed genes would help to elucidate downstream pathways implicated in the diseases associated with NALCN. Finally, we did not apply patch-clamp recordings or in vivo electrophysiological techniques to explore the NALCN currents and activities of PBL glutamatergic neurons under CFA pain modelling.

In summary, our study uncovered a novel ion channel, namely, NALCN, that modulates inflammatory pain via PBL glutamatergic neuron-CeA projections. NALCN in PBL glutamatergic neurons may be an underlying therapeutic target for the control of inflammatory pain.

4. Materials and Methods

4.1. Animals

C57BL/6J male mice (8 weeks, 20–22 g) were given free access to food and water and housed under standard conditions (22–24 °C, 45–55% humidity) with a 12:12 light cycle in this study. All procedures were approved by the Animal Ethics Committee of the West China Hospital of Sichuan University (Approval ID: 2021420A, Chengdu, Sichuan, China) and complied with the Animal Research guidelines: Reporting of In Vivo Experiments (ARRIVE) guidelines. All efforts were made to reduce the number of experimental animals and their suffering as much as possible.

4.2. Stereotaxic Injection

Wild-type mice were randomly grouped and labelled with ear tags. Mice were anaesthetised with 2% isoflurane and fixed in a stereotaxic device (RWD, Life Science, Shenzhen, China). Their body temperatures were maintained by a heating blanket. After cutting the scalp to expose the skull of the mice, two holes were drilled into the skull above the PBL to inject virus into the bilateral PBL (bregma: −5.4 mm, lateral: ±1.2 mm, depth: −3.4 mm). Viruses were injected at a rate of 100 nL/min, and the total volume was 500 nL except for tracer virus (300 nL). In chemogenetic manipulation, AAV2/9-mCaMKIIa-hM3D(Gq)-mCherry-ER2-WPRE-pA,_AAV2/9-mCaMKIIa-hM4D(Gi)-mCherry-ER2-WPRE-pA_and AAV2/9-mCaMKIIa-mCherry-WPRE-pA virus were bilaterally microinjected into the PBL of C57BL/6J mice. To manipulate NALCN, we injected AAV2/9-mCaMKIIa-mIRNAI (NALCN)-mCherry-WPRE-pA_and_AAV2/9-mCaMKIIa-mCherry mIRNAI(NC)-WPRE-pA into the bilateral PBL. To study PBL projections, scAAV2/1-hSyn-EGFP-WPRE-pA anterograde tracer virus was used. After viruses were injected, the glass electrode remained in place for five minutes and then was slowly withdrawn to prevent virus reflux. After this procedure, the scalps of the mice were sutured and the mice were placed on a heating blanket for resuscitation. When the mice awakened, they were put back into their cage.

4.3. Immunofluorescence Staining

The mice were anaesthetised with 2% isoflurane and transcardially perfused with phosphate-buffered saline (PBS, pH 7.4) and 4% paraformaldehyde until their limbs stiffened. The brains were fixed in 4% paraformaldehyde for at least 24 h at 4 °C and then immersed in 30% sucrose solution for another 24 h. A freezing microtome (CM1850, Leica, Hesse-Darmstadt, Germany) was used to prepare 20-μm-thick brain slices.

Then, brain slices were labelled with primary antibodies at 4 °C overnight, including C-Fos (1:500, mouse, GTX16902, Neobioscience, Shenzhen, China), NALCN (1:500, rabbit, ASC-022, Alomone, Jerusalem, Israel), and C-Fos (1:500, rabbit, ab222699, Abcam, Cambridge, UK). After overnight incubation, the primary antibody was washed with PBS and incubated with the secondary antibody, including Alexa Fluor 488 goat anti-mouse (1:500, ZF-0512, ZSGB-BIO, Beijing, China), Alexa Fluor 555 donkey anti-mouse (1:500, A-31570, Thermo Fisher Scientific, Waltham, Massachusetts, USA) and Alexa Fluor 550 goat anti-rabbit (1:500, 550045, ZEN-BioScience, Chengdu, China) for 2 h. All images were taken with a Zeiss Axio Imager Z.2. and analysed using ImageJ software V1.8.0 (National Institutes of Health, Bethesda, MA, USA).

4.4. In Vivo Chemogenetic Manipulation

For behavioural experiments, clozapine N-oxide powder (CNO, Sigma, St. Louis, MO, USA) was diluted to 0.5 mg·mL^{-1} with 0.9% saline and injected intraperitoneally at 2.5 mg·kg^{-1}. In the control group, CNO was replaced with an equal volume of saline. Behavioural experiments began 50–60 min after CNO or saline injection.

4.5. Behavioral Tests for Pain

Pain tests included the von Frey filaments and Hargreaves tests to evaluate mechanical and thermal sensitivities separately. Before the formal test, the mice were acclimated on a wire mesh elevated frame for 2 h a day for a week. On the experimental day, mice were placed individually in a transparent cube Plexiglas chamber (length = 10 cm) and habituated for 1 h until they were immobile but awake. For the von Frey test, an up-down approach was used [38]. In the Hargreaves test, mice lifted their feet after a thermal radiation beam (35% intensity) was emitted by an infrared heat pain tester (37570-001, Ugo Basile, Italy). The latency time between the start of the radiation and the mice lifting their feet was defined as the thermal withdrawal latency. If the withdrawal latency exceeded 20 s or the mouse moved autonomously, the data were discarded.

4.6. Inflammatory Pain Mode

Under 2% isoflurane anaesthesia, 20 μL CFA (1 mg·mL^{-1}, Sigma-Aldrich, St. Louis, MO, USA) was injected subcutaneously with an insulin needle in the left plantar of mice. Control mice were injected with an equal volume of saline.

4.7. Implantation of Local Field Potential (LFP) Electrodes

To record the PBL local field potential, three electrodes were implanted into each mouse, including one recording electrode, one common electrode, and one grounding electrode. The grounding electrode can prevent interference from surrounding information during recording. The common electrode and grounding electrode were anchoring screws (1 mm diameter, 2 mm long) fixed to the skull, and the recording electrode was an insulating silver wire inserted into the PBL. Under 2% isoflurane anaesthesia, the mouse skull was fully exposed after cutting the scalp. When recording the PBL local field potential, the recording electrode was implanted into the left PBL, the common electrode was implanted into the right PBL, and the grounding electrode was implanted next to the bregma point (bregma: −1 mm, lateral side: −1.2 mm). The three electrodes were all connected to a mini plug and fixed with glass ionomer cement. When the mice awakened, they were placed back into the cage.

4.8. LFP Recordings and Analyses

The LFP recordings were conducted one week after surgery, and signals were recorded at a sampling frequency of 500 Hz using a Pinnacle EEG Recording System (Part# 8200-SL; Pinnacle Technology, Salinas, CA, USA). After original signals were obtained using the Sirenia acquisition system, MATLAB (version 2006a; MathWorks, Santa Clara, CA, USA) and GraphPad Prism 9.0 software (GraphPad Software, San Diego, CA, USA) were used to amplify, digitise and further analyse the signals [39]. A single recording duration was 40 min. The data from the 30–35 min period were selected for analysis and sample presentation.

4.9. Study Design

Part 1: First, we randomly and equally divided 24 C57BL/6J mice into two groups and microinjected AAV2/9-mCaMKIIa-hM4D(Gi)-mCherry-ER2-WPRE-pA and AAV2/9-mCaMKIIa-mCherry-WPRE-pA viruses. After approximately 3 weeks, 10 in the control group and 11 in the hM4D group were used for pain behavioural tests. We first tested the basal values of the von Frey threshold and thermal withdrawal latency in both groups. After intraperitoneal injection of saline or CNO, the von Frey threshold and thermal withdrawal latency were tested again (Figure 1A).

Part 2: In this part, we randomly and equally divided 24 C57BL/6J mice into two groups, and then AAV2/9-mCaMKIIa-hM3D(Gq)-mCherry-ER2-WPRE-pA and AAV2/9-mCaMKIIa-mCherry-WPRE-pA viruses were injected into the bilateral PBL of the mice. After waiting for 3 weeks, the basal values of the von Frey threshold and thermal withdrawal latency were tested in 10 mice in the control group and 11 mice in the hM3D group before applying treatment measures. Then, after intraperitoneal injection of saline or CNO, the von Frey threshold and thermal withdrawal latency were tested again in both groups (Figure 2A).

Part 3: First, we randomly and equally divided 20 C57BL/6J mice into saline and CFA groups (10 mice per group). The von Frey threshold and thermal withdrawal latency were tested on days 0 (basal value), 1 (saline or CFA plantar modelling), 3, 5 and 7. On day 0, basal values of the local field potential of PBL were tested in both groups. On day 1, the local field potential of PBL was tested in saline- or CFA-modelled mice (Figure 3A).

Part 4: Twenty C57 BL/6J mice were randomly and equally divided into saline and CFA groups (10 mice per group). The brains of the two groups were extracted 30 min after saline or CFA plantar modelling. Immunofluorescence staining of NALCN was performed in PBL. In addition, 48 C57 BL/6J mice were randomly and equally divided into 4 groups (12 mice per group); 2 groups of mice were injected with AAV2/9-mCaMKIIa-mIRNAI(NALCN)-mCherry-WPRE-pA virus into PBL, and another 2 groups were injected with AAV2/9-mCaMKIIa-mCherry mIRNAI(NC)- WPRE-pA virus. After waiting for 3 weeks, 4 groups of mice were tested for von Frey threshold and thermal withdrawal latency on days 0 (basal values), 1 (saline or CFA plantar modelling), 3, 7, 10, and 14 (Figure 4A).

Part 5: First, scAAV2/1-hSyn-EGFP-WPRE-pA anterograde tracer virus was microinjected into PBL of 5 C57 BL/6J mice. After waiting for 2 weeks, the whole brains of mice were removed and observed for tracer virus projection fluorescence. After determining the projection nuclei, twenty C57 BL/6J mice were randomly and equally divided into saline and CFA groups (10 mice per group). Whole brains of mice were taken 30 min after saline and CFA plantar modelling in the two groups. Immunofluorescence staining for C-Fos was performed on the identified projection nuclei of each mouse. Next, 40 C57

BL/6J mice were randomly and equally divided into 4 groups: 2 groups were injected with AAV2/9-mCaMKIIa-mIRNAI(NALCN)-mCherry-WPRE-pA virus, and another 2 groups were injected with AAV2/9-mCaMKIIa-mCherry mIRNAI(NC)-WPRE-pA virus. After waiting for 3 weeks, the 4 groups underwent saline or CFA plantar modelling followed by C-Fos immunofluorescence staining (Figure 6A).

4.10. Statistical Analysis

Data are expressed as the mean ± standard deviation (SD) or mean ± standard error of the mean (SEM), and statistical analyses were conducted using SPSS version 25.0 (SPSS Inc., Chicago, IL, USA) and GraphPad Prism 9.0 software (GraphPad software, San Diego, CA, USA).

The normality of data distribution was assessed by the Shapiro–Wilk test. Data from two groups were compared using independent samples *t* tests. Repeated measurement data were analysed using repeated-measures multivariate ANOVA. After immunofluorescence staining, data from mice without corresponding viral expression in PBL were excluded. Detailed statistical analysis methods are described in the figure legends. $p < 0.05$ was considered statistically significant.

5. Conclusions

In conclusion, this study demonstrated that NALCN in PBL glutamatergic neurons is a key ion channel involved in the regulation of inflammatory pain. NALCN may serve as a critical molecular target for the control of inflammatory pain.

Supplementary Materials: The following supporting information can be downloaded at: https://www.mdpi.com/article/10.3390/ijms241511907/s1.

Author Contributions: Conceptualization, L.W., C.Z. and D.Z.; methodology, L.W., Y.W. and D.Z.; software, L.W., Y.W. and J.J.; validation, J.L., C.Z. and D.Z.; resources, C.Z. and D.Z.; data curation, L.W., Y.W., J.J. and D.Z.; writing—original draft preparation, L.W; writing—review and editing, J.L, C.Z. and D.Z.; visualization, L.W., Y.W., C.Z. and D.Z.; supervision, J.L, C.Z. and D.Z.; All authors have read and agreed to the published version of the manuscript.

Funding: This work was supported by Grant No. 2022NSFSC1399 (To Donghang Zhang) from the Natural Science Foundation of Sichuan Province; Grant No. 2021M692276 (To Donghang Zhang) from the China Postdoctoral Science Foundation; Grant No. 21PJ014 (To Donghang Zhang) from the Health Commission of Sichuan Province, and No. 20HXBH144 (to Donghang Zhang) from Post-Doctor Research Project, West China Hospital, Sichuan University.

Institutional Review Board Statement: The animal study protocol was approved by the Animal Ethics Committee of the West China Hospital of Sichuan University (Approval ID: 2021420A; date of approval: 26 February 2021).

Informed Consent Statement: Not applicable.

Data Availability Statement: Not applicable.

Conflicts of Interest: The authors declare no conflict of interest.

References

1. Li, Y.; Yang, M.; Wu, F.; Cheng, K.; Chen, H.; Shen, X.; Lao, L. Mechanism of electroacupuncture on inflammatory pain: Neural-immune-endocrine interactions. *J. Tradit. Chin. Med.* **2019**, *39*, 740–749. [PubMed]
2. Varrassi, G.; Alon, E.; Bagnasco, M.; Lanata, L.; Mayoral-Rojals, V.; Paladini, A.; Pergolizzi, J.V.; Perrot, S.; Scarpignato, C.; Tölle, T. Towards an Effective and Safe Treatment of Inflammatory Pain: A Delphi-Guided Expert Consensus. *Adv. Ther.* **2019**, *36*, 2618–2637. [CrossRef] [PubMed]
3. Afridi, B.; Khan, H.; Akkol, E.K.; Aschner, M. Pain Perception and Management: Where do We Stand? *Curr. Mol. Pharmacol.* **2021**, *14*, 678–688. [CrossRef]
4. Li, J.; Chen, Y.; Liu, J.; Zhang, D.; Liang, P.; Lu, P.; Shen, J.; Miao, C.; Zuo, Y.; Zhou, C. Elevated Expression and Activity of Sodium Leak Channel Contributes to Neuronal Sensitization of Inflammatory Pain in Rats. *Front. Mol. Neurosci.* **2021**, *14*, 723395. [CrossRef] [PubMed]

5. Basso, L.; Aboushousha, R.; Fan, C.Y.; Iftinca, M.; Melo, H.; Flynn, R.; Agosti, F.; Hollenberg, M.D.; Thompson, R.; Bourinet, E.; et al. TRPV1 promotes opioid analgesia during inflammation. *Sci. Signal.* **2019**, *12*, eaav0711. [CrossRef]
6. Wang, J.; Xu, D.; Shen, L.; Zhou, J.; Lv, X.; Ma, H.; Li, N.; Wu, Q.; Duan, J. Anti-inflammatory and analgesic actions of bufotenine through inhibiting lipid metabolism pathway. *Biomed. Pharmacother.* **2021**, *140*, 111749. [CrossRef]
7. Hou, Z.; Wen, Q.; Zhou, W.; Yan, P.; Zhang, H.; Ding, J. Topical Delivery of Ketorolac Tromethamine via Cataplasm for Inflammatory Pain Therapy. *Pharmaceutics* **2023**, *15*, 1405. [CrossRef]
8. Turnaturi, R.; Chiechio, S.; Pasquinucci, L.; Spoto, S.; Costanzo, G.; Dichiara, M.; Piana, S.; Grasso, M.; Amata, E.; Marrazzo, A.; et al. Novel N-normetazocine Derivatives with Opioid Agonist/Sigma-1 Receptor Antagonist Profile as Potential Analgesics in Inflammatory Pain. *Molecules* **2022**, *27*, 5135. [CrossRef]
9. O'Brien, T.; Christrup, L.L.; Drewes, A.M.; Fallon, M.T.; Kress, H.G.; McQuay, H.J.; Mikus, G.; Morlion, B.J.; Perez-Cajaraville, J.; Pogatzki-Zahn, E.; et al. European Pain Federation position paper on appropriate opioid use in chronic pain management. *Eur. J. Pain* **2017**, *21*, 3–19. [CrossRef]
10. Smith, J.C. Respiratory rhythm and pattern generation: Brainstem cellular and circuit mechanisms. *Handb. Clin. Neurol.* **2022**, *188*, 1–35.
11. Karthik, S.; Huang, D.; Delgado, Y.; Laing, J.J.; Peltekian, L.; Iverson, G.N.; Grady, F.; Miller, R.L.; McCann, C.M.; Fritzsch, B.; et al. Molecular ontology of the parabrachial nucleus. *J. Comp. Neurol.* **2022**, *530*, 1658–1699. [CrossRef] [PubMed]
12. Liu, P.C.; Yao, W.; Chen, X.Y.; Su, W.K.; Zheng, Z.H.; Yan, X.B.; Deng, Y.L.; Shi, K.G.; Liu, X.; Gao, Y.W.; et al. Parabrachial nucleus astrocytes regulate wakefulness and isoflurane anaesthesia in mice. *Front. Pharmacol.* **2022**, *13*, 991238. [CrossRef] [PubMed]
13. Choi, S.; Hachisuka, J.; Brett, M.A.; Magee, A.R.; Omori, Y.; Iqbal, N.U.; Zhang, D.; DeLisle, M.M.; Wolfson, R.L.; Bai, L.; et al. Parallel ascending spinal pathways for affective touch and pain. *Nature* **2020**, *587*, 258–263. [CrossRef] [PubMed]
14. Deng, J.; Zhou, H.; Lin, J.K.; Shen, Z.X.; Chen, W.Z.; Wang, L.H.; Li, Q.; Mu, D.; Wei, Y.C.; Xu, X.H.; et al. The Parabrachial Nucleus Directly Channels Spinal Nociceptive Signals to the Intralaminar Thalamic Nuclei, but Not the Amygdala. *Neuron* **2020**, *107*, 909–923.e6. [CrossRef]
15. Teuchmann, H.L.; Hogri, R.; Heinke, B.; Sandkühler, J. Anti-Nociceptive and Anti-Aversive Drugs Differentially Modulate Distinct Inputs to the Rat Lateral Parabrachial Nucleus. *J. Pain* **2022**, *23*, 1410–1426. [CrossRef]
16. Liu, S.; Ye, M.; Pao, G.M.; Song, S.M.; Jhang, J.; Jiang, H.; Kim, J.H.; Kang, S.J.; Kim, D.I.; Han, S. Divergent brainstem opioidergic pathways that coordinate breathing with pain and emotions. *Neuron* **2022**, *110*, 857–873.e9. [CrossRef]
17. Yang, H.; de Jong, J.W.; Cerniauskas, I.; Peck, J.R.; Lim, B.K.; Gong, H.; Fields, H.L.; Lammel, S. Pain modulates dopamine neurons via a spinal-parabrachial-mesencephalic circuit. *Nat. Neurosci.* **2021**, *24*, 1402–1413. [CrossRef]
18. Liu, C.Y.; Lee, M.L.; Yang, C.S.; Chen, C.M.; Min, M.Y.; Yang, H.W. Morphological and physiological evidence of a synaptic connection between the lateral parabrachial nucleus and neurons in the A7 catecholamine cell group in rats. *J. Biomed. Sci.* **2015**, *22*, 79. [CrossRef]
19. Zhang, D.; Zhao, W.; Liu, J.; Ou, M.; Liang, P.; Li, J.; Chen, Y.; Liao, D.; Bai, S.; Shen, J.; et al. Sodium leak channel contributes to neuronal sensitization in neuropathic pain. *Prog. Neurobiol.* **2021**, *202*, 102041. [CrossRef]
20. Shi, Y.; Abe, C.; Holloway, B.B.; Shu, S.; Kumar, N.N.; Weaver, J.L.; Sen, J.; Perez-Reyes, E.; Stornetta, R.L.; Guyenet, P.G.; et al. Nalcn Is a "Leak" Sodium Channel That Regulates Excitability of Brainstem Chemosensory Neurons and Breathing. *J. Neurosci.* **2016**, *36*, 8174–8187. [CrossRef]
21. Ford, N.C.; Ren, D.; Baccei, M.L. NALCN channels enhance the intrinsic excitability of spinal projection neurons. *Pain* **2018**, *159*, 1719–1730. [CrossRef] [PubMed]
22. Roeder, Z.; Chen, Q.; Davis, S.; Carlson, J.D.; Tupone, D.; Heinricher, M.M. Parabrachial complex links pain transmission to descending pain modulation. *Pain* **2016**, *157*, 2697–2708. [CrossRef]
23. Liao, Z.; Liu, Y.; Wang, Y.; Lu, Q.; Peng, Y.; Liu, Q. Case Report: A de novo Variant in NALCN Associated With CLIFAHDD Syndrome in a Chinese Infant. *Front. Pediatr.* **2022**, *10*, 927392. [CrossRef]
24. Stenehjem, K.K.; Schweigert, J.; Kumar, P. Atypical Presentation of Viral Gastroenteritis in a Three-year-old Child Due to a UNC80 Mutation. *Cureus* **2019**, *11*, e4395. [CrossRef] [PubMed]
25. Kelesoglu, F.M.; Kaya, M.; Sayili, E.T. Novel nonsense mutation in UNC80 in a Turkish patient further validates the sociable skill and severe gastrointestinal problems as part of disease spectrum. *Am. J. Med. Genet. A* **2023**, *191*, 1959–1962. [CrossRef]
26. Winczewska-Wiktor, A.; Hirschfeld, A.S.; Badura-Stronka, M.; Wojsyk-Banaszak, I.; Sobkowiak, P.; Bartkowska-Śniatkowska, A.; Babak, V.; Steinborn, B. Central Apneas Due to the CLIFAHDD Syndrome Successfully Treated with Pyridostigmine. *Int. J. Environ. Res. Public Health* **2022**, *19*, 775. [CrossRef] [PubMed]
27. Zhang, J.; Xu, W.; Gui, J.; Huang, X. Genetic analysis of a case of chromosome 14q microdeletion. *Zhonghua Yi Xue Yi Chuan Xue Za Zhi* **2020**, *37*, 178–181.
28. Wang, J.; Yang, Y.; Liu, J.; Qiu, J.; Zhang, D.; Ou, M.; Kang, Y.; Zhu, T.; Zhou, C. Loss of sodium leak channel (NALCN) in the ventral dentate gyrus impairs neuronal activity of the glutamatergic neurons for inflammation-induced depression in male mice. *Brain Behav. Immun.* **2023**, *110*, 13–29. [CrossRef]
29. Zieglgänsberger, W. Substance P and pain chronicity. *Cell Tissue Res.* **2019**, *375*, 227–241. [CrossRef]
30. Xu, L.; Jiang, H.; Feng, Y.; Cao, P.; Ke, J.; Long, X. Peripheral and central substance P expression in rat CFA-induced TMJ synovitis pain. *Mol. Pain* **2019**, *15*, 1744806919866340. [CrossRef]

31. Lu, B.; Su, Y.; Das, S.; Wang, H.; Wang, Y.; Liu, J.; Ren, D. Peptide neurotransmitters activate a cation channel complex of NALCN and UNC-80. *Nature* **2009**, *457*, 741–744. [CrossRef] [PubMed]
32. Lu, T.Z.; Feng, Z.P. NALCN: A regulator of pacemaker activity. *Mol. Neurobiol.* **2012**, *45*, 415–423. [CrossRef] [PubMed]
33. Ren, D. Sodium leak channels in neuronal excitability and rhythmic behaviors. *Neuron* **2011**, *72*, 899–911. [CrossRef]
34. Chiang, M.C.; Nguyen, E.K.; Canto-Bustos, M.; Papale, A.E.; Oswald, A.M.; Ross, S.E. Divergent Neural Pathways Emanating from the Lateral Parabrachial Nucleus Mediate Distinct Components of the Pain Response. *Neuron* **2020**, *106*, 927–939.e5. [CrossRef] [PubMed]
35. Yang, C.F.; Kim, E.J.; Callaway, E.M.; Feldman, J.L. Monosynaptic Projections to Excitatory and Inhibitory preBötzinger Complex Neurons. *Front. Neuroanat.* **2020**, *14*, 58. [CrossRef]
36. Katagiri, A.; Kato, T. Multidimensional role of the parabrachial nucleus in regulating pain-related affective disturbances in trigeminal neuropathic pain. *J. Oral Sci.* **2020**, *62*, 160–164. [CrossRef]
37. Burchell, A.; Mansour, Y.; Kulesza, R. Leveling up: A long-range olivary projection to the medial geniculate without collaterals to the central nucleus of the inferior colliculus in rats. *Exp. Brain Res.* **2022**, *240*, 3217–3235. [CrossRef]
38. Ruan, Y.; Ling, J.; Ye, F.; Cheng, N.; Wu, F.; Tang, Z.; Cheng, X.; Liu, H. Paeoniflorin alleviates CFA-induced inflammatory pain by inhibiting TRPV1 and succinate/SUCNR1-HIF-1α/NLPR3 pathway. *Int. Immunopharmacol.* **2021**, *101 Pt B*, 108364. [CrossRef]
39. Jiang, J.; Zhao, Y.; Liu, J.; Yang, Y.; Liang, P.; Huang, H.; Wu, Y.; Kang, Y.; Zhu, T.; Zhou, C. Signatures of Thalamocortical Alpha Oscillations and Synchronization with Increased Anaesthetic Depths under Isoflurane. *Front. Pharmacol.* **2022**, *13*, 887981. [CrossRef]

International Journal of
Molecular Sciences

Article

Sodium-Calcium Exchanger 2: A Pivotal Role in Oxaliplatin Induced Peripheral Neurotoxicity and Axonal Damage?

Elisa Ballarini [1,2,†], Alessio Malacrida [1,2,†], Virginia Rodriguez-Menendez [1,2], Eleonora Pozzi [1,2], Annalisa Canta [1,2], Alessia Chiorazzi [1,2], Laura Monza [1,2], Sara Semperboni [1,2], Cristina Meregalli [1,2], Valentina Alda Carozzi [1,2], Maryamsadat Hashemi [1,2], Gabriella Nicolini [1,2], Arianna Scuteri [1,2], Stephen N. Housley [3], Guido Cavaletti [1,2] and Paola Alberti [1,2,*]

1 School of Medicine and Surgery, University of Milano-Bicocca, 20126 Milan, Italy
2 NeuroMI (Milan Center for Neuroscience), 20126 Milan, Italy
3 Integrated Cancer Research Center, School of Biological Sciences, Georgia Institute of Technology, Atlanta, GA 30332, USA
* Correspondence: paola.alberti@unimib.it; Tel.: +39-02-6448-8154
† These authors contributed equally to this work.

Abstract: Oxaliplatin (OHP)-induced peripheral neurotoxicity (OIPN) is a frequent adverse event of colorectal cancer treatment. OIPN encompasses a chronic and an acute syndrome. The latter consists of transient axonal hyperexcitability, due to unbalance in Na^+ voltage-operated channels (Na^+VOC). This leads to sustained depolarisation which can activate the reverse mode of the Na^+/Ca^{2+} exchanger 2 (NCX2), resulting in toxic Ca^{2+} accumulation and axonal damage (ADa). We explored the role of NCX2 in in vitro and in vivo settings. Embryonic rat Dorsal Root Ganglia (DRG) organotypic cultures treated with SEA0400 (SEA), a NCX inhibitor, were used to assess neuroprotection in a *proof-of-concept* and *pilot* study to exploit NCX modulation to prevent ADa. In vivo, OHP treated mice (7 mg/Kg, i.v., once a week for 8 weeks) were compared with a vehicle-treated group (n = 12 each). Neurophysiological and behavioural testing were performed to characterise acute and chronic OIPN, and morphological analyses were performed to detect ADa. Immunohistochemistry, immunofluorescence, and western blotting (WB) analyses were also performed to demonstrate changes in NCX2 immunoreactivity and protein expression. In vitro, NCX inhibition was matched by ADa mitigation. In the in vivo part, after verifyingboth acute and chronic OIPN had ensued, we confirmed via immunohistochemistry, immunofluorescence, and WB that a significant NCX2 alteration had ensued in the OHP group. Our data suggest NCX2 involvement in ADa development, paving the way to a new line of research to prevent OIPN.

Keywords: NCX2; voltage-operated ion channels; chemotherapy-induced peripheral neurotoxicity; chemotherapy induced peripheral neuropathy; axonal damage; axonal hyperexcitability; immunofluorescence; immunohistochemistry; nerve excitability testing; neuropathology; neuroprotection

1. Introduction

Oxaliplatin (OHP)-induced peripheral neurotoxicity (OIPN) is a frequent toxicity, experienced by a growing population of colorectal cancer survivors that can be long-lasting, or even permanent [1,2], altering patients' quality of life [3,4]. At the state of the art, there is no efficacious preventive or curative treatment for OIPN [5]. One of the reasons for this unmet clinical need is the incomplete knowledge on axonal damage (ADa) mechanisms [6,7]. Therefore, robust experimental evidence is still required to devise novel treatments. OIPN is characterised by two different conditions. A *chronic sensory neuropathy* characterised by ADa, known as chemotherapy-induced peripheral neurotoxicity (CIPN), which is one of the commonest late toxicities of several anticancer drugs (platinum drugs, taxanes, vinca alkaloids, proteasome inhibitors, epothilones, and thalidomide [8–11]). OIPN, however, is also characterised by a *specific acute neurotoxicity syndrome*, as soon as after the first chemotherapy cycle. Acute signs/symptoms occur nearly in all OHP-treated patients mirroring an axonal hyperexcitability state: transient cold-induced paraesthesia at limb extremities, cold-induced dysesthesia at oral cavity/pharynx, jaw spasm, and cramps, lasting mainly the 24–72 h after each OHP administration [12,13]. A possible causative link between acute and chronic OIPN was hypothesised. Even though acute OIPN is transient and never dose-limiting, the underlying axonal hyperexcitability could be potentially linked to cellular stress and, therefore, ADa [14,15]. Several preclinical in vitro observations showed sodium voltage-operated ion channel (Na^+VOC) alterations are involved in acute OIPN. In the literature, alternative mechanisms involving also other channels were hypothesised [16], but there are robust in vitro data, in vivo data, and in silico modelling demonstrating that Na^+VOC alterations are pivotal and other channels/transporters, if involved, are a secondary element [17–19]. We already provided clinical observations in line with this: a more severe chronic OIPN was associated with a more pronounced acute OIPN and specific Na^+VOC polymorphisms [20,21], whereas K^+ voltage-operated ion channels ones were not associated with OIPN [22]. Exploiting nerve excitability testing (NET), Na^+VOC dysfunctions were also demonstrated in OHP-treated patients, before ADa ensued [15,23,24]. On the basis of these observations, we tested topiramate (TPM), a known Na^+VOC modulator approved for clinical use in a refined rat model [25]; complete neuroprotection was demonstrated via a refined set of multiple outcome measures [26]. However, the exact causative link between acute Na^+VOC dysfunctions and chronic ADa is still to be investigated. Na^+VOC alterations are only transient [1,18] and functional, thus OIPN is quite different respect to the condition of "sick-NaV channels" extensively described by Morris et al. [27], in relation to traumatic/ischemic/inherited conditions; in these cases, structural degradation of axolemma bilayer is present. Therefore, we hypothesised that OHP-related ADa was due to a downstream event respect to **Na^+VOC impaired functioning**, which would be a transient, functional, **upstream event**. The Na^+/Ca^{2+} exchanger (NCX) family could play a pivotal role in this regard, in particular its isoform 2 (NCX2); NCX2 is widely expressed along distal parts of the axons together with Na^+VOC 1.6, 1.8, and 1.9 [28], as well as in dorsal root ganglia (DRG) [28], the primary target of OHP neurotoxicity [29]. NCX is a bidirectional transporter dependent on the gradient of the two involved ions, Na^+ and Ca^{2+} [30]. NCX is one of the key regulators of Ca^{2+} homeostasis and, in normal conditions, NCX extrudes Ca^{2+} from cells (*forward mode*), but, in case of ion unbalance (e.g., enhanced Na^+ influx, which determines an aberrant neuronal depolarization [31]) NCX starts to operate in the *reverse mode*, resulting in Ca^{2+} importing [32]. Ca^{2+} overload is a relevant contributor leading to cellular damage due to activation of Ca-sensitive calpain, phospholipases, and nitric oxide synthase [28,33,34]. Ca^{2+} intraneuronal levels should be tightly controlled; Ca^{2+}, in fact, is one the major triggers of neurotransmitter release [35] and is essential for several other functions related to neuronal excitability [36], integration of signals [36], synaptic plasticity [37], gene expression [38], metabolism [39], and programmed cell death [40]. Therefore, an uncontrolled level of this ion can lead to neuronal damage and death [41]. Since NCX *reverse mode* activation leads to alterations in Ca^{2+} balance, it is an intriguing **possible downstream event** leading to ADa. Preliminary in vitro or ex vivo findings suggested that NCX2 has a relevant role in ADa in peripheral nerves [34]. Earlier studies were done on rat optic nerves (that strictly anatomically and embryologically speaking are part of the central and not the peripheral nervous system), showing that metabolic anoxia-induced stress resulted in a persistent Na^+ influx via Na^+VOC, and this was also related to the metabolic impairment of the Na^+/potassium ATP-ase pump, leading to NCX *reverse mode* activation and subsequent Ca^{2+}-related toxicity in axons [42]; similar findings were then replicated in peripheral nervous system myelinated axons exposed to anoxia [43,44]. Ca^{2+} toxicity due to the *reverse mode* activation was related to several neurological diseases, and the possible beneficial role of its modulation has been described in different conditions such as brain ischemia [31], Alzheimer disease [45], and amyotrophic lateral sclerosis [46]. However, further observations are still required. Therefore, we performed in vitro and in vivo experiments to shed light into the potential role of NCX2 in ADa development. In the in vivo study, we provide evidence of potential NCX2 involvement in OHP-related ADa, and in the in vitro part we propose a *proof-of-concept* and *pilot* study of NCX modulation effects.

2. Results

2.1. In Vitro Observations

The role of NCX2 in OIPN was investigated in vitro using the well-established experimental DRG explant model obtained from embryonic Sprague Dawley rats; neurite elongation is the key parameter to assess neurotoxicity and neuroprotection [47]. As shown in Figure 1, DRG exposed to culture medium emitted long neurites, which shortened after exposure to OHP for 24 h (−20%, p-value < 0.05 vs. control) and 48 h (−60%, p-value < 0.001 vs. control). DRG pre-treated with SEA0400 (20 μM), a strong and selective NCX family inhibitor [48], for 3 h did not affect neurite elongation under control conditions, and protected neurite outgrowth of DRG exposed to OHP for 24 h (+40%, p-value < 0.01 vs. OHP) and 48 h (+20%, p-value < 0.05 vs. OHP).

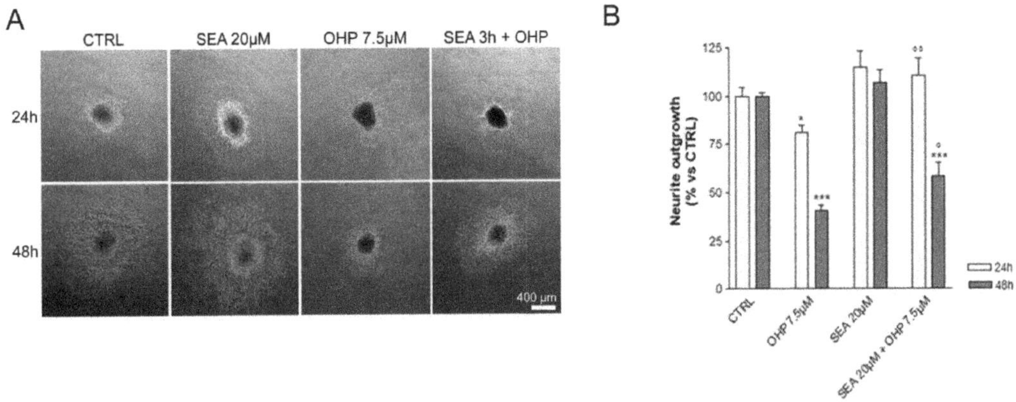

Figure 1. In vitro experiment results. (**A**) shows representative images of neurite elongation at 24 and 48 h (upper and lower panel, respectively); CTRL: control group; SEA: SEA0400-treated; OHP: Oxaliplatin-treated. (**B**) shows neurite outgrowth as a percentage respect to the CTRL group (statistical significance of one-way ANOVA (followed by Tukey's Multiple Comparison Test) is also provided: * $p < 0.05$ vs. CTRL; *** $p < 0.001$ vs. CTRL; ° $p < 0.05$ vs. OHP, °° $p < 0.01$ vs. OHP).

2.2. In Vivo Observations

Study design for the characterization of an in vivo comprehensive OIPN mouse model was as follows. Nerve conduction studies (NCS) and behavioural test parameters were used to verify homogeneity between the experimental groups before 1st chemotherapy administration (no statistically significant difference for all parameters). Acute OPIN was verified after 1st OHP injection monitoring NET changes in the subsequent 72 h. Chronic OIPN and, therefore, ADa development was verified via NCS, neuropathology (caudal nerve morphometry/morphology and intraepidermal nerve fiber density [IENFD]), and behavioural tests after 8 weeks of treatment. Immunohistochemistry, immunofluorescence, and western blotting were used to assess if OIPN was associated with alterations in immunoreactivity and protein expression in DRG of animals treated with OHP for 8 weeks.

2.2.1. NET after the 1st Administration

NET verified acute OIPN ensued after the 1st chemotherapy cycle. Over 72 h of monitoring, the most notable change was related to an upward shift in the recovery cycle curve (p-value < 0.01 at 2 ms and at 2.5 ms refractoriness assessment, Mann–Whitney U-test). These findings are consistent with our previous study and suggest the onset of OHP-related axonal hyperexcitability that alters Na^{+}VOC functioning, leading to an aberrant depolarization [25]. Figure 2 gives an overview of the monitoring over 72 h showing statistically significant parameters. Individual traces of each recording are provided in supplementary Figure S1.

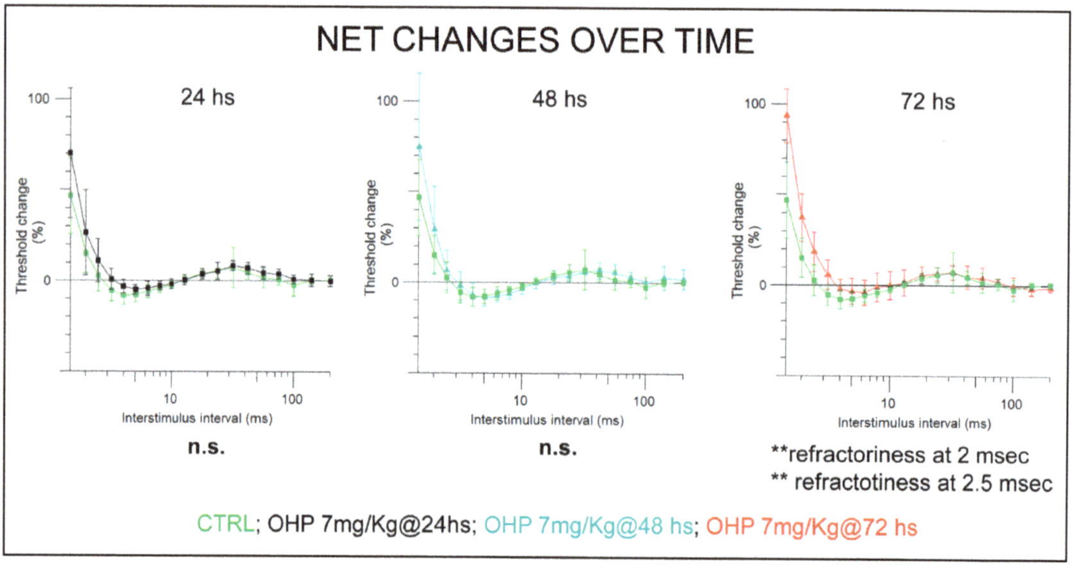

Figure 2. NET findings at recovery cycle at 24, 48, 72 h following 1st Oxaliplatin i.v., administration. Statistical significance of Mann–Whitney U-test is shown. In the graph, individual values are accompanied by SEM bars (** *p*-value < 0.01 vs. CTRL). CTRL: control group (green curves); OHP: OHP-treated group (curves at 24 h are shown in black, at 48 h in cyan, and at 72 h in red).

2.2.2. NCS and Behavioral Tests at the End of Treatment

A pattern compatible with polyneuropathy was observed via NCS: significant decrease in both caudal and digital nerve sensory action potential (SAP) amplitude and digital nerve conduction velocity was observed in OHP-treated animals. These findings demonstrated that relevant OHP-related ADa ensued, able to cause a functional impairment of large myelinated fibers, which are the ones tested with NCS [49]. Similarly, mechanical allodynia (Dynamic test) had ensued in OHP-treated animals ($p < 0.001$), mirroring a functional impairment ensued in small fibers (that convey painful information) [26]. Data are summarised in Figure 3.

2.2.3. Caudal Nerve Morphology and Morphometry

Morphological examination of caudal nerves, harvested after sacrifice at the end of 8 weeks of treatment, demonstrated mild ADa in large myelinated fibers, matching observations obtained via NCS. Morphometrical analyses showed a statistically significant decrease in the fiber density of large myelinated fibers in the OHP treated group. Figure 4 shows representative photographs of caudal nerve morphology and statistical tests (Mann–Whitney U-test).

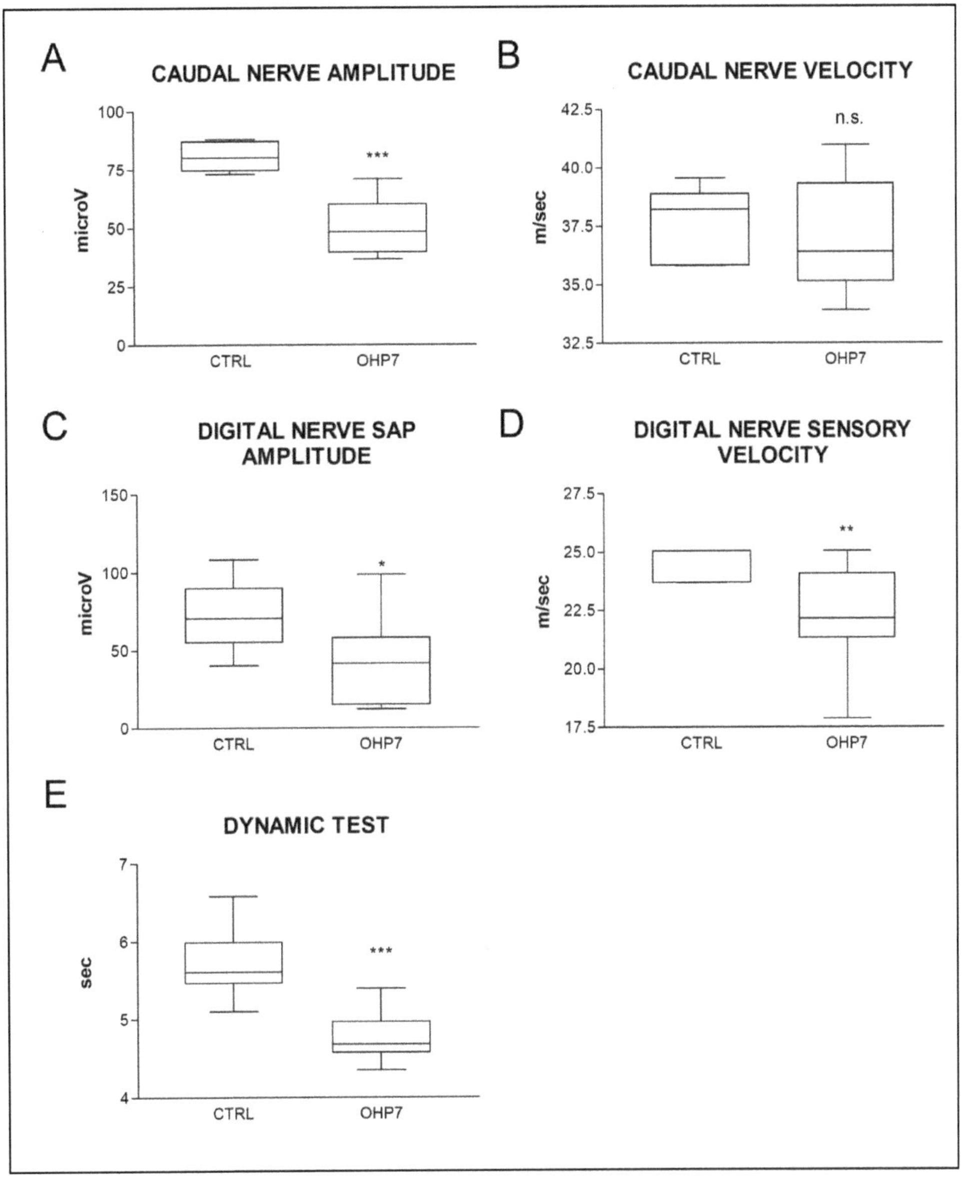

Figure 3. In (**A–D**): nerve conduction study data at the end of treatment is shown. In (**E**) Dynamic (mechanical allodynia) test data at the end of treatment is shown. Statistical significance of Mann–Whitney U-test is shown in all graphs. The box-and-whiskers graphs show median and quartile values, as well as maximum and minimum values. * p-value < 0.05 vs. CTRL, ** p-value < 0.01 vs. CTRL, *** p-value < 0.001 vs. CTRL. CTRL: control group; OHP7: OHP-treated group. SAP: sensory action potential. n.s. means no significance.

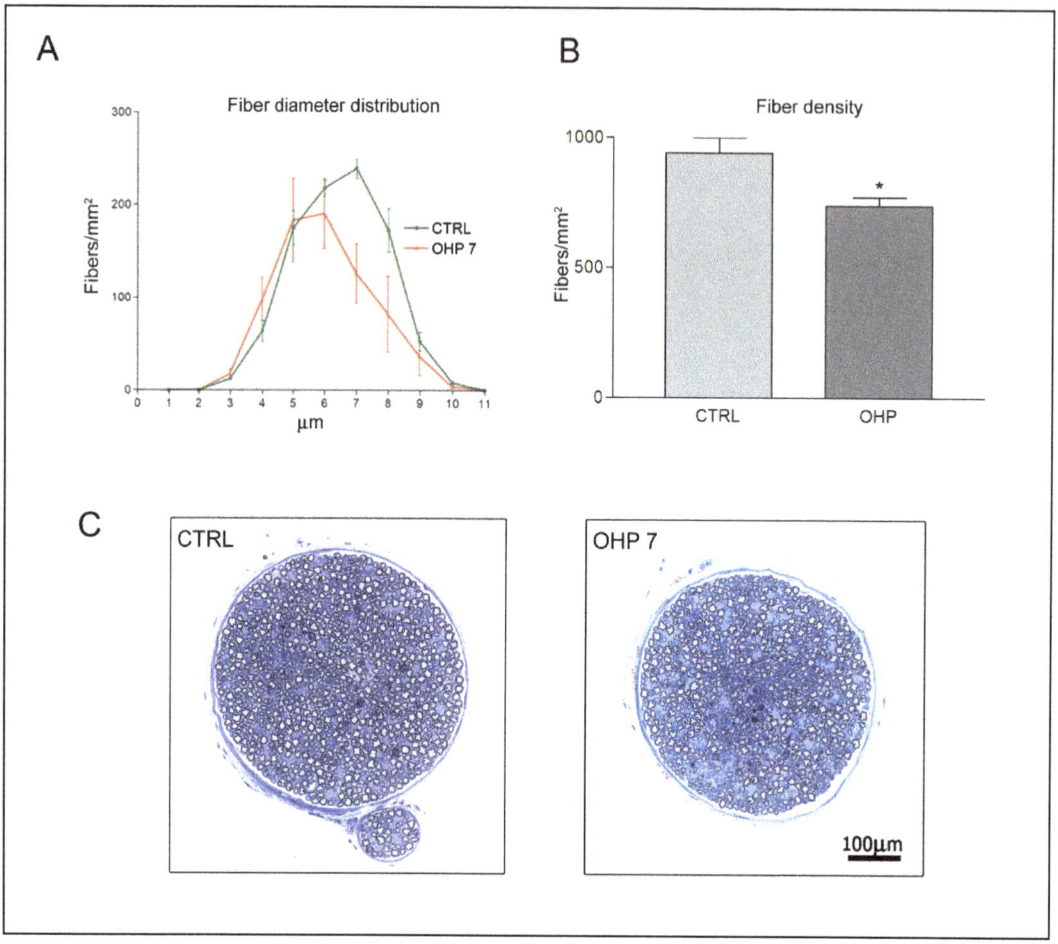

Figure 4. Morphological and morphometrical assessment of caudal nerves at the end of treatment. (**A**) shows the graph of the distribution of fiber diameters (SEM bars are represented). (**B**) shows the box-plot graph (standard deviation bar is shown) of statistical analysis (*t*-test) of morphometry. In (**C**)) representative images of CTRL and OHP animals are shown, to highlight the mild axonal loss in caudal nerves of the OHP group: fiber density is moderately diminished and degenerating fibers are visible in OHP group. CTRL: control group; OHP7: OHP-treated group. * p-value < 0.05 vs. CTRL.

2.2.4. Intraepidermal Nerve Fiber Density (IENFD)

Analysis of IENFD allows to formally count small nerve fibers (and eventually assess their loss), relying on cutaneous terminals. At the end of treatment, IENFD demonstrated a statistically significant decrease in treated animals compared to controls, mirroring the functional impairment that was observed at Dynamic test for mechanical allodynia. Mann–Whitney U-test significance and representative images are shown in Figure 5.

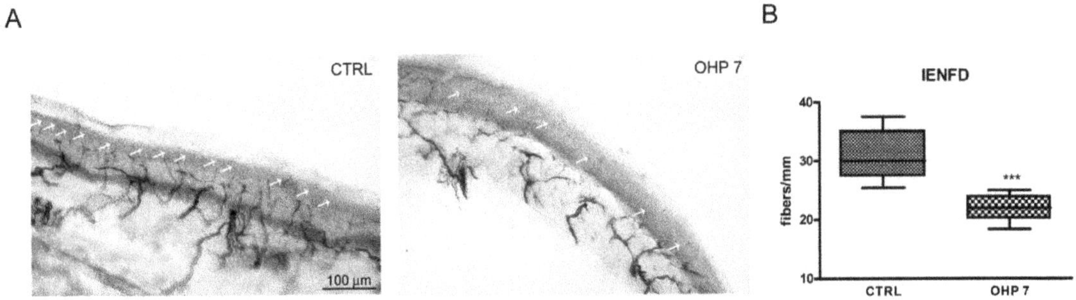

Figure 5. IENFD at the end of treatment. In (**A**), representative images of IEFND in both groups are shown; white arrows point out small fibers visible in each photograph. In (**B**), the graph representing the statistical significance of Mann–Whitney U-test is shown (in the box-and-whiskers graphs median and quartile values, as well as maximum and minimum values, are shown; *** $p < 0.001$ vs. CTRL). CTRL: control group; OHP7: OHP-treated group.

2.2.5. Immunohistochemistry (IHC), Immunofluorescence (IF), and Western Blotting (WB) for NCX2

DRG harvested at sacrifice after 8 weeks of treatment were used to assess whether a difference can be found between control and OHP animals. We conducted analyses with three techniques to verify the robustness of our observations. IHC and IF showed a similar pattern of mainly cytosolic NCX2 immunoreactivity (shown in red in IF images and in dark brown in DAB-stained IHC images); a statistically significant reduction in the OHP group with respect to control was observed with both techniques, exploiting a quantitative measurement of NCX2 immunoreactivity (Figure 1C,D, respectively), and a similar reduction was demonstrated via WB analysis on the DRG pool (Figure 6E,F). This is an indirect confirmation that NCX2 *reverse mode* was activated. If *reverse mode* was activated due to an aberrant depolarization, as already demonstrated by Boscia et al. [50], a repetitive spreading of depolarisation current resulted in a downregulation of NCX2 in neurons (this is an endogenous autoprotective mechanism to prevent Ca^{2+} overload).

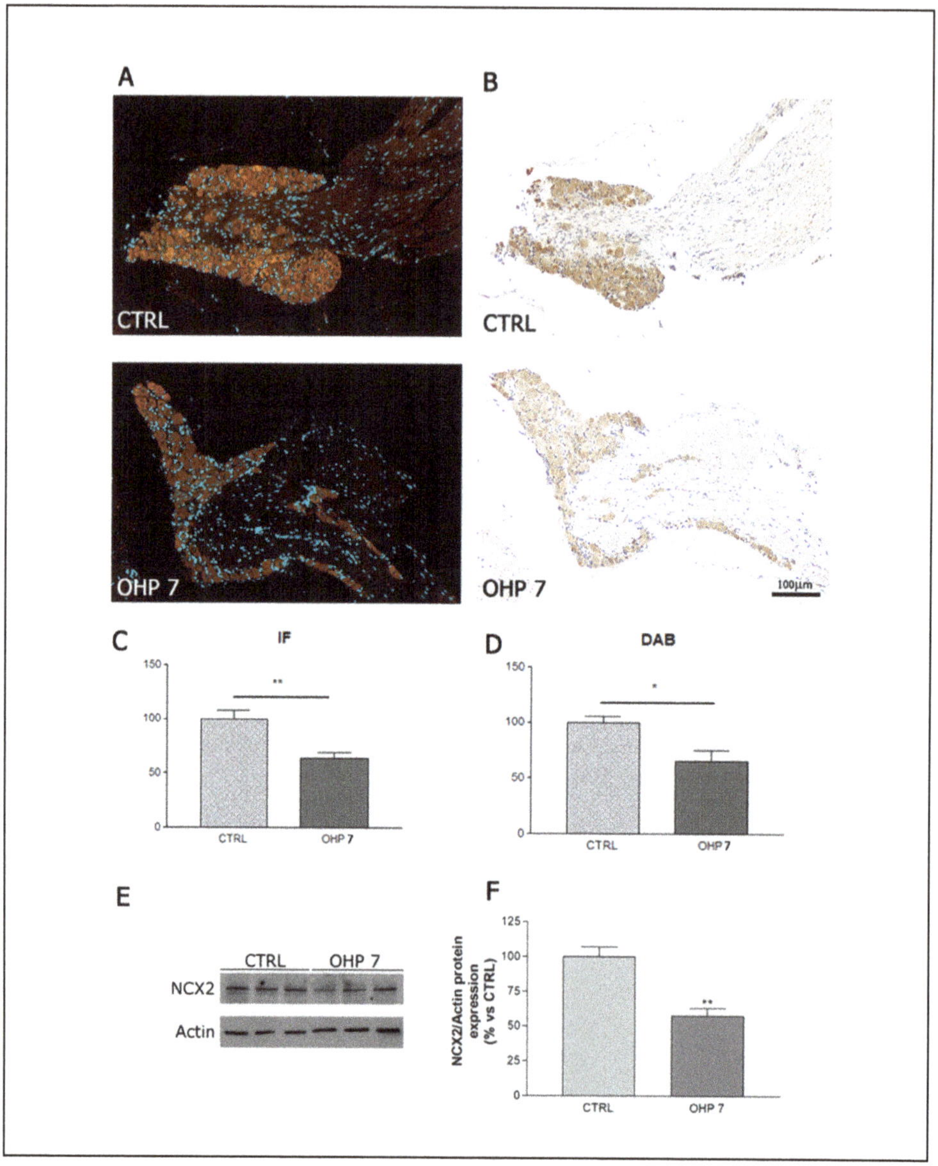

Figure 6. Immunohistochemistry, immunofluorescence, and western blotting for NCX2. On the upper panels representative images of immunofluorescence (**A**) and immunohistochemistry (**B**) are shown, accompanied by graphs showing the statistical significance (*t*-test) for IF and DAB quantification, in (**C**,**D**), respectively). In the bottom panel, the running lane of the western blotting (**E**) is shown as well as the graph (**F**), showing the statistical significance in NCX2 quantification (*t*-test). CTRL: control group; DAB: 3, 3′-diaminobenzidine staining in immunohistochemistry; IF: immunofluorescence; OHP7: OHP-treated group. For each column graph, the standard deviation bar is represented. * p-value < 0.05 vs. CTRL; ** p-value < 0.001 vs. CTRL.

3. Discussion

So far, no mouse model has been described reproducing both acute and chronic OIPN; therefore, we first validated the robustness of the proposed model. Recently, Makker et al., in fact, characterised a mouse model showing only acute OIPN features, exploiting NET [18]. As performed in a previous study from our group in a rat model [25], we relied on a multimodal approach to demonstrated both acute and chronic OIPN had ensued [6,26]. Recovery cycle parameters at NET after the first cycle showed the typical acute OIPN pattern: OHP group showed an upward shift of the recovery cycle curve, compatible with altered Na^{+}VOC functioning as already demonstrated [15,18,25]. We relied on multiple outcome measures to verify the full expression of chronic OIPN (i.e., ADa) at the end of treatment [26]. NCS were used to demonstrate the presence of a sensory axonal neuropathy; consistent with clinical observations, we observed mainly a significant drop in SAP amplitude compatible with sensory ADa [20,51,52]. Morphology of caudal nerves matched neurophysiological results, and morphometry demonstrated axonal loss of myelinated fibers. Larger diameter fibers were the ones with the most evident damage, matching clinical experience; patients mainly show sensory large fibers impairment, up to a disabling sensory ataxia [1,11,53]. Even if large fibers are the main target of OHP ADa, impairment in small fibers was clearly demonstrated in animal models [25,54] and was also observed in a small clinical cohort [55]; our model also reproduced this feature as demonstrated via IENFD reduction. Finally, allodynia was tested to demonstrate a functional impairment in small fibers; mechanical allodynia was tested, instead of temperature-triggered allodynia, to avoid possible overlap between acute OIPN and ADa, given the former is cold triggered [1,11]. Thus, the full OIPN spectrum was reproduced in our animals allowing us to perform NCX2-related observations in a setting mirroring the clinical condition.

We verified NCX2 immunolocalisation in DRG via immunohistochemistry/ immunofluorescence. NCX2, quite notably, is known for being well represented in DRG [33], which are the first target in case of ADa due to OHP [11]. Even if indirectly, our data were compatible with NCX2 *reverse mode* activation. First of all, we demonstrated that the prerequisite of *reverse mode* activation ensued (i.e., Na^{+} dysfunction) via NET; this is the trigger to switch the mode of NCX functioning. If *reverse mode* was activated via an aberrant depolarization, neurons downregulated NCX to avoid a Ca^{2+} overload [50], which is the final effect of a prolonged reverse mode activation. Therefore, the 2 months of treatment and repetitive oxaliplatin administration (each injection is a trigger for reverse mode activation) were able to determine a *reverse mode* activation relevant enough that the downregulation mechanism was established; this is an endogenous and autoprotective mechanism that it is not ultimately strong enough to prevent axonal damage though, as evidenced by ADa development in OHP animals as seen at NCS, Dynamic test for allodynia, and histopathological observations of large and small nerve fibers as discussed above.

Preliminary, mainly in vitro findings suggested that NCX2 modulation might mitigate toxic neuropathies [56,57], pointing out that a pharmacological modulation of this axis can be more efficacious than the endogenous mechanisms. To further investigate this, we conducted a *proof-of-concept* and *pilot* in vitro study to test beneficial effects of NCX modulation. Among possible inhibitors, we selected SEA0400 since it is the most potent and selective NCX inhibitor [58]. Other options, such as the less strong NCX inhibitor KB-R7943, showed a rather low specificity for NCX family even at low dosages, with larger impact on off-targets with respect to SEA0400 [58–60]. Moreover, SEA0400 was already used in studies that investigated nervous system injury due to NCX2 *reverse mode* activation/Ca^{2+} toxicity, caused by alterations of Na^{+} currents: Matsuda et al. [58] showed that SEA0400 attenuated, dose-dependently, damage in models of brain ischemia. Koyama et al. [61] also obtained promising neuroprotection with SEA0400 in a model of brain ischemia. Notably, in a different model of peripheral neuropathy, Yilmaz et al. showed that some neuronal populations of the peripheral nervous system can be resistant to the effect of KB-R7943, but responsive to SEA0400 [57]. Therefore, SE0400 profile was the most appropriate drug for a *proof-of-concept* and *pilot* experiment to test our hypothesis. However, the specific model of disease and system we are dealing with was carefully considered to select the range of dosages to be tested. Literature data reported above, concerning the central nervous system, mostly used nM concentrations [62], but we also considered other published works to better mirror the peripheral nervous system setting; to obtain effects on axons (squid giant axon), a higher—despite still selective—dosage was require in the range of μM [63]. Moreover, in the work by Yilmaz et al., in a different model of peripheral neuropathy [57], SEA0400 was used in a microM concentrations, as well as in neuroprotection experiments relying on SH-SY5Y cells [64]. We also carefully evaluated the most appropriate OHP dose to be used, relying on the extensive review of Calls et al. [65], which presented an overview of the in vitro and in vivo models of OIPN; starting from these, we selected a dosage that was similar to the mean plasma levels in patients treated with OHP [66,67] and able to induce neurite alterations.

To ensure SEA0400 was active against down-stream events, secondary to Na^+VOC alteration, a pre-treatment exposure to SEA0400 was performed (3 h); it was necessary, in fact, to administer SEA0400 before OHP exposure to ensure the block of NCX activity before the *reverse mode* was activated, as a consequence of Na^+VOC unbalance, similarly to what we previously performed in rats with TPM [25]. Exploiting a consolidated approach to assess neurotoxicity in vitro [68], we observed promising data suggesting the neuroprotective role of NCX inhibition against OHP detrimental effects in neurons. This confirms, indirectly, that NCX2 *reverse mode* activation plays a role in ADa. However, some limitations should be acknowledged. Ours is a *proof-of-concept* and *pilot* study that would require further investigation. Despite the fact that our data are promising, of course, it could be argued that a combined administration of Na^+ current and Ca^{2+} current distinct modulators might be effective too in preventing calcium overload as described in models of cardiac ischemia-reperfusion injury and cardiac glycoside toxicity [69]. However, targeting NCX with novel drugs/approaches, instead of these ion channels, is nevertheless a promising strategy since, differently from these ion channels, it is a relatively simpler family; Na^+ and Ca^{2+} ion channels are characterised by several subtypes with relevant functional differences, which makes off-target effects quite easier [70,71]. Of course another point to be considered is that SEA0400 too, despite its specificity, could exert eventual off-target actions, in particular at higher dosages [60]; but, as stated above, the dosages we selected are expected to be quite selective in the peripheral nervous system. Overall, it can be suggested that our data pointed out that NCX2 is potentially a pivotal element in ADa and, therefore, a druggable target to be further investigated. This paves the way to a new line of research to further explore the intriguing possibility that NCX2 modulation might prevent ADa, relying on highly selective approaches to avoid off-target effects (e.g., NCX2 modulation via siRNA in in vivo OIPN models).

4. Materials and Methods

4.1. In Vitro Experiments

4.1.1. Drugs

OHP was purchased from Sigma Chemical Co. (Milan, Italy). It was dissolved in physiological solution to make a stock solution of 1 mg/mL, which was diluted with medium to obtain the working concentration of 7.5 μM. This concentration has been selected in a similar range to the mean plasma concentrations in patients treated with OHP [64,70] and able to induce a relevant reduction in neurite elongation after exposure. The strong and selective NCX inhibitor SEA0400 [58] was purchased from Sigma Chemical Co. (Milan, Italy). It was dissolved in dimethyl sulfoxide (DMSO) to make a stock solution of 1 mM, which was diluted with medium to obtain the 20 μM concentration; this concentration was selected after *proof-of-concept* experiments testing a scaling dose of SEA0400 (1, 10, 20, and 30 μM) and selecting the most promising dosage in terms of neuroprotection. The inhibition effect of SEA0400 was evaluated by a 3 h pre-treatment of SEA0400 then followed by the OHP exposure. Pre-treatment was mandatory on the basis of our previously published data, demonstrating that neuroprotection against OIPN is obtained modulating ion channels/currents before OHP exposure [25].

4.1.2. Dorsal Root Ganglia (DRG) Explants

DRG from 15-day-old embryonic Sprague Dawley rats were aseptically removed and cultured onto a single layer of rat-tail collagen surfaces in 35 mm dishes (4 ganglia/dish). DRG were incubated in media (MEM plus 10% calf bovine serum, 50 µg/mL ascorbic acid, 1.4 mM L-glutamine, 0.6% glucose) with 5 ng/mL NGF for 2 h in a 5% CO_2 humidified incubator at 37 °C. DRG were pre-treated with SEA0400 (20 µM) and then also exposed to OHP 7.5 µM for 24 h and 48 h. Phase contrast micrographs of all DRG were made after 24 h and 48 h of OHP exposure. For each DRG, the longest neurite was measured. These magnified measurements were compared with a calibration grating photographed under identical conditions. Each experiment was performed three times to validate the results.

4.2. In Vivo Observations

4.2.1. Animal and Housing

Experiments were performed on male balb/c mice weighing 18–20 g on arrival (Envigo, Bresso, Italy). Care and husbandry of animals were in conformity with the institutional guidelines in compliance with national (D.L. n. 26/2014) and international laws and policies (EEC Council Directive 86/609, OJ L 358, 1, Dec. 12, 1987; Guide for the Care and Use of Laboratory Animals, U.S. National Research Council, 1996; be carried out in accordance with the U.K. Animals (Scientific Procedures) Act, 1986 and associated guidelines, EU Directive 2010/63/EU for animal experiments, or the National Institutes of Health guide for the care and use of Laboratory animals (NIH Publications No. 8023, revised 1978). Animals were maintained under standard animal housing conditions, thus with a 12 h light–dark cycle and a room temperature and relative humidity at 20 ± 2 °C and 55 ± 10%, respectively. Drug- and vehicle-treated mice were housed separately with free access to water and food.

4.2.2. Study Design

Animals were divided into 2 groups (n = 12 each) and treated as follows: control (CTRL) group was composed of vehicle-treated animals (5% glucose solution); OHP group was treated with OHP 7 mg/kg i.v., once a week per 8 weeks (1 qw 8 ws). Sample size for each experiment was calculated on the basis of nerve conduction velocity (NCV) reference values of our laboratory [72], assuming that the relevant difference between CTRL and OHP groups is 5 m/s (standard deviation = 7); thus, if a 2-sided 5% alpha and a 80% power is set, the sample size is 7 animals/group (www.dssresearch.com/KnowledgeCenter/toolkitcalculators/samplesizecalculators.aspx (accessed on 12 December 2019)). In each experiment the sample size was increased above the defined number (12 animals/group) in order to have enough animals to be tested at each time point, taking into account the different duration of treatment between experiments and avoiding underpowered statistical analysis in case of animal loss due to treatment. To ensure groups homogeneity, animals were randomised based on nerve conduction studies (NCS) values at baseline. Dynamic test and NCS were performed at baseline to ensure homogeneity among groups and then repeated and the end of treatment to ensure neuropathy induction on all animals/group. Nerve excitability testing was performed at 24, 48, 72 h after the 1st OHP administration to monitor changes in axonal excitability on all animals/group. At the end of the observational period (after the execution of NCS and behavioural tests at the end of treatment) animals were sacrificed and specimen for histological analysis and western blotting were harvested (3 animals/group).

4.2.3. Drug Administration

OHP (Oxaliplatin 5 mg/mL solution, Accord Healthcare Limited, Durham, NC, USA), 7 mg/kg or the vehicle solution (5% glucose solution) was administered i.v. in the tail vein once a week for a period of 8 weeks.

4.2.4. Dynamic Test

The mechanical nociceptive threshold was assessed using a Dynamic Aesthesiometer Test (model 37450, Ugo Basile Biological Instruments, Comerio, Italy), which generated a linearly increasing mechanical force. At each time point, after the acclimatization period, a pointed metallic filament (0.5-mm diameter) was applied to the plantar surface of the hind paw, which exerted a progressively increasing punctuate pressure, reaching up to 15 g within 15 s. The pressure evoking a clear voluntary hind-paw withdrawal response was recorded automatically and taken as representing the mechanical nociceptive threshold index. Results represented the maximal pressure (expressed in grams) tolerated by animals. There was an upper limit cut-off of 15 s, after which the mechanical stimulus was automatically terminated.

4.2.5. Caudal Nerve Morphology and Morphometry

Caudal nerves were isolated and dissected out to avoid stretching. Nerves were immediately fixed by immersion in 3% glutaraldehyde in 0.12 M phosphate buffer solutions pH 7.4, post-fixed in OsO4 and embedded in epoxy resin. Then, 1.5-µm semithin sections were obtained and stained with toluidine blue. Morphometric analysis was performed on a 60x stitched image of a single nerve section per animal (3 animals/group) with Nexcope NE 920 light microscope (TIEsseLab S.r.l., Milan, Italy). The images were acquired in stitching mode using Capture V2.0 Software (Revolutionary Computational Imaging Software, TIEsseLab S.r.l., Milan, Italy) and processed by an automatic image analyser (Image-Pro Plus Software, Immagini e Computer SNC, Milan, Italy). The fiber diameter and the frequency distribution of myelinated fibers were determined, and data were analysed with GraphPad Prism Software (GraphPad Software, San Diego, CA, USA).

4.2.6. IENFD

Hind paws were collected after sacrifice, and 3-mm round shapes were taken and immediately fixed in PLP 2% (paraphormaldehyde-lysine-periodate sodium). Three sections (20 µm thick) were randomly obtained from each footpad and immunostained with rabbit polyclonal antiprotein gene product 9.5 (UCHL1/PGP 9.5 (protein gene product 9.5) Rabbit Polyclonal antibody, Proteintech, Illinois, Rosemont, IL, USA) using a free-floating protocol. The total number of PGP-positive fibers IENFs were counted under a light microscope (Nexcope NE 920 light microscopel TIEsseLab S.r.L., Milan, Italy) at 40X magnification with the assistance of a microscope-mounted video camera. Individual fibers were counted (blind examiner) as they crossed the dermal–epidermal junction, and secondary branching within the epidermis will be excluded. A computerised system measured epidermidis length to calculate the linear density of IENF/mm.

4.2.7. Immunohistochemistry for NCX2

DRG of three animals per group were dissected, fixed in 10% formalin for 3 h at RT and paraffin embedded. Three-µm-thick slices were cut with a Leica RM2265 microtome (Microsystems GmbH, Wetzlar, Germany). Immunohistochemistry was performed using a rabbit polyclonal anti-Na^+/Ca^{2+} Exchanger 2 antibody (NCX-2, TA328916, OriGene, Rockville, MD, USA). Paraffin sections were deparaffinised with xylene, rehydrated, and heated in a steamer for 20 min (1 mM EDTA pH 7.4) to retrieve antigens. Endogenous peroxidase activity was quenched by incubation in 3% H_2O_2 for 10 min at RT. The slides were washed in PBS and incubated in 5% NGS for 1 h at RT. Then, the sections were incubated with anti-NCX-2 antibody (1:200 in 1% NGS) o/n at 4 °C. The following day, the slides were washed and incubated with a biotinylated secondary antibody to rabbit IgG for 1 h at RT (1:100, Vector Laboratories, Peterborough, UK) followed by incubation with streptavidin-conjugated horseradish peroxidase for 1 h at RT (1:100, ABC kit Vectastain, Vector Laboratories, Peterborough, UK). The antigen–antibody complex was visualised by incubating the sections with 3,3-diaminobenzidine hydrochloride (DAB) (Sigma, St. Louis, MO, USA) dissolved in PBS with 10 µL of 3% H_2O_2. Negative controls were incubated only with the secondary antibody. DAB intensity of images was then quantified using ImageJ software and Colour Deconvolution plugin.

4.2.8. Immunofluorescence for NCX2

Three-µm-thick paraffin DRG sections were deparaffinised with xylene and rehydrated and non-specific tissue binding was blocked with 5% NGS and 5% BSA in PBS for 1 h. Samples were then incubated overnight at 4 °C with NCX2 primary rabbit antibody (1:200, OriGene, Rockville, MD, USA) and with a goat anti-rabbit secondary antibody Alexa red 546 (1:200, Invitrogen Waltham, MA, USA) for 1 h at room temperature. Sections were then examined, and stitched images were acquired using an epifluorescence microscope (Cell observer. Zeiss, Germany). Fluorescence intensity of images was then quantified using ImageJ software.

4.2.9. Western Blotting for NCX2

DRG specimens were harvested at sacrifice at end of treatment, and proteins were extracted after chemical and mechanical lysis. Lysis solution was made up of 10% glycerol, 25 mM TrisHCl pH 7.5, 1% Triton X-100, 5 mM EDTA pH 8.0, 1 mM EGTA Ph 8.0, 10 mM sodium orthovanadate, 4 mM PMSF, 1% aprotinin, and 20 mM sodium pyrophosphate. Protein concentration was quantified by Bradford method, and 10 µg were denatured and loaded onto 10% SDS-PAGE. After electrophoresis, proteins were transferred to nitrocellulose membranes and immunoblotting analysis was performed following manufacturer instructions. Briefly, membranes were blocked with 5% non-fat milk blocking solution and then incubated with primary antibodies against NCX2 (1:500, OriGene, Rockville, MD, USA) and beta actin (1:1000, Santa Cruz Biotechnology, Dallas, TX, USA). After incubation, membranes were washed and then incubated with appropriate horseradish peroxidase-conjugated secondary antibodies (anti-rabbit, 1:1000, PerkinElmer, Waltham, MA, USA; anti-goat, 1:2000 Santa Cruz Biotechnology, Dallas, TX, USA). Immunoreactive proteins were visualised using an ECL chemiluminescence system (Amersham, Sullivan County, TE, USA).

4.2.10. NCS

Nerve conduction studies were performed with the electromyography apparatus Matrix Light (Micromed, Mogliano Veneto, Italy) and performed as previously described [72]. Briefly, subdermal needle electrodes were used (Ambu Neuroline (Ambu, Ballerup, Denmark)). Recordings were performed under deep isofluorane anesthesia and animal body temperature was monitored and kept constant at 37 ± 0.5 °C with a thermal pad (Harvard Apparatus, Holliston, MA, USA). Caudal nerve and digital nerves were assessed. Caudal nerve conduction study was performed positioning a pair of recording needle electrodes at the base of the tail (inter-electrode distance: 0.5 cm) and a pair of stimulating needle electrodes (inter-electrode distance: 0.5 cm) was placed 3.5 cm distally with respect to the active recording electrode; the ground electrode was placed 1 cm distally to the active recording electrode. Digital nerve conduction study was obtained positioning the positive recording electrode in front of the patellar bone, the negative recording electrode close to ankle bone, the positive and negative stimulating electrodes close to the fourth toe near the digital nerve and under the paw, respectively; the ground electrode was placed in the sole. Intensity, duration, and frequency of stimulation were set up in order to obtain optimal results. Averaging technique was applied carefully. Recordings filters were kept between 20 Hz and 3 KHz. Sweep was kept at 0.5 ms.

4.2.11. NET

During recordings, mice were under deep isoflurane anaesthesia and body temperature was kept constant at 37 ± 0.5 °C, as stated above. The montage to perform caudal nerve recordings was adapted slightly, modifying the previously devised protocol by Boërio et al. [73]. Briefly, disposable, non-polarizable, subdermal platinum iridium needle electrodes (MedCat Supplies, Klazienaveen, The Netherlands) were used for stimulation; as ground and recording electrodes, subdermal needle electrodes were instead used (Ambu Neuroline (Ambu, Ballerup, Denmark)). The stimulating cathode was placed at the base of the tail and the anode was placed in the base of the rear foot, on the same side of the caudal nerve tested (left). Recording needle electrodes were inserted subcutaneously in the left side of the tail (interelectrode distance: 1 cm), 3.5 cm distally respect to the cathode. Ground electrode was inserted in the right side of the tail, in mid between the cathode and the active recording electrode. As a stimulator, an isolated linear bipolar constant current stimulant (maximal output ±10 mA, DS4, Digitimer, Welwyn Garden City, UK); the Xcell3 Microelectrode Amplifier (FHC, Bowdoin, ME) was connected to the recording electrodes via a customised probe/adapter specifically designed by FHC for our needs; National Instrument USB-6221-BNC Acquisition Device (National Instrument Italy, Assago, Italy) was used to connect all these instruments. For NET recordings, Qtrac© software (Institute of Neurology, Queen Square, London, UK) and TROND protocol were used. For the purposes of this study, Recovery Cycle is the curve of interest. Recovery Cycle was performed using paired pulse stimuli: a supramaximal stimulus followed by a conditioning one at different interstimulus intervals (2–200 ms) [74]. Threshold changes at 2 and 2.5 ms interstimulus intervals were used to determine refractoriness. The minimum mean threshold change of three adjacent points defined superexcitability; instead, subexcitability was determined as the minimum mean threshold change obtained after a 10 ms interstimulus intervals.

4.2.12. Statistical Analyses

Descriptive statistics were generated for all variables. Normally distributed data were analyzed with parametric tests (t-test, one-way Anova followed by Dunnet test) and non-normally distributed with non-parametric tests (Mann–Whitney U-test, Kruskal–Wallis test, followed by pairwise comparison, adjusted by the Bonferrroni correction for multiple tests). Two-sided tests were used. A p-value < 0.05 was set as significant. All analyses were conducted in GraphPad environment (v4.0), apart from NET recordings. The latter were analyzed with QtraS© (Institute of Neurology, Queen Square, London, UK), specifically designed to dialogue with the recording software QtraC© (Institute of Neurology, Queen Square, London, UK).

5. Conclusions

Our journey from the bench side to bed side and vice versa allowed us to pave the way to a new line of research exploiting NCX2 modulation to prevent OIPN-related ADa, highlighting a novel druggable target to cure this condition and, potentially, other conditions in which ADa ensues, as consequences of Na^{+}VOC/NCX family alterations.

Supplementary Materials: The following supporting information can be downloaded at: https://www.mdpi.com/article/10.3390/ijms231710063/s1.

Author Contributions: Conceptualization, P.A.; methodology, P.A., G.C., E.B., and A.M.; formal analysis, P.A., E.B., A.M., E.P., and V.R.-M.; investigation, P.A., E.B., A.M., S.S., E.P., L.M., A.C. (Annalisa Canta), A.C. (Alessia Chiorazzi), V.R.-M., C.M., V.A.C., G.N., A.S., and M.H.; resources, P.A., S.N.H. and G.C.; data curation, P.A., E.B., A.M., and V.R.-M.; writing—original draft preparation, P.A.; writing—review and editing of the first draft, G.C. and S.N.H.; supervision, P.A.; project administration, P.A.; funding acquisition, P.A. and G.C. All authors have read and agreed to the published version of the manuscript.

Funding: PA is recipient of *Premio Giovani Talenti UNIMIB/Accademia Dei Lincei* for year 2020 and 2021. G.C. is recipient of MUR/PRIN 2017ZFJCS3 unrestricted research grant. S.N.H. is recipient of NIH Grants R01CA221363 and R01HD090642 and the Northside Hospital Foundation, Inc.

Institutional Review Board Statement: The study was conducted in compliance with the Declaration of Helsinki, and the procedures were in accordance with the authorizations received by the Italian Ministry of Health (919/2015-PR).

Informed Consent Statement: Not applicable.

Data Availability Statement: Data will be made available upon request to the corresponding author.

Conflicts of Interest: The authors declare no conflict of interest. The funders had no role in the design of the study; in the collection, analyses, or interpretation of data; in the writing of the manuscript; or in the decision to publish the results.

References

1. Alberti, P. Platinum-drugs induced peripheral neurotoxicity: Clinical course and preclinical evidence. *Expert. Opin. Drug Metab. Toxicol.* **2019**, *15*, 487–497. [CrossRef]
2. Briani, C.; Argyriou, A.A.; Izquierdo, C.; Velasco, R.; Campagnolo, M.; Alberti, P.; Frigeni, B.; Cacciavillani, M.; Bergamo, F.; Cortinovis, D.; et al. Long-term course of oxaliplatin-induced polyneuropathy: A prospective 2-year follow-up study. *J. Peripher. Nerv. Syst.* **2014**, *19*, 299–306. [CrossRef]
3. Cavaletti, G.; Alberti, P.; Argyriou, A.A.; Lustberg, M.; Staff, N.P.; Tamburin, S.; Toxic Neuropathy Consortium of the Peripheral Nerve Society. Chemotherapy-induced peripheral neurotoxicity: A multifaceted, still unsolved issue. *J. Peripher. Nerv. Syst.* **2019**, *24*, S6–S12. [CrossRef]
4. Alberti, P.; Cavaletti, G.; Cornblath, D.R. Toxic neuropathies: Chemotherapy-induced peripheral neurotoxicity. *Curr. Opin. Neurol.* **2019**, *35*, 676–683. [CrossRef]
5. Loprinzi, C.L.; Lacchetti, C.; Bleeker, J.; Cavaletti, G.; Chauhan, C.; Hertz, D.L.; Kelley, M.R.; Lavino, A.; Lustberg, M.B.; Paice, J.A.; et al. Prevention and Management of Chemotherapy-Induced Peripheral Neuropathy in Survivors of Adult Cancers: ASCO Guideline Update. *J. Clin. Oncol.* **2020**, *28*, 3325–3348. [CrossRef]
6. Alberti, P.; Lehmann, H.C. Chemotherapy induced peripheral neurotoxicity: Six essential articles for effective future research. *Exp. Neurol.* **2021**, *337*, 113555. [CrossRef]
7. Alberti, P.; Cavaletti, G. Management of side effects in the personalized medicine era: Chemotherapy-induced peripheral neuropathy. *Methods Mol. Biol.* **2014**, *1175*, 301–322. [CrossRef]
8. Velasco, R.; Alberti, P.; Bruna, J.; Psimaras, D.; Argyriou, A.A. Bortezomib and other proteosome inhibitors-induced peripheral neurotoxicity: From pathogenesis to treatment. *J. Peripher. Nerv. Syst.* **2019**, *24*, S52–S62. [CrossRef]
9. Tamburin, S.; Park, S.B.; Alberti, P.; Demichelis, C.; Schenone, A.; Argyriou, A.A. Taxane and epothilone-induced peripheral neurotoxicity: From pathogenesis to treatment. *J. Peripher. Nerv. Syst.* **2019**, *24*, S40–S51. [CrossRef]
10. Islam, B.; Lustberg, M.; Staff, N.P.; Kolb, N.; Alberti, P.; Argyriou, A.A. Vinca alkaloids, thalidomide and eribulin-induced peripheral neurotoxicity: From pathogenesis to treatment. *J. Peripher. Nerv. Syst.* **2019**, *24*, S63–S73. [CrossRef]
11. Staff, N.P.; Cavaletti, G.; Islam, B.; Lustberg, M.; Psimaras, D.; Tamburin, S. Platinum-induced peripheral neurotoxicity: From pathogenesis to treatment. *J. Peripher. Nerv. Syst.* **2019**, *24*, S26–S39. [CrossRef]
12. Lucchetta, M.; Lonardi, S.; Bergamo, F.; Alberti, P.; Velasco, R.; Argyriou, A.A.; Briani, C.; Bruna, J.; Cazzaniga, M.; Cortinovis, D.; et al. Incidence of atypical acute nerve hyperexcitability symptoms in oxaliplatin-treated patients with colorectal cancer. *Cancer Chemother. Pharmacol.* **2012**, *70*, 899–902. [CrossRef]
13. Argyriou, A.A.; Velasco, R.; Briani, C.; Cavaletti, G.; Bruna, J.; Alberti, P.; Cacciavillani, M.; Lonardi, S.; Santos, C.; Cortinovis, D.; et al. Peripheral neurotoxicity of oxaliplatin in combination with 5-fluorouracil (FOLFOX) or capecitabine (XELOX): A prospective evaluation of 150 colorectal cancer patients. *Ann. Oncol.* **2012**, *23*, 3116–3122. [CrossRef]
14. Grolleau, F.; Gamelin, L.; Boisdron-Celle, M.; Lapied, B.; Pelhate, M.; Gamelin, E. A possible explanation for a neurotoxic effect of the anticancer agent oxaliplatin on neuronal voltage-gated sodium channels. *J. Neurophysiol.* **2001**, *85*, 2293–2297. [CrossRef]
15. Park, S.B.; Lin, C.S.; Krishnan, A.V.; Goldstein, D.; Friedlander, M.L.; Kiernan, M.C. Oxaliplatin-induced neurotoxicity: Changes in axonal excitability precede development of neuropathy. *Brain* **2009**, *132*, 2712–2723. [CrossRef]
16. Wu, S.N.; Chen, B.S.; Wu, Y.H.; Peng, H.; Chen, L.T. The mechanism of the actions of oxaliplatin on ion currents and action potentials in differentiated NG108-15 neuronal cells. *Neurotoxicology* **2009**, *30*, 677–685. [CrossRef]
17. Dimitrov, A.G.; Dimitrova, N.A. A possible link of oxaliplatin-induced neuropathy with potassium channel deficit. *Muscle Nerve* **2012**, *45*, 403–411. [CrossRef]
18. Makker, P.G.S.; White, D.; Lees, J.G.; Parmar, J.; Goldstein, D.; Park, S.B.; Howells, J.; Moalem-Taylor, G. Acute changes in nerve excitability following oxaliplatin treatment in mice. *J. Neurophysiol.* **2020**, *124*, 232–244. [CrossRef]
19. Adelsberger, H.; Quasthoff, S.; Grosskreutz, J.; Lepier, A.; Eckel, F.; Lersch, C. The chemotherapeutic oxaliplatin alters voltage-gated Na^+ channel kinetics on rat sensory neurons. *Eur. J. Pharmacol.* **2000**, *406*, 25–32. [CrossRef]

20. Velasco, R.; Bruna, J.; Briani, C.; Argyriou, A.A.; Cavaletti, G.; Alberti, P.; Frigeni, B.; Cacciavillani, M.; Lonardi, S.; Cortinovis, D.; et al. Early predictors of oxaliplatin-induced cumulative neuropathy in colorectal cancer patients. *J. Neurol. Neurosurg. Psychiatry* **2014**, *85*, 392–398. [CrossRef]
21. Argyriou, A.A.; Cavaletti, G.; Antonacopoulou, A.; Genazzani, A.A.; Briani, C.; Bruna, J.; Terrazzino, S.; Velasco, R.; Alberti, P.; Campagnolo, M.; et al. Voltage-gated sodium channel polymorphisms play a pivotal role in the development of oxaliplatin-induced peripheral neurotoxicity: Results from a prospective multicenter study. *Cancer* **2013**, *119*, 3570–3577. [CrossRef] [PubMed]
22. Argyriou, A.A.; Antonacopoulou, A.G.; Alberti, P.; Briani, C.; Bruna, J.; Velasco, R.; Anastopoulou, G.G.; Park, S.B.; Cavaletti, G.; Kalofonos, H.P. Liability of the voltage-gated potassium channel KCNN3 repeat polymorphism to acute oxaliplatin-induced peripheral neurotoxicity. *J. Peripher. Nerv. Syst.* **2019**, *24*, 298–303. [CrossRef] [PubMed]
23. Krishnan, A.V.; Goldstein, D.; Friedlander, M.; Kiernan, M.C. Oxaliplatin and axonal Na^+ channel function in vivo. *Clin. Cancer Res.* **2006**, *12*, 4481–4484. [CrossRef]
24. Heide, R.; Bostock, H.; Ventzel, L.; Grafe, P.; Bergmans, J.; Fuglsang-Frederiksen, A.; Finnerup, N.B.; Tankisi, H. Axonal excitability changes and acute symptoms of oxaliplatin treatment: In vivo evidence for slowed sodium channel inactivation. *Clin. Neurophysiol.* **2018**, *129*, 694–706. [CrossRef] [PubMed]
25. Alberti, P.; Canta, A.; Chiorazzi, A.; Fumagalli, G.; Meregalli, C.; Monza, L.; Pozzi, E.; Ballarini, E.; Rodriguez-Menendez, V.; Oggioni, N.; et al. Topiramate prevents oxaliplatin-related axonal hyperexcitability and oxaliplatin induced peripheral neurotoxicity. *Neuropharmacology* **2020**, *164*, 107905. [CrossRef]
26. Bruna, J.; Alberti, P.; Calls-Cobos, A.; Caillaud, M.; Damaj, M.I.; Navarro, X. Methods for in vivo studies in rodents of chemotherapy induced peripheral neuropathy. *Exp. Neurol.* **2020**, *325*, 113154. [CrossRef]
27. Morris, C.E.; Boucher, P.A.; Joós, B. Left-shifted nav channels in injured bilayer: Primary targets for neuroprotective nav antagonists? *Front. Pharmacol.* **2012**, *3*, 19. [CrossRef]
28. Persson, A.K.; Black, J.A.; Gasser, A.; Cheng, X.; Fischer, T.Z.; Waxman, S.G. Sodium-calcium exchanger and multiple sodium channel isoforms in intra-epidermal nerve terminals. *Mol. Pain* **2010**, *6*, 84. [CrossRef]
29. Cavaletti, G.; Tredici, G.; Petruccioli, M.G.; Dondè, E.; Tredici, P.; Marmiroli, P.; Minoia, C.; Ronchi, A.; Bayssas, M.; Etienne, G.G. Effects of different schedules of oxaliplatin treatment on the peripheral nervous system of the rat. *Eur. J. Cancer* **2001**, *37*, 2457–2463. [CrossRef]
30. Annunziato, L.; Pignataro, G.; Di Renzo, G.F. Pharmacology of brain Na^+/Ca^{2+} exchanger: From molecular biology to therapeutic perspectives. *Pharmacol. Rev.* **2004**, *56*, 633–654. [CrossRef]
31. Annunziato, L.; Pignataro, G.; Boscia, F.; Sirabella, R.; Formisano, L.; Saggese, M.; Cuomo, O.; Gala, R.; Secondo, A.; Viggiano, D.; et al. ncx1, ncx2, and ncx3 gene product expression and function in neuronal anoxia and brain ischemia. *Ann. NY Acad. Sci.* **2007**, *1099*, 413–426. [CrossRef] [PubMed]
32. Blaustein, M.P.; Lederer, W.J. Sodium/calcium exchange: Its physiological implications. *Physiol. Rev.* **1999**, *79*, 763–854. [CrossRef] [PubMed]
33. Persson, A.K.; Kim, I.; Zhao, P.; Estacion, M.; Black, J.A.; Waxman, S.G. Sodium channels contribute to degeneration of dorsal root ganglion neurites induced by mitochondrial dysfunction in an in vitro model of axonal injury. *J. Neurosci.* **2013**, *33*, 19250–19261. [CrossRef]
34. Persson, A.K.; Hoeijmakers, J.G.J.; Estacion, M.; Black, J.A.; Waxman, S.G. Sodium Channels, Mitochondria, and Axonal Degeneration in Peripheral Neuropathy. *Trends Mol. Med.* **2016**, *22*, 377–390. [CrossRef] [PubMed]
35. Neher, E.; Sakaba, T. Multiple roles of calcium ions in the regulation of neurotransmitter release. *Neuron* **2008**, *59*, 861–872. [CrossRef] [PubMed]
36. Marty, A.; Zimmerberg, J. Diffusion into the patch-clamp recording pipette of a factor necessary for muscarinic current response. *Cell Signal.* **1989**, *1*, 259–268. [CrossRef]
37. Malenka, R.C.; Kauer, J.A.; Zucker, R.S.; Nicoll, R.A. Postsynaptic calcium is sufficient for potentiation of hippocampal synaptic transmission. *Science* **1988**, *242*, 81–84. [CrossRef]
38. Szekely, A.M.; Costa, E.; Grayson, D.R. Transcriptional program coordination by N-methyl-D-aspartate-sensitive glutamate receptor stimulation in primary cultures of cerebellar neurons. *Mol. Pharmacol.* **1990**, *38*, 624–633.
39. McCormack, J.G.; Denton, R.M. Intracellular calcium ions and intramitochondrial Ca^{2+} in the regulation of energy metabolism in mammalian tissues. *Proc. Nutr. Soc.* **1990**, *49*, 57–75. [CrossRef]
40. Chalfie, M.; Wolinsky, E. The identification and suppression of inherited neurodegeneration in Caenorhabditis elegans. *Nature* **1990**, *345*, 410–416. [CrossRef]
41. Nikoletopoulou, V.; Tavernarakis, N. Calcium homeostasis in aging neurons. *Front. Genet.* **2012**, *3*, 200. [CrossRef]
42. Stys, P.K.; Waxman, S.G.; Ransom, B.R. Ionic mechanisms of anoxic injury in mammalian CNS white matter: Role of Na^+ channels and Na^+-Ca^{2+} exchanger. *J. Neurosci.* **1992**, *12*, 430–439. [CrossRef]
43. Lehning, E.J.; Doshi, R.; Isaksson, N.; Stys, P.K.; LoPachin, R.M. Mechanisms of injury-induced calcium entry into peripheral nerve myelinated axons: Role of reverse sodium-calcium exchange. *J. Neurochem.* **1996**, *66*, 493–500. [CrossRef]
44. Lehning, E.J.; Doshi, R.; Stys, P.K.; LoPachin, R.M. Mechanisms of injury-induced calcium entry into peripheral nerve myelinated axons: In vitro anoxia and ouabain exposure. *Brain Res.* **1995**, *694*, 158–166. [CrossRef]

45. Pannaccione, A.; Piccialli, I.; Secondo, A.; Ciccone, R.; Molinaro, P.; Boscia, F.; Annunziato, L. The Na. *Cell Calcium* **2020**, *87*, 102190. [CrossRef]
46. Anzilotti, S.; Valsecchi, V.; Brancaccio, P.; Guida, N.; Laudati, G.; Tedeschi, V.; Petrozziello, T.; Frecentese, F.; Magli, E.; Hassler, B.; et al. Prolonged NCX activation prevents SOD1 accumulation, reduces neuroinflammation, ameliorates motor behavior and prolongs survival in a ALS mouse model. *Neurobiol. Dis.* **2021**, *159*, 105480. [CrossRef]
47. Scuteri, A.; Ravasi, M.; Monfrini, M.; Milano, A.; D'Amico, G.; Miloso, M.; Tredici, G. Human Mesenchymal Stem Cells Protect Dorsal Root Ganglia from the Neurotoxic Effect of Cisplatin. *Anticancer. Res.* **2015**, *35*, 5383–5389.
48. Amran, M.S.; Hashimoto, K.; Homma, N. Effects of sodium-calcium exchange inhibitors, KB-R7943 and SEA0400, on aconitine-induced arrhythmias in guinea pigs in vivo, in vitro, and in computer simulation studies. *J. Pharmacol. Exp. Ther.* **2004**, *310*, 83–89. [CrossRef]
49. Alberti, P. Chemotherapy-induced peripheral neurotoxicity - outcome measures: The issue. *Expert. Opin. Drug. Metab. Toxicol.* **2017**, *13*, 241–243. [CrossRef]
50. Boscia, F.; Gala, R.; Pignataro, G.; de Bartolomeis, A.; Cicale, M.; Ambesi-Impiombato, A.; Di Renzo, G.; Annunziato, L. Permanent focal brain ischemia induces isoform-dependent changes in the pattern of Na^+/Ca^{2+} exchanger gene expression in the ischemic core, periinfarct area, and intact brain regions. *J. Cereb. Blood Flow Metab.* **2006**, *26*, 502–517. [CrossRef]
51. Argyriou, A.A.; Cavaletti, G.; Briani, C.; Velasco, R.; Bruna, J.; Campagnolo, M.; Alberti, P.; Bergamo, F.; Cortinovis, D.; Cazzaniga, M.; et al. Clinical pattern and associations of oxaliplatin acute neurotoxicity: A prospective study in 170 patients with colorectal cancer. *Cancer* **2013**, *119*, 438–444. [CrossRef]
52. Alberti, P.; Rossi, E.; Argyriou, A.A.; Kalofonos, H.P.; Briani, C.; Cacciavillani, M.; Campagnolo, M.; Bruna, J.; Velasco, R.; Cazzaniga, M.E.; et al. Risk stratification of oxaliplatin induced peripheral neurotoxicity applying electrophysiological testing of dorsal sural nerve. *Support. Care Cancer* **2018**, *26*, 3143–3151. [CrossRef]
53. Grisold, W.; Cavaletti, G.; Windebank, A.J. Peripheral neuropathies from chemotherapeutics and targeted agents: Diagnosis, treatment, and prevention. *Neuro. Oncol.* **2012**, *14*, iv45–iv54. [CrossRef]
54. Boyette-Davis, J.; Dougherty, P.M. Protection against oxaliplatin-induced mechanical hyperalgesia and intraepidermal nerve fiber loss by minocycline. *Exp. Neurol.* **2011**, *229*, 353–357. [CrossRef]
55. Koskinen, M.J.; Kautio, A.L.; Haanpää, M.L.; Haapasalo, H.K.; Kellokumpu-Lehtinen, P.L.; Saarto, T.; Hietaharju, A.J. Intraepidermal nerve fibre density in cancer patients receiving adjuvant chemotherapy. *Anticancer Res.* **2011**, *31*, 4413–4416.
56. Brenneman, D.E.; Kinney, W.A.; Ward, S.J. Knockdown siRNA Targeting the Mitochondrial Sodium-Calcium Exchanger-1 Inhibits the Protective Effects of Two Cannabinoids Against Acute Paclitaxel Toxicity. *J. Mol. Neurosci.* **2019**, *68*, 603–619. [CrossRef]
57. Yilmaz, E.; Gold, M.S. Paclitaxel-induced increase in NCX activity in subpopulations of nociceptive afferents: A protective mechanism against chemotherapy-induced peripheral neuropathy? *Cell Calcium* **2016**, *60*, 25–31. [CrossRef]
58. Matsuda, T.; Arakawa, N.; Takuma, K.; Kishida, Y.; Kawasaki, Y.; Sakaue, M.; Takahashi, K.; Takahashi, T.; Suzuki, T.; Ota, T.; et al. SEA0400, a novel and selective inhibitor of the Na^+-Ca^{2+} exchanger, attenuates reperfusion injury in the in vitro and in vivo cerebral ischemic models. *J. Pharmacol. Exp. Ther.* **2001**, *298*, 249–256.
59. Arakawa, N.; Sakaue, M.; Yokoyama, I.; Hashimoto, H.; Koyama, Y.; Baba, A.; Matsuda, T. KB-R7943 inhibits store-operated Ca^{2+} entry in cultured neurons and astrocytes. *Biochem. Biophys. Res. Commun.* **2000**, *279*, 354–357. [CrossRef]
60. Lee, C.; Hryshko, L.V. SEA0400: A novel sodium-calcium exchange inhibitor with cardioprotective properties. *Cardiovasc. Drug. Rev.* **2004**, *22*, 334–347. [CrossRef]
61. Koyama, Y.; Matsui, S.; Itoh, S.; Osakada, M.; Baba, A.; Matsuda, T. The selective Na^+-Ca^{2+} exchange inhibitor attenuates brain edema after radiofrequency lesion in rats. *Eur. J. Pharmacol.* **2004**, *489*, 193–196. [CrossRef] [PubMed]
62. Shenoda, B. The role of Na^+/Ca^{2+} exchanger subtypes in neuronal ischemic injury. *Transl. Stroke. Res.* **2015**, *6*, 181–190. [CrossRef] [PubMed]
63. Beaugé, L.; DiPolo, R. SEA-0400, a potent inhibitor of the Na^+/Ca^{2+} exchanger, as a tool to study exchanger ionic and metabolic regulation. *Am. J. Physiol. Cell Physiol.* **2005**, *288*, C1374–C1380. [CrossRef]
64. Nashida, T.; Takuma, K.; Fukuda, S.; Kawasaki, T.; Takahashi, T.; Baba, A.; Ago, Y.; Matsuda, T. The specific Na^+/Ca^{2+} exchange inhibitor SEA0400 prevents nitric oxide-induced cytotoxicity in SH-SY5Y cells. *Neurochem. Int.* **2011**, *59*, 51–58. [CrossRef]
65. Calls, A.; Carozzi, V.; Navarro, X.; Monza, L.; Bruna, J. Pathogenesis of platinum-induced peripheral neurotoxicity: Insights from preclinical studies. *Exp. Neurol.* **2020**, *325*, 113141. [CrossRef] [PubMed]
66. Graham, M.A.; Lockwood, G.F.; Greenslade, D.; Brienza, S.; Bayssas, M.; Gamelin, E. Clinical pharmacokinetics of oxaliplatin: A critical review. *Clin. Cancer Res.* **2000**, *6*, 1205–1218. [PubMed]
67. Lersch, C.; Schmelz, R.; Eckel, F.; Erdmann, J.; Mayr, M.; Schulte-Frohlinde, E.; Quasthoff, S.; Grosskreutz, J.; Adelsberger, H. Prevention of oxaliplatin-induced peripheral sensory neuropathy by carbamazepine in patients with advanced colorectal cancer. *Clin. Colorectal. Cancer* **2002**, *2*, 54–58. [CrossRef]
68. Cavaletti, G.; Nicolini, G.; Marmiroli, P. Neurotoxic effects of antineoplastic drugs: The lesson of pre-clinical studies. *Front. Biosci.* **2008**, *13*, 3506–3524. [CrossRef]
69. Soliman, D.; Wang, L.; Hamming, K.S.; Yang, W.; Fatehi, M.; Carter, C.C.; Clanachan, A.S.; Light, P.E. Late sodium current inhibition alone with ranolazine is sufficient to reduce ischemia- and cardiac glycoside-induced calcium overload and contractile dysfunction mediated by reverse-mode sodium/calcium exchange. *J. Pharmacol. Exp. Ther.* **2012**, *343*, 325–332. [CrossRef]

70. Catterall, W.A.; Swanson, T.M. Structural Basis for Pharmacology of Voltage-Gated Sodium and Calcium Channels. *Mol. Pharmacol.* **2015**, *88*, 141–150. [CrossRef]
71. de Lera Ruiz, M.; Kraus, R.L. Voltage-Gated Sodium Channels: Structure, Function, Pharmacology, and Clinical Indications. *J. Med. Chem.* **2015**, *58*, 7093–7118. [CrossRef] [PubMed]
72. Monza, L.; Fumagalli, G.; Chiorazzi, A.; Alberti, P. Addressing the Need of a Translational Approach in Peripheral Neuropathy Research: Morphology Meets Function. *Brain Sci.* **2021**, *11*, 139. [CrossRef] [PubMed]
73. Boërio, D.; Greensmith, L.; Bostock, H. Excitability properties of motor axons in the maturing mouse. *J. Peripher. Nerv. Syst.* **2009**, *14*, 45–53. [CrossRef] [PubMed]
74. Kiernan, M.C.; Bostock, H. Effects of membrane polarization and ischaemia on the excitability properties of human motor axons. *Brain* **2000**, *123 Pt 12*, 2542–2551. [CrossRef]

International Journal of
Molecular Sciences

MDPI

Review

Ion Channel and Transporter Involvement in Chemotherapy-Induced Peripheral Neurotoxicity

Eleonora Pozzi [1,†], Giulia Terribile [2,†], Laura Cherchi [1], Sara Di Girolamo [1], Giulio Sancini [2] and Paola Alberti [1,3,*]

1 Experimental Neurology Unit, School of Medicine and Surgery, University of Milano-Bicocca, 20900 Monza, Italy; eleonora.pozzi@unimib.it (E.P.); l.cherchi1@campus.unimib.it (L.C.); sara.digirolamo@unimib.it (S.D.G.)
2 Human Physiology Unit, School of Medicine and Surgery, University of Milano-Bicocca, 20900 Monza, Italy; giulia.terribile@unimib.it (G.T.); giulio.sancini@unimib.it (G.S.)
3 Fondazione IRCCS San Gerardo dei Tintori, 20900 Monza, Italy
* Correspondence: paola.alberti@unimib.it; Tel.: +39-02-6448-8154
† These authors contributed equally to this work.

Abstract: The peripheral nervous system can encounter alterations due to exposure to some of the most commonly used anticancer drugs (platinum drugs, taxanes, vinca alkaloids, proteasome inhibitors, thalidomide), the so-called chemotherapy-induced peripheral neurotoxicity (CIPN). CIPN can be long-lasting or even permanent, and it is detrimental for the quality of life of cancer survivors, being associated with persistent disturbances such as sensory loss and neuropathic pain at limb extremities due to a mostly sensory axonal polyneuropathy/neuronopathy. In the state of the art, there is no efficacious preventive/curative treatment for this condition. Among the reasons for this unmet clinical and scientific need, there is an uncomplete knowledge of the pathogenetic mechanisms. Ion channels and transporters are pivotal elements in both the central and peripheral nervous system, and there is a growing body of literature suggesting that they might play a role in CIPN development. In this review, we first describe the biophysical properties of these targets and then report existing data for the involvement of ion channels and transporters in CIPN, thus paving the way for new approaches/druggable targets to cure and/or prevent CIPN.

Keywords: chemotherapy-induced peripheral neurotoxicity; chemotherapy-induced peripheral neuropathy; neuropathic pain; ion channels; neuropathy; NCX; sodium voltage-operated channels; axonal damage; potassium channels

1. Introduction

Chemotherapy-induced peripheral neurotoxicity (CIPN) is a relevant, potentially persistent adverse event of the most commonly used drugs in cancer treatment: platinum drugs, taxanes, vinca alkaloids, proteasome inhibitors and thalidomide [1]. CIPN features are mainly ones of sensory polyneuropathy, even if motor and autonomic impairment is also reported [1]. Sensory disturbances at limb extremities consist of paresthesia/dysesthesia and neuropathic pain at limb extremities and a sensory loss that can be so pronounced that it impairs fine manipulation and gait [2].

In the state of the art, there is no curative or symptomatic treatment for this condition, even though a moderate effect of duloxetine as a symptomatic treatment was demonstrated [3]. An incomplete knowledge of CIPN mechanisms is one of the main reasons for this unmet clinical need. In the last few years, (an) intriguing target(s) emerged as potentially relevant for preventing/treating CIPN: modulation of ion channels/transporters. In this review, we provide first a description of the potential target of interest, and we report studies describing their potential involvement in CIPN.

2. Biophysical Properties of Transporters/Ion Channels of Interest

In Figure 1, the general features of each class of transporters/ion channels addressed are presented.

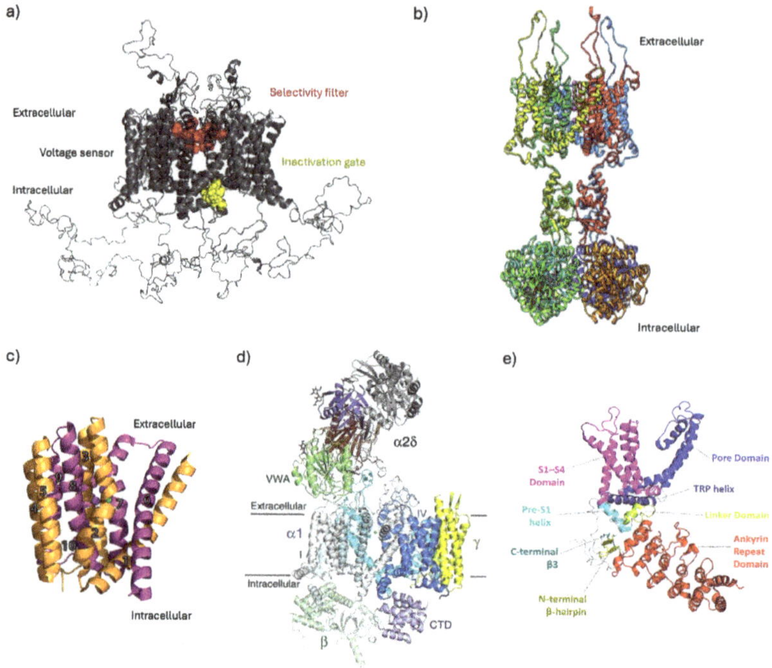

Figure 1. Channels' 3D structures. (**a**) 3D structure of voltage-gated Na^+ 1-type channel (Homo sapiens). The scheme includes all domains: the selectivity filter (red), the inactivation gate (yellow) and the voltage sensor (gray). Modified from Romanova et al., 2022 [4]. (**b**) 3D structure of voltage-gated K^+ 1-type channel. View from the membrane (side view). Two of four domains are shown, with front and back domains deleted to allow the structure to be seen. Purple spheres represent K^+ ions. Modified from Kariev et al., 2024 [5]. (**c**) 3D structure of sodium–calcium exchanger (Methanococcus jannaschii). Helices 1–5 (TM1–5) are orange, and helices 6–10 (TM6–10) are purple. Purple and green spheres represent Na^+ and Ca^{2+} ions, respectively. Modified from Giladi et al., 2016 [6]. (**d**) Three-dimensional structure of high-voltage-activated Ca^{2+} channel (rabbit). The structure model is color-coded for distinct subunits. Green sphere represents Ca^{2+} ion. Modified from Mochida et al., 2018 [7]. (**e**) 3D structure of TRPV3 (mouse). Side view of the tetramer protein. The structure model is color-coded for distinct subunits. Modified from Kalinovskii et al., 2023 [8].

2.1. Voltage-Gated Na^+ Channels (VGSCs)

Voltage-gated Sodium (Na^+) channels (VGSCs) play a strategical role in pain mechanisms as they are involved in the cellular excitability and, not only in the generation, but also in the propagation, of action potential (AP). AP is the main unit for conducting information, including painful stimuli from the peripheral receptors to the higher centers [9]. VGSCs are considered key determinants of nociceptor excitability [10]. Clinical and experimental data suggest that changes in VGSC expression [11], trafficking or kinetic properties [12] could play a key role in the pathogenesis of neuropathic pain. Indeed, they have been studied to investigate the aetiology of pathological pain sensation, and they have been considered important pharmacological targets [13].

2.1.1. Topology and Gating

VGSCs are heteromeric transmembrane complexes consisting of a principal highly glycosylated [14] pore-forming α subunit (260 kDa), sufficient on its own to form a functional ion-conducting voltage-gated channel, and an additional β subunit (33–45 kDa) involved (one or more) in the modulation of channel gating. The α subunit is constituted by four homologous (with amino acid homology > 75% [14]) domains (D1–D4) connected to each other by three intracellular loops, respectively (L1–L3). Each domain consists of six transmembrane-spanning hydrophobic α-helical segments (S1–S6), and each segment is connected by two intracellular (between S2–S3 and S4–S5) and three extracellular linkers (between S1–S2, S3–S4 and S5–S6; the last ones are called P-loops). Both the N-terminus and C-terminus are intracellular [9].

Among segments, S4 is highly conserved and acts in each domain as a voltage sensor for the channel. This region contains a net electrical charge, called a gating charge, due to the presence of basic amino acids (positively charged, i.e., lysine and arginine). The movement of these positive charges through the electrical field, in response to a change in the membrane potential, imparts a change in free energy, needed for the transition of the channel between its functional states (called gating) [15]. Indeed, the channel can assume three different functional states: (i) closed and activatable (resting), (ii) open (active) and (iii) closed and non-activatable (inactivated or refractory) [13]. VGSCs stay in the closed state under membrane resting potentials, they open on depolarization and they close rapidly (fast inactivation) on repolarization, or more slowly (slow inactivation) on prolonged depolarization. This process leaves the channel refractory for some time after repolarization, and it has to recover back to the closed state to open again on the next depolarization. Recovery after inactivation is called repriming [16]. This inactivation is known as open-state inactivation. There is another type of inactivation, known as closed-state inactivation, that occurs when a channel can inactivate directly from a closed state before it opens. The two types of inactivation are not mutually exclusive, although many channels inactivate mainly through one of these [17]. Some VGSCs (i.e., most of the VGSCs of Purkinje cells) open with depolarization, but then, rather than inactivating, they become blocked due to an open channel blocker. In this case, blocked channels reopen upon repolarization as the blocker unbinds, producing the so-called resurgent current [18].

Although resurgent current flows through the same channels as transient (in the voltage clamp: brief macroscopic Na^+ current evoked by a depolarization step) and persistent current (the proportionately tiny, residual Na^+ current that lasts throughout the step), it shows different voltage dependence and kinetics. It is dynamically gating (differently from persistent current), and, at any given potential, it has a rising phase, a peak and a falling phase [18].

According to the molecular mechanism underlying the gating properties of VGSCs, on depolarization, positive charges within S4 move outward along a spiral path, initiating a morphological change that opens the pore, allowing the inward passage of Na^+ (rising phase of the AP). The movement toward the extracellular side is accompanied by the movement of the linker L3 (DIII–DIV), which acts as an "inactivation gate". This movement is probably due to a reduction in electrostatic repulsion between the charges in S4 and those in the linker, and it allows the process known as fast inactivation (on a timescale of ms) of the channel (involved in the termination of AP and in the regulation of the refractory period). On sustained depolarization, VGSCs undergo a process known as slow inactivation (on a timescale of s), involved in the regulation of membrane excitability by increasing the AP threshold and limiting their burst duration and their propagation within dendrites. After the refractory period, the recovery occurs when S4 goes back into the membrane, and the inactivation gate moves away from the pore [19].

According to the molecular mechanism underlying the Na^+ influx upon repolarization, it has been suggested that endogenous factors, functioning as an open channel blocking particles (i.e., small intracellular peptides), can enter the channel while it is open. Upon repolarization, the blocker is expelled due to its positive charge, giving rise to resurgent currents [20].

Regarding the channel structure, among the linkers of the four domains, the most important are L3 (DIII–DIV) and L1 (DI–II). The first one contains three hydrophobic residues (isoleucine, phenylalanine and methionine) that are involved in the fast inactivation of the channel, and it is phosphorylated by protein kinase C (PKC); the second one is phosphorylated by protein kinase A (PKA) [21]. The phosphorylation may be a key element in the modulation of channel activity since PKC and PKA are involved in neuropathic and inflammatory pain, respectively [22].

Another important part of the receptor is the portion that connects the segments S5 and S6 (called P-loops), which contains four conserved amino acids, aspartate, glutamate, lysine and alanine (DEKA motif), that represent the selective filter which determines the channel's Na^+ permeability over the other cations [23].

2.1.2. Classification and Isoforms Involved in Neuropathic Pain

There are nine functionally characterized α isoforms (Nav 1.1–1.9) and four β isoforms (β_1–β_4). The different isoforms are characterized by a different level of homology, as discussed in detail by Isom, an aspect that should be carefully weighted since [24] these isoforms display different kinetics and voltage-dependent properties, and they are associated with auxiliary protein that regulates channel trafficking and gating. These characteristics allow a cell-type-specific modulation of the channel [25].

In the absence of subtype-selective Na^+ channel blockers, the isoforms are classified according to their sensitivity to tetrodotoxin (TTX) in TTX-sensitive ones (Na_V 1.1–Na_V 1.4, Nav 1.6 and Nav 1.7) that have an IC50 in the nanomolar range and TTX-resistant ones (Nav 1.5, Nav 1.8, Nav 1.9) that have an IC50 in the micromolar range [14].

Among the nine functionally expressed isoforms, four are considered to have a key role in the pathogenesis of neuropathic pain: Nav 1.3, Nav 1.7, Nav 1.8 and Nav 1.9 [13,25].

Nav 1.3

The Nav 1.3 isoform is encoded by the SCN3A gene, and it is a TTX-sensitive isoform (Kd = 1.8–4 nM) [25]. It is highly expressed in the central nervous system (CNS) [9]. This isoform produces a fast inactivation and activation current, and it is characterized by rapid repriming. It shows slow closed-state inactivation that leads to a large ramp current in response to small, slow depolarization. The regulation is cell dependent [26]. Nav 1.3 is upregulated in several pain disorders, and the resulting hyperexcitability may explain its involvement; the fast kinetic that characterizes this channel supports its role in allowing peripheral nerves to fire at high frequencies [27,28].

Nav 1.7

The Nav 1.7 isoform is encoded by the SCN9A gene, and it is a TTX-sensitive isoform (Kd = 4.3–25 nM) [25]. It is preferentially expressed in peripheral neurons and normally highly expressed in small-diameter dorsal root ganglion (DRG) neurons with unmyelinated and slow conduction axons (C-fibers) [13]. This channel is involved in the regulation of sensory neurons' excitability, and it is one of the main contributors to human pain disorders [29].

It is characterized by a fast activation and inactivation but slow repriming; these biophysical properties make it well suited for low-frequency firing in C-fibers. Moreover, Nav 1.7 is characterized by slow closed-state inactivation, a mechanism that allows the channel to produce a large ramp current in response to small and slow depolarizations [30]. The ability of this isoform to boost subthreshold stimuli increases the probability that neurons can reach their threshold for firing APs. Based on these characteristics, Nav 1.7 is thought to act as a threshold channel [31]. This isoform produces resurgent currents in a subset of DRG neurons (their production crucially depends on cell background). These currents are triggered by repolarization following a strong depolarization and support burst firing in, for example, cerebellar Purkinje neurons [32]. Nav 1.7 channels might also regulate neurotransmitter release at the nociceptors' central terminals [33].

The key role of the Nav 1.7 isoform in pain mechanisms is supported by the evidence that different mutations affecting the SCN9A gene can modify its biophysical properties in a pro-excitatory manner compared to the wild-type channel by (a) causing a hyperpolarization shift in activation, allowing the channel to open after weaker depolarization [34]; (b) causing wider ramp currents after the same small, slow depolarization [35]; (c) impairing slow activation with the result of increased firing rate [36]; (d) causing a depolarizing shift in fast inactivation, resulting in fewer inactivated channels at any given potential and a persistent current [37]; and (e) causing an increase in resurgent currents [37]. Moreover, some mutations cause a decrease in the single AP threshold and an increase in the firing frequency in small DRG [38].

Nav 1.8

The Na_V 1.8 isoform is encoded by the SCN10A gene, and it is a TTX-resistant isoform (Kd = 40–60 μM) [25]. It could represent an ideal therapeutic target because it is selectively expressed in sensory neurons and mostly in small-diameter DRG neurons [13]. This isoform is characterized by a slow activation and inactivation [13] but rapid repriming [25] and depolarized voltage dependency for activation and inactivation. It produces the majority of the Na^+ current during the AP depolarizing phase in neurons in which it is expressed [39]. It can support repetitive firing in response to depolarizing input [10]. All these biophysical properties and its localization in free nerve endings suggest that the Nav 1.8 isoform may have an important role in nociceptor excitability [29] and in nociceptive information transmission [10].

Nav 1.9

The Nav 1.9 isoform is encoded by the SCN11A gene, and it is a TTX-resistant isoform (Kd = 40 μM) [25]. It is selectively and highly expressed in small-diameter DRG neurons with unmyelinated and slow conduction axons (C-fibers) [13], but it is downregulated in injured neurons [40]. This isoform is characterized by hyperpolarized voltage dependency of activation, close to the resting membrane potentials of neurons (−60/−70 mV), and it is characterized by ultraslow inactivation [30]. These properties allow it to produce a persistent Na^+ current since its activation and inactivation curves allow it to be activated at resting potentials [41]. A downregulation in its expression and the consequent associated decrease in the persistent current could lead to more hyperpolarized membrane potentials, allowing recovery of TTX-sensitive Na^+ channels from inactivation [40].

2.2. *Voltage-Gated K^+ Channels (Kv)*

Since the first extraordinary landmark studies by Hodgkin and Huxley using the patch clamp technique, knowledge about voltage-dependent potassium channels is becoming ever deeper. K^+ channels are probably the largest and most diverse family of ion channels [42–44], represented by more than 80 known loci, which encode multiple pore-forming subunits in the mammalian genome. To account for the expansion of identified K^+ channel genes, a parallel—KCN—nomenclature was initiated by the Human Genome Organization (HUGO) [45], which complements the standardized—Kv—nomenclature.

2.2.1. Topology and Gating

Among K^+ channels, the most important contributors to neuronal excitability are the voltage-dependent Kv channels that regulate resting membrane potential, membrane repolarization, action potential shape, firing frequency and adaptation in both the central and the peripheral nervous system (CNS and PNS, respectively) [42,46].

The Kv channels are tetramers of α subunits, each with S1–S6 α-helical transmembrane segments. The S4 segment is the voltage sensor that allows the opening of the channel at a precise membrane potential (high- or low-threshold kinetics) depending on the specific channel type. This leads to the activation of an outward K^+ ion current across the cell membrane via the S5–S6 pore (P-region) [47–50]. There are approximatively 40 mammalian genes encoding α subunits, which are divided into 12 families (Kv1–12) [43,47]. Different genes within a family are denoted with an additional number after the decimal point, such as Kv1.1 and Kv1.2, roughly in order of their molecular characterization [43]. Both the N- and C-termini are located intracellularly, and often link auxiliary citoplasmatic β subunits (Kvβ) that further modify the gating properties, the neuronal distribution and the functions of Kv channels [42,45,48]. An important example is represented by the Kvβ2 subunit, which strongly modifies the K1.1 and K1.5 channels' activation threshold both directly and indirectly [43,48]. Moreover, the β subunit of Kvβ2.1 allows the distribution of these channels in the juxtaparanodal zones adjacent to nodes of Ranvier in large-diameter myelinated axons in both the CNS and the PNS [46,51].

It is now widely known that Kv channels form an exceedingly diverse group, much more so than one would predict simply based on the number of distinct genes that encode them. This diversity arises from several factors [44]. The main one is the ability to form both homo- and hetero-tetramers between different subunits within the same family. Each subunit's composition could dictate distinct biophysical and functional properties, different interactions with second messengers, variable spatial and temporal expression and diverse regulation by pathophysiological processes [42,44,52]. Furthermore, the presence of a variety of modulatory partners can also critically modify Kv functions. Notably, members of the Kv5, Kv6, Kv8 and Kv9 families encode 10 "silent" subunits (KvS), which do not form conducting channels on their own but co-assemble with other Kv subunits with significant physiological consequences [44,52]. An important example is the association of Kv9.1 or Kv9.3 with Kv2.1, which leads to increased currents compared with Kv2.1 homomers. Moreover, the interplay of Kv2.1/Kv9.1 in A-fiber neurons allows the channel involvement in pain signaling and a direct participation in nociceptive pathways [52,53].

2.2.2. Channel Kinetics

Inactivating K^+ Channels

Among many, one reliable way of distinguishing the huge variety of K^+ voltage-dependent channels is by current kinetics and, in particular, the presence or absence of inactivation. Regarding the inactivating group, many neurons express two main classes of currents: slow inactivating, commonly called D-current (I_D), and fast inactivating or A-current (I_A) [54–56]. I_D has been reported to differ from I_A in so far as it shows slower inactivation rates and enhanced sensitivity to K^+ channel selective blockers [56,57]. Moreover, I_D has been identified in different types of neurons with a fast-conducting axon. Here, it provides a secure conduction, synchronizing rapid synaptic inputs and facilitating rapid membrane recovery [56,58].

Like I_D, I_A is activated transiently at a low-threshold level but inactivated rapidly during both large and small depolarizations from rest. Thus, I_A has a fine mechanism that modulates its amplitude even with small voltage changes around the resting potential [55,59]. I_A is mediated primarily by Kv4 family α subunits, which are widely expressed in both the CNS and PNS. In particular, Kv4 mainly localizes in cell bodies and dendrites of cortical and hippocampal pyramidal neurons, in small nociceptive type neurons and in the larger mechanoreceptor type [56,60]. This specific neuronal distribution allows I_A not only to regulate the integration and the propagation of the excitatory synaptic potentials, but also to modulate the back-propagating potentials in dendritic branches [60,61]. Among the five A-channels in mammals, recently, several authors have focused their attention on two specific subunits: Kv4.2 and Kv4.3 [43,55,62,63]. The latter is often associated with different β subunits and multiple K^+ channel interacting proteins, such as KChIPs, which can modulate Kv4 current properties, neuronal trafficking and, in turn, its functions [63,64]. Kv4.2 subunits are specifically localized in the dendritic membrane, fine-tuning the back-propagation of signals and dendritic excitability, whereas Kv4.3 resides principally on the soma [43,62]. Highlighting Kv4's fundamental functions, different defects in trafficking, expression or kinetics are observed in several disorders, such as memory deficits or peripheral neuropathic pain [55,57,62].

Delayed Rectifier Currents

Regarding the non-inactivating K^+ voltage-dependent channels, the landmarks of this group are the channels characterized by a delayed rectifier current (K_{DR}). The "classical" K_{DR} shows fast-activating kinetics at a low-threshold level and is mediated by Kv1/KCNA family α subunits [42,65]. In particular, Kv1.1 activates rapidly upon small membrane depolarizations, while Kv1.2 requires stronger ones [66,67]. These latter are predominantly localized at the axon initial segment (AIS), the presynaptic terminal sites and juxtaparanodal regions of the nodes of Ranvier of medium and large myelinated axons [46,47,68,69]. Consistent with their neuronal distribution and kinetics, Kv1.1 and Kv1.2 have a pivotal role not only in limiting the action potential generation and propagation, but also in modulating the shape and the rate of action potentials and the neurotransmitter release [43,60,70]. Kv1.1 and Kv1.2 often combine with β subunits, such as Kvβ2, to form functional heteromeric K^+ channels, whose trafficking and functions depend on the association with other accessory proteins [60,65,70]. Notably, some proteins with great relevance are the cell adhesion molecules Caspr2 and TAG-1, the cytoskeletal scaffold 4.1B and multiple members of the ADAM family that allow a different distribution along the axon areas [70,71].

In addition to the "classical" K_{DR}, the M-type current (I_M) has a certain relevance too. I_M shows slow-activating kinetics at a low-threshold level and is principally mediated by Kv7/KCNQ family α subunits [67,72,73]. Like the Kv1.1 subunit, K7 activation requires only small depolarizations, but it can even be activated at the resting potential. Moreover, Kv7 channel activity can be modulated by G-proteins associated with Muscarinic acetylcholine receptors, hence the origin of the I_M name [73,74]. Kv7 kinetics mirror its functions as it aims to maintain the resting potential and reduce neuronal excitability [43,52,72]. Consistent with their fundamental role, Kv7 subunits are selectively localized to the AIS and the nodes of Ranvier, where homomers of the Kv7.2 subunit or heteromers with Kv7.3 exclusively form functional channels [42,60,67]. As well as in the CNS, Kv7.2 and Kv7.3 are widely present in the PNS, where they play an important role in the nociceptive pathway. Indeed, an alteration of their currents may cause a strong neuronal hyperexcitability, leading to different physiopathologies such as epilepsy, peripheral sensitization and neuropathic pain [52,73,75,76].

2.3. Sodium–Calcium Exchanger (NCX) Family

The NCX exchangers (Ca^{2+}/Na^+) are members of a much larger family of transport proteins, the CaCA (Ca^{2+}/cation antiporter) superfamily, which play a hallmark role in controlling the Ca^{2+} flux across the plasma membrane or between intracellular compartments [77,78]. The CaCA proteins possess a conserved sequence and share a similar topology that has been extensively studied in many organisms [79–81]. In particular, mammals express three different SLC8 (A1–A3) genes, which, respectively, encode three NCX (1–3) exchangers that share about 70% overall amino acid identity [77,81].

The NCX family has a general topology composed of 10 transmembrane domain segments (TMS) and a large intracellular regulatory f-loop between TMS5 and TMS6. The latter contains two calcium-binding domains (CBD1-2) and an XIP domain. CBD1-2 are regulatory domains required for intracellular ion sensing and binding; otherwise, the XIP domain consists of a small auto-inhibitory sequence that confers Na^+-dependent inactivation. Moreover, there are two highly conserved repeats, α-1 and α-2, localized between TMS2-TMS3 and TMS7-TMS8, respectively, which form the ion transport regions [79,82–84]. At the post-transcriptional level, only NCX1 and NCX3 undergo alternative splicing of the primary nuclear SLC8A1 and SLC8A3 transcripts, and each variant exhibits distinct properties for Ca^{2+} sensing and fluxes [77,79].

NCX is a low-affinity high-capacity transporter that shows an electrogenic coupling ratio of 1:3 Ca^{2+}/Na^+ ions [85]. In particular, NCX mediates Ca^{2+} extrusion by combining the latter with the influx of Na^+ ions, depending on the electrochemical gradient of each one under physiological conditions [85,86]. In addition to the forward mode just described, in some cases, NCX can contribute to Ca^{2+} influx into cells by operating in the reverse mode, coupling Ca^{2+} influx with Na^{+} efflux [86–88]. The latter mode is involved in the regulatory process of glutamatergic gliotransmission between astrocytes and neurons and in the NMDA/AMPA receptors' activity, which produce Ca^{2+} entry [86,89]. Thus, NCX dysregulation plays a central role in subsequent Ca^{2+}-related toxicity, which is strongly involved in the development of several diseases [86,87,90]. Furthermore, during both normal and pathophysiological conditions, NCX has emerged as a dominant mechanism for the Ca^{2+} efflux pathway and the protective regulation of cell homeostasis [77,91].

Consistent with its crucial role, NCX is present in both excitable and non-excitable tissues, such as brain, heart, kidney, pancreas and liver [79,92]. It is relevant that each splice variant is expressed in a tissue-specific manner, and only the brain shows significant expression of all three isoforms. Moreover, each variant has a unique cellular and subcellular distribution and probably a specialized functional role in Ca^{2+} homeostasis [83]. Although the NCX was discovered and extensively studied in cardiac myocytes, it has particularly important functions in both the CNS and the PNS, where excitable cells experience transient Ca^{2+} fluxes [85,93]. In particular, NCX1 is expressed ubiquitously, and it shows a great abundance in heart and brain, whereas NCX2 and NCX3 are more highly expressed in the PNS and skeletal muscle than NCX1 [92,94].

NCX2

In the CNS, NCX2 is involved in neurotransmitter release in both neuronal and glial cells with a relevant distribution in astrocytes where the NCX2 antibody strongly cross-reacted with a glial fibrillar protein [93,95]. Moreover, the overall high expression of NCX2 in brain could partly be the result of the ratio of astrocytes to neurons (4:1) [95]. Consistent with this distribution, it is broadly clear that NCX2 exerts its influence at the framework of a neuron–glia network [79]. Moreover, NCX2 is highly detected in the membranes of neuronal cell bodies and at the presynaptic level in a cerebral structure-dependent manner [93,94,96]. For example, NCX2 is the major isoform involved in the clearance of Ca^{2+} in presynaptic terminals of CA1 hippocampal pyramidal neurons, where it is essential for the control of synaptic plasticity, memory, learning and cognition [83,96]. In the PNS, NCX2 is localized within the cell bodies of small-diameter DRG neurons and throughout the entire length of neurites and neuritic tips [97]. More precisely, NCX2 is localized in epidermal free nerve endings and in mechanosensory nerve endings, which include nociceptors [97,98]. Here, NCX2 is co-expressed with Nav 1.6, Nav 1.7, Nav 1.8 and Nav 1.9, whose physiological activities influence each other in both normal and pathological conditions [86,91,98]. In this regard, Persson and colleagues demonstrated that the presence of NCX2 together with Na^{+} channels within epidermal nociceptive terminals may thus make these fibers especially sensitive to injury when energetically challenged [99]. Overall, NCX2 has an important role not only in regulating Ca^{2+} homeostasis but also in noxious stimulus transmission [86,91,97,99]. Thus, NCX2 exerts important antinociceptive effects, and its alteration is strongly related to peripheral sensitization and neuropathic pain [86].

NCX3

In the CNS, NCX3 is expressed at the lowest level of the three isoforms and may play a highly specialized role in Ca^{2+} homeostasis in certain cell types, such as small subfields of neurons involved in Long-Term Potentiation (LTP) [93,100]. Consistent with its distribution in hippocampal neuronal cell bodies as well as in the associated dendritic network, Molinaro and colleagues demonstrated that NCX3 impairment has important consequences for basal synaptic transmission, LTP regulation, spatial learning and memory performance [100]. Moreover, this isoform shows a particularly low expression level in astrocytes too [95]. Interestingly, NCX3 is the main isoform expressed in oligodendrocyte progenitor cells (OPC), where it has a critical role in oligodendrocytes' maturation and in consequent neuronal myelination [90]. Thus, NCX3 dysfunction causes a relevant reduced size of hypo-myelinated spinal cord, where it is highly localized in both white and gray matter [83]. Unlike NCX1-2, NCX3 has the peculiar capability of maintaining Ca^{2+} homeostasis even when ATP levels are reduced significantly, highlighting its major role in neuronal preservation and protection during different pathophysiological conditions [90,100].

2.4. *Voltage-Gated Ca^{2+} Channels (VGCCs or Cav)*

Voltage-gated Ca^{2+} channels (VGCCs or Cav) are some of the most important regulators of Ca^{2+} concentration [101] that are under fine regulation in order to maintain it lower inside the cell (≈50–100 nM) than in the extracellular milieu (≈2 mM) [102]. Indeed, free Ca^{2+} is an important intracellular messenger in all cells, where it controls several cellular functions but can become toxic and cause cell death [103]. The homeostatic control of intracellular calcium concentration ($[Ca^{2+}]_i$) is maintained through the action of the plasma membrane Ca^{2+} transport ATPase (PMCA) and Na^+/Ca^{2+} exchanger (NCX) in a resting cell. Upon elevated $[Ca^{2+}]_i$, Ca^{2+} is sequestered intracellularly by the sarcoendoplasmic reticulum Ca^{2+}-ATPase (SERCA) and by the mitochondrial Ca^{2+} uniporter (mtCU), which are activated by Ca^{2+}. Elevation of $[Ca^{2+}]_i$ can result from either the influx of extracellular Ca^{2+} or the release of Ca^{2+} from intracellular stores after various cell stimuli such as membrane depolarization, extracellular signaling molecules or intracellular messengers [104]. Since VGCCs are some of the main regulators of $[Ca^{2+}]_i$, they have an important role, especially in neurons, where, in addition to functions common to all cells, they also regulate cell excitability, and they have an important role in any neuronal function, including physiologic nociception and neuropathic pain [101].

2.4.1. Topology and Subunits

VGCCs are heteromultimeric complexes consisting of a principal transmembrane pore-forming α_1 subunit (190 kDa) that includes the structural and functional machinery required to conduct Ca^{2+} (Ca^{2+}-selective pore, voltage sensor and gating mechanism) and a combination of three auxiliary subunits, a disulfide-linked $\alpha_2\delta$ dimer (170 kDa), an intracellular phosphorylated β subunit (55 kDa) and a transmembrane γ subunit (33 kDa), involved in the modulation of channel properties [105].

The α_1 subunit is constituted by four homologous domains (D_1–D_4) connected to each other by three intracellular loops, respectively (L_1–L_3). Each domain consists of six transmembrane-spanning hydrophobic α-helical segments (S_1–S_6), and each segment is connected by two intracellular (between S_2–S_3 and S_4–S_5) and three extracellular linkers (between S_1–S_2, S_3–S_4 and S_5–S_6; the last ones are called P-loops). Both the N-terminus and C-terminus are intracellular [15].

The α_1 subunit is the key molecule of the channel complex because it is capable of Ca^{2+} conduction, while the $\alpha_2\delta$, β and γ subunits are auxiliary. It is known that some types of VGCCs are formed through the assembly of α_1 subunits and ancillary ones (i.e., the high-voltage-activated channel; see below), while others appear to lack these ancillary subunits (i.e., the low-voltage-activated channel; see below) [9].

The $\alpha_2\delta$ subunit increases the maximum current density by increasing the α_1 expression in plasmatic membrane together with reducing its turnover (exerts its effect on Ca^{2+} channel trafficking) and increasing also the inactivation rate [106]. It might have a role in modulating the excitability of DRG; indeed, the $\alpha_2\delta_1$ overexpression resulted in enhanced currents and altered biophysical properties in sensory neurons, supporting its contribution in neuropathic pain [107].

Similarly, the β subunit increases the maximum current density by causing a hyperpolarization shift in activation and by increasing the channel opening probability. It is not yet clear if its effect is also due to an enhancement of α_1 trafficking [106].

Instead, the γ subunit, despite being known to bind the α_1 D_4 domain, has a controversial role as auxiliary subunit, and its functions are largely unknown [108].

VGCCs generally activate, inactivate and deactivate slower than VGSCs and can therefore be distinguished on a temporal basis [109].

2.4.2. Classification, Physiological and Pharmacological Properties and Isoforms Involved in Neuropathic Pain

VGCCs are classified according to (i) their activation threshold or (ii) their amino acid sequence homology of the α_1 subunit (Cav). In particular, VGCCs are classified in high-voltage-activated (HVA) and in low-voltage-activated (LVA) channels, and, at the molecular genetic level, are classified in three families (Cav1, Cav2 and Cav3) [109]. α_1 subunits are currently classified into 10 subtypes (Cav1.1–Cav1.4, Cav2.1–Cav2.3 and Cav3.1–Cav3.3), of which 9 are expressed in the nervous system.

All 10 α_1 subunits subtypes share a common transmembrane topology, but they have different biophysical properties (activation, inactivation, conductance and deactivation), expression patterns and pharmacology [102].

There is another classification according to Ca^{2+} currents that show distinct biophysical properties and that are differently modulated by pharmacological agents. For the different classes of Ca^{2+} currents, an alphabetical nomenclature has been adopted (L, N, P/Q, R and T type) [110].

HVA channels activate in response to strong depolarizations, and they generate long-lasting calcium influxes. They are further classified as L type (Cav1.1–Cav1.4; L for large and long-lasting), P/Q type (Cav2.1; P for Purkinje cells), N type (Cav2.2; N for neurons) and R type (Cav2.1; R for resistant) according to their different pharmacological sensitivity. L-type Ca^{2+} is blocked by dihydropyridines, phenylalkylamines and benzothiazepines (organic L-type Ca^{2+} channel antagonists). N-type, P/Q-type and R-type Ca^{2+} are blocked by specific polypeptide toxins from snail and spider venoms.

LVA channels, also classified as T type (Cav3.1–Cav3.3; T for tiny and transient), activate in response to weak depolarizations (between −65 mV and −50 mV), and they open transiently (under both brief and long depolarizations). They are resistant to the other VGCCs' blockers. They are expressed in a wide variety of cell types, where they are involved in shaping the AP and controlling patterns of repetitive firing. T-type channels also generate the so-called window current, defined as small tonic inward current around resting membrane potentials, as a result of activation and inactivation curve overlap [109,110].

It has been shown that some VGCCs have greater involvement in pain pathways than others; among these, there are N-type and some T-type channels (Cav3.2) [111,112].

Cav2.2 (N-Type Channel)

Cav2.2 is exclusively expressed in the central and peripheral nervous systems including the brain, spinal cord and primary sensory neurons [102]. Its expression is particularly high in the superficial layer (laminae I and II) of the dorsal horn, which includes the major termination zone of nociceptive primary afferents [113]. Several studies demonstrated the implication of Cav2.2 channels in the transmission of pain signals at the spinal level. Indeed, it has been demonstrated that the pharmacological block of N-type channels stops the release of pro-nociceptive neurotransmitters such as glutamate and substance P [114]. Also, genetic models support its involvement, since Cav2.2$^{-/-}$ mice showed markedly reduced neuropathic pain symptoms after spinal nerve ligation [115]. The N-type channel inhibitors have been suggested in the treatment of some forms of pain [116].

Cav3.2 (T-Type Channel)

Cav3.2 is expressed in all parts of the sensory neurons involved in the transmission of the nociceptive signal, including peripheral nerve endings, axons, soma and dorsal horn synapses [116]. It is known that T-type channels modulate cellular excitability and rhythmic activity, and that they are involved in pathophysiological conditions related to neuronal hyperexcitability [113]. Cav3.2 might have a role in the nociceptive pathway or in lowering the threshold for AP generation [117] or in enhancing Ca^{2+}-dependent neurotransmitter release, which results in synaptic facilitation [118]. Its role in pain states is supported by findings that showed that antisense knockdown of Cav3.2, but not Cav3.1 and Cav3.3, channels in DRG neurons, resulted in marked antinociceptive, anti-hyperalgesia and anti-allodynia effects [119].

2.5. Transient Receptor Potential Family (TRPA1, TRPM8 and TRPV1)

Transient receptor potential (TRP) receptors are involved in the development of chemotherapy-induced peripheral neuropathic pain, which is a common side effect of selected chemotherapeutic agents such as oxaliplatin [120]. TRP channels have six transmembrane-spanning domains (S1–S6), with a pore-forming loop between S5 and S6, and both the C- and N-termini are located intracellularly [121]. TRPC channels are nonselective cation channels expressed in excitable and non-excitable cells [122]. TRPV channels are a part of the TRP channel superfamily and named for their sensitivity to vanilloid and capsaicin [123]. In most tissues, TRPV channels serve as sensors for different pain stimuli (heat, pressure and pH) and contribute to the homeostasis of electrolytes, the maintenance of barrier functions and the development of macrophages [124]. TRPA1 is the only TRPA protein present in humans. TRPA1 is a sensor for diverse noxious external stimuli such as intense cold, irritating compounds, mechanical stimuli, reactive chemicals and endogenous signals associated with cellular damage [125]. The TRPM subfamily consists of eight members. TRPM1–TRPM8.42 TRPMs are involved in several physiological and pathological processes. TRPM channels possess a large cytosolic domain, making them the largest members of the TRP superfamily [126]. The mucolipin family of the ion channel TRP superfamily (TRPML) includes three members: TRPML1, TRPML2 and TRPML3. Defects in TRPML function are predicted to have important effects on organelle acidification, vesicle fusion, endosome maturation and signaling, thus suggesting that this protein family plays a key role in normal and pathological conditions [127].

3. Ion Channels/Transporters in Chemotherapy-Induced Peripheral Neurotoxicity Models

First of all, studies are characterized by a vast heterogeneity in dosages/schedules of chemotherapy drugs as well as the outcome measures elected to detect/grade neurotoxicity, making comparing different studies complex. Moreover, a clear distinction should be made in particular for in vivo studies; not all studies actually demonstrated the onset of neuropathy but just a neuropathic pain/nocifensive behavior, making it arguable, in particular in the preventive setting, that an actual effect on neuropathy onset can be expected in a clinical setting. Keeping in mind these aspects, the following key information can be retained. Tables 1 and 2 summarize in vitro and in vivo findings, respectively, to provide the reader with a broader overview of what is available in the literature; however, the majority of the data presented cannot be easily translated to the clinical setting since the dosage/schedule does not mirror the actual phenomenon observed in clinical practice. Therefore, while planning a study in this field, a careful evaluation of this aspect should be performed, as recently suggested by Cavaletti et al. in 2024 [128].

Table 1. In vitro studies of Ion Channels/Transporters in Chemotherapy-Induced Peripheral Neurotoxicity Models.

Authors	Target	Cell Culture and Treatment	Neurotoxicity Assessment	Observations
Ballarini et al. [87]	NCX2	OHP 7.5 μM. Rat DRG neurons	Neurite elongation	Protection of neurite outgrowth with a selective NCX blocker
Adelsberger et al. [129]	VGSCs	OHP 250 μM. Rat DRG and hippocampal neurons	Patch clamp recordings	Increase in the Na^+ current on DRG neurons but not on hippocampal neurons
Chang et al. [130]	VGSCs	PTX 0.1–1 μM. Human DRG neurons	Immunohistochemistry, qRT-PCR, transient Na^+ currents and action potential frequency	Increase in Nav 1.7 mRNA expression but not Nav 1.8. Increased transient Na^+ currents amplitude and action potential firing frequency

Table 1. *Cont.*

Authors	Target	Cell Culture and Treatment	Neurotoxicity Assessment	Observations
Lee et al. [131]	VGSCs	OHP 10 and 100 μM. Rat DRG neurons	Patch clamp recordings	Alteration of VGSC conductance towards negative membrane potentials in A-fibers of DRGs
Verma et al. [132]	VGSCs Nav 1.7 and Nav 1.8, KDR, KA, leak channel	PTX 250 nM. Rat DRG neurons	Micro/multielectrode array recordings	Decrease in PTX-induced hyperexcitability by a Nav 1.8 blocker and a KDR agonist treatments
Tomaszewski et al. [133]	VGCCs, VGKCs, VGSCs	CDDP 1, 5, 10, 50 and 100 μM. Rat DRG neurons	Patch clamp recordings	Decrease in Ca^{2+} and K^+ currents in small DRG neurons but only a trend toward reduction in Na^+ currents
Brenneman et al. [134]	mNCX-1	PTX 3 μM. Rat DRG neurons	Cell viability assays, IR cell bodies and neuritic areas	mNCX-1 siRNA decreases CBD protection from PTX toxicity, decrease in IR neuronal cell bodies and neuritic IR areas
Li et al. [135]	VGCC T-type	PTX 1 μM. Human DRG neurons	Patch clamp recordings, immunohistochemistry	Increase in Ca^{2+} current, increase in DRG excitability
Leo et al. [136]	VGCCs	CDDP 0.5 or 5 μM. Rat DRG neurons	Patch clamp recordings, immunostaining, calpain activity assay	Decrease in Ca^{2+} current in L-type, P-/Q-type and T-type channels but increase in N-type VGCC currents. Increased expression of N-type VGCC proteins. DRG neuroprotection by N-type VGCC blocker
Schmitt et al. [137]	VGCCs	OHP 1, 10, 100, 250 and 500 μM. Rat DRG neurons	Patch clamp recordings, immunocytochemistry, Western blot, calpain activity assay	Decrease in L-type, P/Q-type and T-type VGCCs currents. Prolonged treatment increased current density. Increase in L-type and T-type VGCCs protein expression. Increase in the action potential amplitude through modulation of T-type and L-type VGCCs
Tomita et al. [138]	VGCCs T type	BTZ 0.1 nM. Mouse neuroblastoma x rat DRG neuron hybrid cells	Western blot, qRT-PCR, patch clamp recordings	Increase in T-type VGCC protein expression. Increase in Ca^{2+} currents
Materazzi et al. [139]	TRPA1, TRPV4	PTX 10, 30 and 50 μM. Mice DRG neurons or esophagus slices	Ca^{2+} imaging, neuropeptides release assay	Modulation of TRPA1 and TRPV4 by Ca^{2+}-dependent CGRP secretion
Nassini et al. [140]	TRPA1	OHP or CDDP 100 μM, guinea pig pulmonary artery. TRPA1+ CHO cells expressing mouse (10 to 300 μM OHP/CDDP)	Guinea pig pulmonary artery assay of neurogenic relaxation. DRG and CHO Ca^{2+} response to OHP or CDDP	OHP and CDDP activate TRPA1 channel on nociceptive nerve terminals. The activation of TRPA1 is mediated by oxidative stress
Sanchez et al. [141]	TRPV4	PTX 1 μM. Human SH-SY5Y cells	qRT-PCR, Western blot, patch clamp recordings, cytosolic Ca^{2+} measurement	Increase in TRPV4 protein and mRNA expression. Increase in outward and inward current density. Increase in cytosolic Ca^{2+} concentrations
Ta et al. [142]	TRPV1, TRPM8, TRPA1	CDDP or OHP 6.7 μM. Rat DRG neurons	qRT-PCR	TRPV1, TRPM8 and TRPA1 mRNA expressions are differently upregulated by CDDP and OHP
Trevisan et al. [143]	TRPA1	BTZ 10 or 100 μM. Mouse DRG neurons	Ca^{2+} imaging	BTZ did not evoke Ca^{2+} responses in TRPA1+ neurons
Ertilav et al. [144]	TRPV1	DT 10 nM. TRPV1 transfected SH-SY5Y cells	Ca^{2+} fluorescence, Western blot	Activation of TRPV1

Table 1. *Cont.*

Authors	Target	Cell Culture and Treatment	Neurotoxicity Assessment	Observations
Anand et al. [145]	TRPV1, TRPA1, TRPM8	OHP 12–120 μM. Rat DRG neurons	Neurite elongation and density, cell viability assay, cAMP assay, Ca^{2+} imaging	TRPV1 and TRPA1 sensitization but not for TRPM8
Leo et al. [146]	TRPA1, TRPV1	CDDP and OHP 10 μM. Rat DRG neurons	Cell viability assay, immunocytochemical staining, cytosolic and intramitochondrial Ca^{2+} measurement	Increase in cytosolic Ca^{2+} concentration and decrease in intramitochondrial Ca^{2+} concentration in TRPA1+ and TRPV1+ DRG neurons
Sanchez et al. [147]	TRPA1	PTX 1 μM. Human SH-SY5Y cells	qRT-PCR, Western blot, patch clamp recordings, cytosolic Ca^{2+} measurement	Increase in TRPA1 protein expression, TRPA1 current density and TRPA1-mediated Ca^{2+} concentrations

AMP: cyclic adenosine monophosphate; ASIC: acid-sensing ion channel; BTZ: bortezomib; CAP: compound action potential. CBD: cannabidiol; CDDP: cisplatin; CGRP: calcitonin gene-related peptide; CHO: Chinese hamster ovary; CMAP: compound muscle action potential; DT: docetaxel; HCN: hyperpolarization-activated cyclic nucleotide gated; IENFD: intraepidermal nerve fiber density; GBP: gabapentin; GIRK: G-protein-gated inward rectifier K+ channel; IR: immunoreactive; K2p1.1: potassium channel subfamily K member 1; KDR: delayed rectifier potassium channel; KA: A-type transient potassium channel; MOR: μ-opioid receptor; NCS: nerve conduction studies; NCV: nerve conduction velocity; NCX: sodium–calcium exchanger; NET: neuronal excitability testing; OHP: oxaliplatin; P2 × 3: purinergic receptor; PTX: paclitaxel; SNAP: sensory nerve action potential. TG: trigeminal ganglia; TREK: TWIK-related K+ channel; TRP: transient receptor potential channels, vanilloid subtype; TRPM: transient receptor potential melastatin; TRPA: ankyrin-type transient receptor potential; TTX: tetrodotoxin; VCR: vincristine; VGCCs: voltage-gated calcium channels; VGSCs: voltage-gated sodium channels; VGKC: voltage-gated potassium channel.

Table 2. In vivo studies of Ion Channels/Transporters in Chemotherapy-Induced Peripheral Neurotoxicity Models.

Authors	Target	Animal Model	Neurotoxicity Assessment	Observations
Ballarini et al. [87]	NCX2	OHP 7 mg/kg in mice, i.v., once a week for 8 weeks	NCS and NET recordings, mechanical allodynia test, immunohistochemistry, Western blot, caudal nerve morphology and morphometry, IENFD	Decrease in NCX2 protein expression in DRGs
Chukyo et al. [120]	TRPV1, TRPA1, TRPM8	OHP 6 mg/kg in rats, single i.p.	Acetone spray test, immunohistochemistry, in situ hybridization	Increase in TRPA1, TRPV1 and TRPM8 protein expression in DRGs. Increase in TRPA1 and TRPV1 mRNA coexpression in DRGs
Caudle and Neubert [148]	HCN, VGSCs, menthol, TRPM8	OHP 10 mg/kg in mice, i.p., two administrations; PTX 26 mg/kg in mice, i.p., four administrations. Dissociated TRG neurons *	Orofacial Pain Assessment Devices, patch clamp recordings	Increase in HCN, VGSCs and menthol evoked TRPM8 currents but not of VGKCs
Nieto et al. [149]	VGSCs TTX sensitive	PTX 2 mg/kg in mice, i.p., 5 days	Heat hyperalgesia test, acetone cold allodynia test, mechanical allodynia test, rotarod test	Decrease in heat hyperalgesia, mechanical and cold allodynia by TTX administration
Makker et al. [150]	VGSCs and VGKCs	OHP 10 mg/kg i.p. or 7.5 and 15 mg/kg i.m. in mice, single dose; 5 mg/kg i.p. on days 0, 2, 4, 6	CMAP and SNAP recording, mathematical modeling of axonal excitability	Change of the depolarization phase and creation of afterdischarges, inactivation of VGCCs, reduction in fast K^+ conductance in motor axons. Increase in hyperpolarization and decrease in peak amplitude in sensory axons

Table 2. *Cont.*

Authors	Target	Animal Model	Neurotoxicity Assessment	Observations
Alberti et al. [151]	VGSC	OHP 5 mg/kg in rats, twice a week for 4 weeks	NCS and NET recordings, mechanical allodynia test, caudal nerve morphology and morphometry, IENFD	Modulating VGSC with topiramate (100 mg/kg per os, daily, starting 5 days before first OHP administration and continuing up to chemotherapy completion) complete neurotoxicity prevention was observed via neurophysiology, neuropathology and behavioral tests
Braden et al. [152]	VGSC Nav 1.7	OHP 3 mg/kg in mice, i.p., on days 0–4 and 10–14	Von Frey test	Decrease in mechanical allodynia through indirect inhibition of Nav 1.7
Di Cesare Mannelli et al. [153]	VGKCs Kv7	PTX 2 mg/kg in mice, i.p., on days 1, 3, 5 and 7; OHP 2.4 mg/kg in mice, i.p., on days 1–2, 5–9, 12–14	Cold plate test, Von Frey test, hot plate test	Kv7 channel blocker XE991 antagonized the pain-relieving activity of H2S donors, demonstrating the role of Kv7 in neuropathic pain
Jia et al. [154]	K^+ channel 1.1 (K2p 1.1)	PTX 4 mg/kg in rats, i.p., every other day for a total of four injections, on days 0, 2, 4, and 6	Mechanical allodynia heat, heat hyperalgesia test and cold hyperalgesia test	Reduction in K^+ channel 1.1
Kagiava et al. [155]	VGKCs	OHP 25, 100 and 500 µM. Rat isolated sciatic nerve *	Evoked CAP recordings	Induce alterations in CAP waveform, firing frequency and repolarization phase through VGKCs but not VGSCs
Kanbara et al. [156]	GIRK1	OHP 2 mg/kg in rats, i.p., twice a week for 4 weeks	Von Frey test	GIRK1 activation contributes to MOR antinociception
Lucarini et al. [157]	VGKC Kv7	OHP 2.4 mg/kg in mice, i.p., on days 1–2, 5–9 and 12–14	Cold plate test	Modulating Kv7 channels, a reduction in painful features is observed
Yilmaz et al. [158]	NCX	PTX 2 mg/kg in rats, on days 0, 2, 4 and 6. Dissociated DRG neurons *	Ca^{2+} imaging	PTX-induced inhibition of Ca^{2+} transients is not modulated by NCX activity
Li et al. [135]	VGCC T type	PTX 2 mg/kg in rats, i.p., on days 0, 2, 4 and 6	Von Frey test, patch clamp recordings, Ca^{2+} imaging, immunohistochemistry, Western blot	Increase in Ca^{2+} current, increase of DRG excitability, increase in T-type VGCC expression in DRGs and spinal cord. Decrease in mechanical allodynia by T-type VGCC blocker
Leo et al. [136]	VGCCs	CDDP 1.5 mg/kg in rats, i.p., two cycles of four daily administrations with four days rest	Von Frey test, hot plate test, rotarod test, Western blot, qRT-PCR	Increased expression of N-type VGCC proteins, but not mRNA in DRGs. Decrease in thermal hyperalgesia and mechanical allodynia by N-type VGCC blocker
Tomita et al. [138]	VGCCs T type	BTZ 0.4 mg/kg in mice, i.p., six administrations in 2 weeks	Western blot, Von Frey test	Increase in T-type VGCCs' protein expression in DRGs. Decrease in mechanical hyperalgesia through T-type VGGC blockers and gene silencing
Nodera et al. [159]	Kv7 VDKCs	CDDP 2.3 mg/kg in mice, i.p., on days 1–5 and 13–17	SNAP recording, NET recording, NCS recording	Axonal protection, preserved membrane potential through increase in K^+ currents with treatment Kv7 agonist retigabine
Kawakami et al. [160]	VGCCs	PTX 2 and 4 mg/kg in rats, i.p., on days 0, 2, 4 and 6. Dissociated DRG neurons	Von Frey test, patch clamp recordings	Increase in Ca^{2+} currents. GBP, a Ca^{2+} channel blocker, reverses mechanical hyperalgesia

Table 2. *Cont.*

Authors	Target	Animal Model	Neurotoxicity Assessment	Observations
Matsumoto et al. [161]	VGCCs α2δ-1 subunit	PTX 4 mg/kg in mice, single i.p. or i.v., or i.p. on days 0, 2, 4 and 6	Heat hyperalgesia test, electrical hyperalgesia test, qRT-PCR, Western blot, immunohistochemistry	Increase in DRGs' expression of α2δ-1 subunit. GBP blockade of VGCCs decreases mechanical allodynia and sensitization of myelinated A-fibers
Okubo et al. [162]	VGCCs T type	PTX 2 mg/kg in rats, i.p., on days 0, 2, 4 and 6	Paw pressure test	Decrease in hyperalgesia through administration of T-type VGCCs selective blockers
Sekiguchi et al. [163]	T-type VGCCs	PTX 4 mg/kg in mice or 2 mg/kg in rats, i.p., on days 0, 2, 4 and 6	Von Frey test, paw pressure test, open field test, rotarod test	T-type VGCC blockers reduce neuropathic mechanical allodynia
Meregalli et al. [164]	VGCCs T type	BTZ 0.2 mg/kg in rats, i.v., three times a week for 4 weeks	NCV measurement, mechanical allodynia test, β-tubulin polymerization assay, IENFD, proteasome inhibition assay	Suvecaltamide modulation of T-type VGCCs reverses NCV and IENFD neuropathy, reverses β-tubulin polymerization increase but does not affect proteasome inhibition by BTZ
Sharma et al. [165]	VGCCs L type	VCR 50 µg/kg in rats, i.p., 10 days administration	acetone drop test, pin-prick test, hot plate test	Decrease in the protective effect of VCR pretreatment on allodynia and hyperalgesia following treatment with T-type VGCC blocker
Materazzi et al. [139]	TRPA1, TRPV4	PTX 6 mg/kg in WT and TRPA1 KO mice, single i.p.	Von Frey test, acetone cold stimulation test	Decrease in mechanical allodynia by TRPA1 and TRPV4 blockers. Decrease in cold hypersensitivity by TRPA1 but not TRPV4 blocker
Nassini et al. [140]	TRPA1	OHP 2 mg/kg i.v., CDDP 2 mg/kg i.p. in C57/BL6, Trpa1$^{+/+}$ or Trpa1$^{-/-}$ mice	Von Frey test, cold plate test, qRT-PCR	TRPA1 modulation decreases painful features related to OHP and CDDP single administration
Ta et al. [142]	TRPV1, TRPM8, TRPA1	CDDP 2.3 mg/kg or OHP 3 mg/kg in WT or TRPV1 KO mice, i.p., 5 days, 5 days rest and 5 days treatment	Von Frey test, radiant heat test, cold plate test, tail immersion test, qRT-PCR, immunohistochemistry	Upregulation of TRPV1 and TRPA1 mRNA following CDDP treatment, but only TRPA1 upregulation following OHP treatment in TGs. Decrease in mechanical allodynia following CDDP and OHP treatment in TRPV1 KO mice. Decrease in CDDP-induced thermal hypersensitivity in TRPV1 KO mice
Trevisan et al. [143]	TRPA1	BTZ 0.2, 0.5 or 1 mg/kg in WT or TRPA1 KO mice, single i.p.	Von Frey test, hot plate test, cold stimulation, chemical hyperalgesia test, rotarod test, Western blot	BTZ did not modify TRPA1 expression level in DRGs. Decrease in mechanical and cold hyperalgesia through TRPA1 agonist treatment and in TRPA1 KO mice
Chen et al. [166]	TRPV1, TRPV4, TRPA1	PTX 1 mg/kg in mice, i.p., on days 0, 2, 4 and 6	Von Frey test, hot plate test, cold hyperalgesia test	Reduction in heat hyperalgesia, but not mechanical allodynia and cold hyperalgesia, through TRPV1 blocking. Reduction in mechanical and heat, but not cold, hyperalgesia through TRPV4 blocking. Reduction in mechanical allodynia, heat and cold hyperalgesia through TRPA1 blocking
Ertilav et al. [144]	TRPV1	DT 30 mg/kg in mice, single i.p. Dissociated DRG neurons *	Von Frey test, hot plate test, Western blot, patch clamp recordings, Ca^{2+} fluorescence, cell viability assay	Increase in cytosolic Ca^{2+} concentration through TRPV1 channel agonist stimulation. Increase in TRPV1 expression level

Table 2. *Cont.*

Authors	Target	Animal Model	Neurotoxicity Assessment	Observations
Hori et al. [167]	TRPV1, TRPV2, P2 × 3 and ASIC3	CDDP 3 mg/kg in rats, i.p., once per week for five weeks	Von Frey test, pin-prick test, mechanical allodynia test, grid force test, rotarod test and immunohistochemistry	Increase in TRPV2, P2 × 3 and ASIC3 expression, but not in TRPV1 in DRGs
Quartu et al. [168]	TRPV1	BTZ 0.20 mg/kg in rats, single i.v., or three administrations for 8 weeks	Caudal NCV recordings, mechanical allodynia test, thermal hyperalgesia test, immunohistochemistry, morphometry, qRT-PCR and Western blot	Reduction in caudal NCV, increase in mechanical allodynia but not of thermal hyperalgesia. Increase in TRPV1 protein expression, but decrease in TRPV1 and CGRP mRNA level, in DRGs and spinal cord
Mao et al. [169]	K2p1.1 channel	PTX 4 mg/kg in mice, i.p., on days 0, 2, 4 and 6	Von Frey test, heat hyperalgesia, conditioned place preference, patch clamp recordings, qRT-PCR, Western blot, immunohistochemistry	PTX induces a decrease of K2P1.1 expression, contributing to chemotherapy-induced neuropathic pain
Pereira et al. [170]	TREK2	OHP 6 mg/kg in WT and TREK2 KO mice, single i.p.	Von Frey test, flinch test, immersion tests, hot plate test, thermal preference test, dynamic cold plate test; qRT-PCR; single C-fibers recordings	Decrease in TREK2 expression in DRGs. TREK2 mediates neuropathic hyperalgesia, regulates heat sensitivity of C-fibers, but does not play a role in noxious thermal hypersensitivity
Rapacz et al. [171]	VGSCs and L-type VGCCs	OHP 10 mg/kg in mice, i.p.	Von Frey test, hot plate test, formalin test	Decrease in mechanical allodynia by VGSCs and VGCCs blocking
Salat et al. [172]	VGSCs	OHP 10 mg/kg in mice, single i.p.	Von Frey test, cold plate tests, rotarod test	Reduced mechanical allodynia following a VGSC blocker

AMP: cyclic adenosine monophosphate; ASIC: acid-sensing ion channel; BTZ: bortezomib; CAP: compound action potential. CBD: cannabidiol; CDDP: cisplatin; CGRP: calcitonin gene-related peptide; CHO: Chinese hamster ovary; CMAP: compound muscle action potential; DT: docetaxel; HCN: hyperpolarization-activated cyclic nucleotide gated; IENFD: intraepidermal nerve fiber density; GBP: gabapentin; GIRK: G-protein-gated inward rectifier K^+ channel; IR: immunoreactive; K2p1.1: potassium channel subfamily K member 1; KDR: delayed rectifier potassium channel; KA: A-type transient potassium channel; MOR: μ-opioid receptor; NCS: nerve conduction studies; NCV: nerve conduction velocity; NCX: sodium–calcium exchanger; NET: neuronal excitability testing; OHP: oxaliplatin; P2 × 3: purinergic receptor; PTX: paclitaxel; SNAP: sensory nerve action potential. TG: trigeminal ganglia; TREK: TWIK-related K^+ channel; TRP: transient receptor potential channels, vanilloid subtype; TRPM: transient receptor potential melastatin; TRPA: ankyrin-type transient receptor potential; TTX: tetrodotoxin; VCR: vincristine; VGCC: voltage-gated calcium channels; VGSC: voltage-gated sodium channels; VGKC: voltage-gated potassium channel. * ex vivo studies.

3.1. Voltage-Gated Sodium Channels

VGSCs were linked to peripheral neurotoxicity in in vitro studies after exposure to oxaliplatin (OHP). Adelsberger et al. [129] first observed that dorsal root ganglia (DRG) neurons exposed to OHP have an increase in Na^+ currents by exploiting patch clamp recordings. Exploiting the same technique, alternations in this channel were demonstrated by multiple other authors, confirming the robustness of this observation [130,131,148] even though dosages/models vary among different studies. Verma et al. [132] tested VGSCs with micro/multielectrode array recordings, showing a possible hyperexcitability in neurons also after exposure to paclitaxel (PTX), and Nieto et al. [149] suggested that they might be involved in painful manifestations in PTX-treated animals.

In vivo studies confirmed involvement of VGSCs in OHP-related hyperexcitability [150], showing that, by decreasing this condition, neuroprotection can be established [151,152]. In particular, preliminary data suggest, in fact, that increased Na^+ currents due to OHP exposure are able to activate the sodium–calcium exchanger (NCX) reverse mode, leading to Ca^{2+} neurotoxicity, as we will discuss in the subsequent sections [87].

3.2. Voltage-Gated Potassium Channels

A potential involvement of VGkCs was observed in the in vitro setting, in particular after exposure to cisplatin CDDP, leading to a reduction in K^+ currents in DRG neurons, which were actually explored in a broader setting aiming to assess changes in Ca^{2+} homeostasis and related voltage-operated ion channels which were found to be altered [133].

In the in vivo setting, they were linked to painful manifestations due to PTX exposure [153,154]. Some authors also suggested that they might play a role in the neurotoxicity manifestations observed in rodent models exposed to OHP [155–157].

3.3. Sodium–Calcium Exchanger

As stated above, the NCX, and, specifically, the isoform NCX2, was suggested to be the actual link between acute OHP-related transient neurotoxicity (i.e., a state of axonal hyperexcitability that mainly lasts in patients 24–72 h after OHP exposure [173–175]) and the actual OHP-induced axonal damage. The hypothesis, corroborated by preliminary findings by Ballarini et al. [87], is that Na^+ currents increased due to OHP exposure trigger NCX2 reverse mode, leading to Ca^{2+} toxicity. Both in vitro [134] and in vivo [158] studies suggested that this family could also play a role in PTX-related neurotoxicity.

3.4. Voltage-Gated Ca^{2+} Channels

Alterations in Ca^{2+}-related currents were suggested in in vitro models after exposure to PTX [135], CDDP [136], OHP [137] and bortezomib (BTZ) [138]. In the in vivo setting, similar observations were made in relation to CDDP [136,159], PTX [135,160–163], BTZ [138,164] and vincristine (VCR) [165].

3.5. Transient Receptor Potential Family (TRPA1, TRPM8 and TRPV1)

In vitro findings suggest that this family could play a role in PTX-, OHP-, CDDP- and BTZ-related CIPN [139–143]. In vivo data support its role mostly in painful phenomena related to CIPN due to OHP, PTX, CDDP and BTZ [120,139,140,142–144,148,166–168].

4. Possible Clinical Translation

The mechanism underlying CIPN remains unclear though mechanistic studies have reported that Na^+, K^+ and Ca^{2+}channels and different types of transient receptor potential family are suggested to be involved [176]. As stated above, preclinical data should be carefully weighted to translate inferences from bench to bed side. As already stated, each study should be carefully evaluated considering schedule/dosages and outcome measures both in the in vitro and in vivo setting. For in vivo studies in particular, it should be pointed out that a robust CIPN animal model cannot rely just on behavioral tests; neuropathological and neurophysiological studies are warranted to test both large and small nerve fibers [177,178], which can be differently affected by different anticancer drugs [175]. Moreover, to translate data on voltage-operated ion channels from the in vitro to the in vivo setting, and then to the clinical setting, a specific method can be used: nerve excitability testing (NET) [179]. NET was first described in humans and then adapted to animal models [180,181]. A virtuous example of this is given by studies performed on OHP. Both in patients [182,183] and animal models [87,150,151], NET was able to demonstrate that OHP can induce a transient alteration that perfectly mirrors the expected pattern of a VGSC's channelopathy. Starting from this observation, a pathogenetic hypothesis was then built, as already discussed, linking these phenomena (i.e., aberrant persistent Na^+ currents) to NCX2 reverse mode activation. Rather promising neuroprotection data were obtained in animal models, targeting the modulation of this double Na^+/Ca^{2+} axis, on the basis of strong outcome measures, as previously stated (i.e., neurophysiology and neuropathology to assess and grade axonal damage) [87], exploiting a selective modulator for the NCX family, such as SEA0400 (see Figure 2). The administration of sodium channel blockers was reported to be effective against oxaliplatin-induced neuropathic pain in humans [184] and animals [185], suggesting that concentrating on the voltage-gated sodium channel may be an effective treatment approach [151].

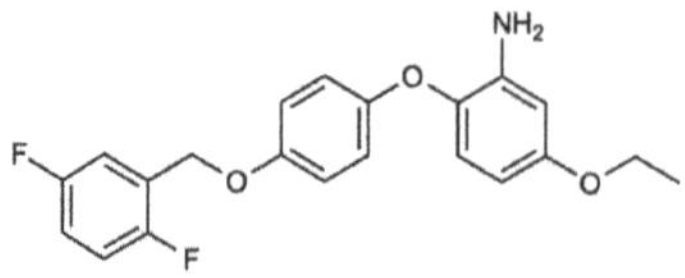

Figure 2. Molecular structure of SEA0400. SEA0400 is a novel and selective inhibitor of the Na^{+}-Ca^{2+} exchanger (NCX), inhibiting Na^{+}-dependent Ca^{2+} uptake.

5. Concluding Remarks

Voltage-operated ion channels/transporters are clearly pivotal components of neurons, and it is not illogical to expect quite a vast involvement of these in CIPN axonal damage; therefore, our hypothesis is that, by targeting these specific elements, it is possible to potentially detect drugs/mechanisms to actually prevent axonal damage (see Figure 3). Already known or even known modulators can be tested, also exploiting the possibility of designing new drugs exploiting innovative approaches such as computational biology, as summarized in a clear-cut review by Azad et al. in 2023 [186].

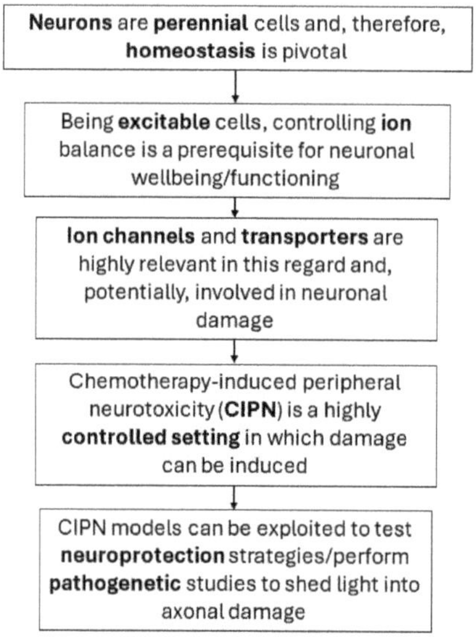

Figure 3. Graphical representation of the overarching hypothesis: targeting ion channels/transporters to prevent axonal damage.

Promising in vitro preclinical data should be carefully weighed based on the specific model used, as well as in vivo studies. Once robust data are available, they could be transferred to the bed side for neuroprotection trials, relying on highly translational outcome measures such as NET. Voltage-operated ion channels are an intriguing option in this setting since, when targeting them, the eventual neuroprotectant drug is unlikely to target the same mechanism that the anticancer drug exploits to obtain its oncological efficacy.

Author Contributions: Conceptualization, P.A.; methodology, P.A.; data curation, E.P., L.C., S.D.G. and G.T.; writing—original draft preparation, E.P., L.C., S.D.G. and G.T.; writing—review and editing, P.A. and G.S.; supervision, P.A.; project administration, P.A.; funding acquisition, P.A. All authors have read and agreed to the published version of the manuscript.

Funding: This research received no external funding.

Acknowledgments: P.A. is supported by the Fondazione Cariplo research grant for Biomedical Research Conducted by Young Researchers.

Conflicts of Interest: The authors declare no conflicts of interest.

References

1. Park, S.B.; Cetinkaya-Fisgin, A.; Argyriou, A.A.; Höke, A.; Cavaletti, G.; Alberti, P. Axonal degeneration in chemotherapy-induced peripheral neurotoxicity: Clinical and experimental evidence. *J. Neurol. Neurosurg. Psychiatry* **2023**, *94*, 962–972. [CrossRef] [PubMed]
2. Cavaletti, G.; Cornblath, D.R.; Merkies, I.S.J.; Postma, T.J.; Rossi, E.; Alberti, P.; Bruna, J.; Argyriou, A.A.; Briani, C.; Velasco, R.; et al. Patients' and physicians' interpretation of chemotherapy-induced peripheral neurotoxicity. *J. Peripher. Nerv. Syst.* **2019**, *24*, 111–119. [CrossRef] [PubMed]
3. Loprinzi, C.L.; Lacchetti, C.; Bleeker, J.; Cavaletti, G.; Chauhan, C.; Hertz, D.L.; Kelley, M.R.; Lavino, A.; Lustberg, M.B.; Paice, J.A.; et al. Prevention and Management of Chemotherapy-Induced Peripheral Neuropathy in Survivors of Adult Cancers: ASCO Guideline Update. *J. Clin. Oncol.* **2020**, *38*, 3325–3348. [CrossRef]
4. Romanova, D.; Balaban, P.; Nikitin, E. Sodium Channels Involved in the Initiation of Action Potentials in Invertebrate and Mammalian Neurons. *Biophysica* **2022**, *2*, 184–193. [CrossRef]
5. Kariev, A.M.; Green, M.E. Water, Protons, and the Gating of Voltage-Gated Potassium Channels. *Membranes* **2024**, *14*, 37. [CrossRef]
6. Giladi, M.; Shor, R.; Lisnyansky, M.; Khananshvili, D. Structure-Functional Basis of Ion Transport in Sodium-Calcium Exchanger (NCX) Proteins. *Int. J. Mol. Sci.* **2016**, *17*, 1949. [CrossRef] [PubMed]
7. Mochida, S. Presynaptic calcium channels. *Neurosci. Res.* **2018**, *127*, 33–44. [CrossRef] [PubMed]
8. Kalinovskii, A.P.; Utkina, L.L.; Korolkova, Y.V.; Andreev, Y.A. TRPV3 Ion Channel: From Gene to Pharmacology. *Int. J. Mol. Sci.* **2023**, *24*, 8601. [CrossRef]
9. Catterall, W.A.; Goldin, A.L.; Waxman, S.G. International Union of Pharmacology. XLVII. Nomenclature and structure-function relationships of voltage-gated sodium channels. *Pharmacol. Rev.* **2005**, *57*, 397–409. [CrossRef]
10. Bennett, D.L.; Woods, C.G. Painful and painless channelopathies. *Lancet Neurol.* **2014**, *13*, 587–599. [CrossRef] [PubMed]
11. Amir, R.; Argoff, C.E.; Bennett, G.J.; Cummins, T.R.; Durieux, M.E.; Gerner, P.; Gold, M.S.; Porreca, F.; Strichartz, G.R. The role of sodium channels in chronic inflammatory and neuropathic pain. *J. Pain* **2006**, *7*, S1–S29. [CrossRef]
12. Devor, M. Sodium channels and mechanisms of neuropathic pain. *J. Pain* **2006**, *7*, S3–S12. [CrossRef] [PubMed]
13. Hargus, N.J.; Patel, M.K. Voltage-gated Na^+ channels in neuropathic pain. *Expert Opin. Investig. Drugs* **2007**, *16*, 635–646. [CrossRef] [PubMed]
14. Wang, J.; Ou, S.W.; Wang, Y.J. Distribution and function of voltage-gated sodium channels in the nervous system. *Channels* **2017**, *11*, 534–554. [CrossRef] [PubMed]
15. Catterall, W.A. Ion channel voltage sensors: Structure, function, and pathophysiology. *Neuron* **2010**, *67*, 915–928. [CrossRef]
16. Ulbricht, W. Sodium channel inactivation: Molecular determinants and modulation. *Physiol. Rev.* **2005**, *85*, 1271–1301. [CrossRef] [PubMed]
17. Ye, W.; Zhao, H.; Dai, Y.; Wang, Y.; Lo, Y.H.; Jan, L.Y.; Lee, C.H. Activation and closed-state inactivation mechanisms of the human voltage-gated K. *Mol. Cell* **2022**, *82*, 2427–2442.e4. [CrossRef] [PubMed]
18. Lewis, A.H.; Raman, I.M. Resurgent current of voltage-gated Na^+ channels. *J. Physiol.* **2014**, *592*, 4825–4838. [CrossRef] [PubMed]
19. Catterall, W.A. From ionic currents to molecular mechanisms: The structure and function of voltage-gated sodium channels. *Neuron* **2000**, *26*, 13–25. [CrossRef] [PubMed]
20. Lampert, A.; Eberhardt, M.; Waxman, S.G. Altered sodium channel gating as molecular basis for pain: Contribution of activation, inactivation, and resurgent currents. *Handb. Exp. Pharmacol.* **2014**, *221*, 91–110. [CrossRef]
21. Cantrell, A.R.; Tibbs, V.C.; Yu, F.H.; Murphy, B.J.; Sharp, E.M.; Qu, Y.; Catterall, W.A.; Scheuer, T. Molecular mechanism of convergent regulation of brain Na^+ channels by protein kinase C and protein kinase A anchored to AKAP-15. *Mol. Cell. Neurosci.* **2002**, *21*, 63–80. [CrossRef] [PubMed]
22. Scheuer, T. Regulation of sodium channel activity by phosphorylation. *Semin. Cell Dev. Biol.* **2011**, *22*, 160–165. [CrossRef] [PubMed]
23. Naylor, C.E.; Bagnéris, C.; DeCaen, P.G.; Sula, A.; Scaglione, A.; Clapham, D.E.; Wallace, B.A. Molecular basis of ion permeability in a voltage-gated sodium channel. *EMBO J.* **2016**, *35*, 820–830. [CrossRef] [PubMed]
24. Isom, L.L. Sodium channel beta subunits: Anything but auxiliary. *Neuroscientist* **2001**, *7*, 42–54. [CrossRef] [PubMed]
25. Dib-Hajj, S.D.; Cummins, T.R.; Black, J.A.; Waxman, S.G. Sodium channels in normal and pathological pain. *Annu. Rev. Neurosci.* **2010**, *33*, 325–347. [CrossRef] [PubMed]
26. Cummins, T.R.; Aglieco, F.; Renganathan, M.; Herzog, R.I.; Dib-Hajj, S.D.; Waxman, S.G. Nav1.3 sodium channels: Rapid repriming and slow closed-state inactivation display quantitative differences after expression in a mammalian cell line and in spinal sensory neurons. *J. Neurosci.* **2001**, *21*, 5952–5961. [CrossRef] [PubMed]

27. Hains, B.C.; Klein, J.P.; Saab, C.Y.; Craner, M.J.; Black, J.A.; Waxman, S.G. Upregulation of sodium channel Nav1.3 and functional involvement in neuronal hyperexcitability associated with central neuropathic pain after spinal cord injury. *J. Neurosci.* **2003**, *23*, 8881–8892. [CrossRef] [PubMed]
28. Li, S.; Yan, G.J.; Tan, Y.X.; Xue, L.L.; Wang, T.H.; Zhao, H.R.; Lu, M.N.; Zhang, H.X.; Mei, R.; Dong, X.H.; et al. Reduced Expression of Voltage-Gated Sodium Channel Beta 2 Restores Neuronal Injury and Improves Cognitive Dysfunction Induced by A. *Neural Plast.* **2022**, *2022*, 3995227. [CrossRef] [PubMed]
29. Hameed, S. Na_V1.7 and Na_V1.8: Role in the pathophysiology of pain. *Mol. Pain* **2019**, *15*, 1744806919858801. [CrossRef] [PubMed]
30. Cummins, T.R.; Howe, J.R.; Waxman, S.G. Slow closed-state inactivation: A novel mechanism underlying ramp currents in cells expressing the hNE/PN1 sodium channel. *J. Neurosci.* **1998**, *18*, 9607–9619. [CrossRef] [PubMed]
31. Rush, A.M.; Cummins, T.R.; Waxman, S.G. Multiple sodium channels and their roles in electrogenesis within dorsal root ganglion neurons. *J. Physiol.* **2007**, *579*, 1–14. [CrossRef] [PubMed]
32. Theile, J.W.; Jarecki, B.W.; Piekarz, A.D.; Cummins, T.R. Nav1.7 mutations associated with paroxysmal extreme pain disorder, but not erythromelalgia, enhance Navbeta4 peptide-mediated resurgent sodium currents. *J. Physiol.* **2011**, *589*, 597–608. [CrossRef] [PubMed]
33. Minett, M.S.; Nassar, M.A.; Clark, A.K.; Passmore, G.; Dickenson, A.H.; Wang, F.; Malcangio, M.; Wood, J.N. Distinct Nav1.7-dependent pain sensations require different sets of sensory and sympathetic neurons. *Nat. Commun.* **2012**, *3*, 791. [CrossRef] [PubMed]
34. Cheng, X.; Dib-Hajj, S.D.; Tyrrell, L.; Wright, D.A.; Fischer, T.Z.; Waxman, S.G. Mutations at opposite ends of the DIII/S4-S5 linker of sodium channel Na V 1.7 produce distinct pain disorders. *Mol. Pain* **2010**, *6*, 24. [CrossRef] [PubMed]
35. Cheng, X.; Dib-Hajj, S.D.; Tyrrell, L.; Waxman, S.G. Mutation I136V alters electrophysiological properties of the Na_V1.7 channel in a family with onset of erythromelalgia in the second decade. *Mol. Pain* **2008**, *4*, 1. [CrossRef] [PubMed]
36. Han, C.; Hoeijmakers, J.G.; Ahn, H.S.; Zhao, P.; Shah, P.; Lauria, G.; Gerrits, M.M.; te Morsche, R.H.; Dib-Hajj, S.D.; Drenth, J.P.; et al. Nav1.7-related small fiber neuropathy: Impaired slow-inactivation and DRG neuron hyperexcitability. *Neurology* **2012**, *78*, 1635–1643. [CrossRef]
37. Faber, C.G.; Hoeijmakers, J.G.; Ahn, H.S.; Cheng, X.; Han, C.; Choi, J.S.; Estacion, M.; Lauria, G.; Vanhoutte, E.K.; Gerrits, M.M.; et al. Gain of function Nav1.7 mutations in idiopathic small fiber neuropathy. *Ann. Neurol.* **2012**, *71*, 26–39. [CrossRef] [PubMed]
38. Dib-Hajj, S.D.; Yang, Y.; Black, J.A.; Waxman, S.G. The Na_V1.7 sodium channel: From molecule to man. *Nat. Rev. Neurosci.* **2013**, *14*, 49–62. [CrossRef] [PubMed]
39. Blair, N.T.; Bean, B.P. Roles of tetrodotoxin (TTX)-sensitive Na^+ current, TTX-resistant Na^+ current, and Ca^{2+} current in the action potentials of nociceptive sensory neurons. *J. Neurosci.* **2002**, *22*, 10277–10290. [CrossRef] [PubMed]
40. Sleeper, A.A.; Cummins, T.R.; Dib-Hajj, S.D.; Hormuzdiar, W.; Tyrrell, L.; Waxman, S.G.; Black, J.A. Changes in expression of two tetrodotoxin-resistant sodium channels and their currents in dorsal root ganglion neurons after sciatic nerve injury but not rhizotomy. *J. Neurosci.* **2000**, *20*, 7279–7289. [CrossRef] [PubMed]
41. Lin, Z.; Santos, S.; Padilla, K.; Printzenhoff, D.; Castle, N.A. Biophysical and Pharmacological Characterization of Nav1.9 Voltage Dependent Sodium Channels Stably Expressed in HEK-293 Cells. *PLoS ONE* **2016**, *11*, e0161450. [CrossRef]
42. Moldovan, M.; Alvarez, S.; Romer Rosberg, M.; Krarup, C. Axonal voltage-gated ion channels as pharmacological targets for pain. *Eur. J. Pharmacol.* **2013**, *708*, 105–112. [CrossRef] [PubMed]
43. Lai, H.C.; Jan, L.Y. The distribution and targeting of neuronal voltage-gated ion channels. *Nat. Rev. Neurosci.* **2006**, *7*, 548–562. [CrossRef] [PubMed]
44. Gutman, G.A.; Chandy, K.G.; Grissmer, S.; Lazdunski, M.; McKinnon, D.; Pardo, L.A.; Robertson, G.A.; Rudy, B.; Sanguinetti, M.C.; Stühmer, W.; et al. International Union of Pharmacology. LIII. Nomenclature and molecular relationships of voltage-gated potassium channels. *Pharmacol. Rev.* **2005**, *57*, 473–508. [CrossRef] [PubMed]
45. Gutman, G.A.; Chandy, K.G.; Adelman, J.P.; Aiyar, J.; Bayliss, D.A.; Clapham, D.E.; Covarriubias, M.; Desir, G.V.; Furuichi, K.; Ganetzky, B.; et al. International Union of Pharmacology. XLI. Compendium of voltage-gated ion channels: Potassium channels. *Pharmacol. Rev.* **2003**, *55*, 583–586. [CrossRef]
46. Rasband, M.N.; Park, E.W.; Vanderah, T.W.; Lai, J.; Porreca, F.; Trimmer, J.S. Distinct potassium channels on pain-sensing neurons. *Proc. Natl. Acad. Sci. USA* **2001**, *98*, 13373–13378. [CrossRef] [PubMed]
47. Fan, L.; Guan, X.; Wang, W.; Zhao, J.Y.; Zhang, H.; Tiwari, V.; Hoffman, P.N.; Li, M.; Tao, Y.X. Impaired neuropathic pain and preserved acute pain in rats overexpressing voltage-gated potassium channel subunit Kv1.2 in primary afferent neurons. *Mol. Pain* **2014**, *10*, 8. [CrossRef] [PubMed]
48. Sokolova, O.; Accardi, A.; Gutierrez, D.; Lau, A.; Rigney, M.; Grigorieff, N. Conformational changes in the C terminus of Shaker K^+ channel bound to the rat Kvbeta2-subunit. *Proc. Natl. Acad. Sci. USA* **2003**, *100*, 12607–12612. [CrossRef] [PubMed]
49. MacKinnon, R. Potassium channels. *FEBS Lett.* **2003**, *555*, 62–65. [CrossRef] [PubMed]
50. Dolly, J.O.; Parcej, D.N. Molecular properties of voltage-gated K^+ channels. *J. Bioenerg. Biomembr.* **1996**, *28*, 231–253. [CrossRef] [PubMed]
51. Rasband, M.N.; Trimmer, J.S.; Peles, E.; Levinson, S.R.; Shrager, P. K^+ channel distribution and clustering in developing and hypomyelinated axons of the optic nerve. *J. Neurocytol.* **1999**, *28*, 319–331. [CrossRef] [PubMed]
52. Busserolles, J.; Tsantoulas, C.; Eschalier, A.; López García, J.A. Potassium channels in neuropathic pain: Advances, challenges, and emerging ideas. *Pain* **2016**, *157* (Suppl. S1), S7–S14. [CrossRef] [PubMed]

53. Tsantoulas, C.; Zhu, L.; Shaifta, Y.; Grist, J.; Ward, J.P.; Raouf, R.; Michael, G.J.; McMahon, S.B. Sensory neuron downregulation of the Kv9.1 potassium channel subunit mediates neuropathic pain following nerve injury. *J. Neurosci.* **2012**, *32*, 17502–17513. [CrossRef] [PubMed]
54. Johnston, J.; Forsythe, I.D.; Kopp-Scheinpflug, C. Going native: Voltage-gated potassium channels controlling neuronal excitability. *J. Physiol.* **2010**, *588*, 3187–3200. [CrossRef] [PubMed]
55. Chien, L.Y.; Cheng, J.K.; Chu, D.; Cheng, C.F.; Tsaur, M.L. Reduced expression of A-type potassium channels in primary sensory neurons induces mechanical hypersensitivity. *J. Neurosci.* **2007**, *27*, 9855–9865. [CrossRef] [PubMed]
56. Everill, B.; Rizzo, M.A.; Kocsis, J.D. Morphologically identified cutaneous afferent DRG neurons express three different potassium currents in varying proportions. *J. Neurophysiol.* **1998**, *79*, 1814–1824. [CrossRef] [PubMed]
57. Takeda, M.; Tsuboi, Y.; Kitagawa, J.; Nakagawa, K.; Iwata, K.; Matsumoto, S. Potassium channels as a potential therapeutic target for trigeminal neuropathic and inflammatory pain. *Mol. Pain* **2011**, *7*, 5. [CrossRef] [PubMed]
58. Brew, H.M.; Forsythe, I.D. Two voltage-dependent K^+ conductances with complementary functions in postsynaptic integration at a central auditory synapse. *J. Neurosci.* **1995**, *15*, 8011–8022. [CrossRef] [PubMed]
59. Carrasquillo, Y.; Nerbonne, J.M. IA channels: Diverse regulatory mechanisms. *Neuroscientist* **2014**, *20*, 104–111. [CrossRef] [PubMed]
60. Jensen, C.S.; Rasmussen, H.B.; Misonou, H. Neuronal trafficking of voltage-gated potassium channels. *Mol. Cell. Neurosci.* **2011**, *48*, 288–297. [CrossRef] [PubMed]
61. Cai, X.; Liang, C.W.; Muralidharan, S.; Kao, J.P.; Tang, C.M.; Thompson, S.M. Unique roles of SK and Kv4.2 potassium channels in dendritic integration. *Neuron* **2004**, *44*, 351–364. [CrossRef] [PubMed]
62. Hu, J.H.; Liu, Y.; Hoffman, D.A. Identification of Kv4.2 protein complex and modifications by tandem affinity purification-mass spectrometry in primary neurons. *Front. Cell. Neurosci.* **2022**, *16*, 1070305. [CrossRef] [PubMed]
63. Norris, A.J.; Foeger, N.C.; Nerbonne, J.M. Interdependent roles for accessory KChIP2, KChIP3, and KChIP4 subunits in the generation of Kv4-encoded IA channels in cortical pyramidal neurons. *J. Neurosci.* **2010**, *30*, 13644–13655. [CrossRef] [PubMed]
64. An, W.F.; Bowlby, M.R.; Betty, M.; Cao, J.; Ling, H.P.; Mendoza, G.; Hinson, J.W.; Mattsson, K.I.; Strassle, B.W.; Trimmer, J.S.; et al. Modulation of A-type potassium channels by a family of calcium sensors. *Nature* **2000**, *403*, 553–556. [CrossRef] [PubMed]
65. Yang, E.K.; Takimoto, K.; Hayashi, Y.; de Groat, W.C.; Yoshimura, N. Altered expression of potassium channel subunit mRNA and alpha-dendrotoxin sensitivity of potassium currents in rat dorsal root ganglion neurons after axotomy. *Neuroscience* **2004**, *123*, 867–874. [CrossRef] [PubMed]
66. D'Adamo, M.C.; Liantonio, A.; Rolland, J.F.; Pessia, M.; Imbrici, P. Kv1.1 Channelopathies: Pathophysiological Mechanisms and Therapeutic Approaches. *Int. J. Mol. Sci.* **2020**, *21*, 2935. [CrossRef] [PubMed]
67. Schwarz, J.R.; Glassmeier, G.; Cooper, E.C.; Kao, T.C.; Nodera, H.; Tabuena, D.; Kaji, R.; Bostock, H. KCNQ channels mediate IKs, a slow K^+ current regulating excitability in the rat node of Ranvier. *J. Physiol.* **2006**, *573*, 17–34. [CrossRef] [PubMed]
68. Devaux, J.J.; Kleopa, K.A.; Cooper, E.C.; Scherer, S.S. KCNQ2 is a nodal K^+ channel. *J. Neurosci.* **2004**, *24*, 1236–1244. [CrossRef] [PubMed]
69. Wang, H.; Kunkel, D.D.; Schwartzkroin, P.A.; Tempel, B.L. Localization of Kv1.1 and Kv1.2, two K channel proteins, to synaptic terminals, somata, and dendrites in the mouse brain. *J. Neurosci.* **1994**, *14*, 4588–4599. [CrossRef]
70. Rasband, M.N. Clustered K^+ channel complexes in axons. *Neurosci. Lett.* **2010**, *486*, 101–106. [CrossRef] [PubMed]
71. Hivert, B.; Marien, L.; Agbam, K.N.; Faivre-Sarrailh, C. ADAM22 and ADAM23 modulate the targeting of the Kv1 channel-associated protein LGI1 to the axon initial segment. *J. Cell Sci.* **2019**, *132*, jcs219774. [CrossRef]
72. Zheng, Y.; Liu, H.; Chen, Y.; Dong, S.; Wang, F.; Wang, S.; Li, G.L.; Shu, Y.; Xu, F. Structural insights into the lipid and ligand regulation of a human neuronal KCNQ channel. *Neuron* **2022**, *110*, 237–247.e234. [CrossRef] [PubMed]
73. Abd-Elsayed, A.; Jackson, M.; Gu, S.L.; Fiala, K.; Gu, J. Neuropathic pain and Kv7 voltage-gated potassium channels: The potential role of Kv7 activators in the treatment of neuropathic pain. *Mol. Pain* **2019**, *15*, 1744806919864256. [CrossRef] [PubMed]
74. Battefeld, A.; Tran, B.T.; Gavrilis, J.; Cooper, E.C.; Kole, M.H. Heteromeric Kv7.2/7.3 channels differentially regulate action potential initiation and conduction in neocortical myelinated axons. *J. Neurosci.* **2014**, *34*, 3719–3732. [CrossRef] [PubMed]
75. Ying, Y.; Gong, L.; Tao, X.; Ding, J.; Chen, N.; Yao, Y.; Liu, J.; Chen, C.; Zhu, T.; Jiang, P. Genetic Knockout of TRPM2 Increases Neuronal Excitability of Hippocampal Neurons by Inhibiting Kv7 Channel in Epilepsy. *Mol. Neurobiol.* **2022**, *59*, 6918–6933. [CrossRef] [PubMed]
76. Cooper, E.C.; Aldape, K.D.; Abosch, A.; Barbaro, N.M.; Berger, M.S.; Peacock, W.S.; Jan, Y.N.; Jan, L.Y. Colocalization and coassembly of two human brain M-type potassium channel subunits that are mutated in epilepsy. *Proc. Natl. Acad. Sci. USA* **2000**, *97*, 4914–4919. [CrossRef] [PubMed]
77. Lytton, J. Na^+/Ca^{2+} exchangers: Three mammalian gene families control Ca^{2+} transport. *Biochem. J.* **2007**, *406*, 365–382. [CrossRef] [PubMed]
78. Golovina, V.A.; Bambrick, L.L.; Yarowsky, P.J.; Krueger, B.K.; Blaustein, M.P. Modulation of two functionally distinct Ca^{2+} stores in astrocytes: Role of the plasmalemmal Na/Ca exchanger. *Glia* **1996**, *16*, 296–305. [CrossRef]
79. Khananshvili, D. The SLC8 gene family of sodium-calcium exchangers (NCX)—Structure, function, and regulation in health and disease. *Mol. Asp. Med.* **2013**, *34*, 220–235. [CrossRef] [PubMed]
80. On, C.; Marshall, C.R.; Chen, N.; Moyes, C.D.; Tibbits, G.F. Gene structure evolution of the Na^+-Ca^{2+} exchanger (NCX) family. *BMC Evol. Biol.* **2008**, *8*, 127. [CrossRef] [PubMed]

81. Cai, X.; Lytton, J. The cation/Ca^{2+} exchanger superfamily: Phylogenetic analysis and structural implications. *Mol. Biol. Evol.* **2004**, *21*, 1692–1703. [CrossRef] [PubMed]
82. Khananshvili, D. Basic and editing mechanisms underlying ion transport and regulation in NCX variants. *Cell Calcium* **2020**, *85*, 102131. [CrossRef] [PubMed]
83. Sharma, V.; O'Halloran, D.M. Recent structural and functional insights into the family of sodium calcium exchangers. *Genesis* **2014**, *52*, 93–109. [CrossRef] [PubMed]
84. Hilge, M.; Aelen, J.; Vuister, G.W. Ca^{2+} regulation in the Na^{+}/Ca^{2+} exchanger involves two markedly different Ca^{2+} sensors. *Mol. Cell* **2006**, *22*, 15–25. [CrossRef] [PubMed]
85. Bode, K.; O'Halloran, D.M. NCX-DB: A unified resource for integrative analysis of the sodium calcium exchanger super-family. *BMC Neurosci.* **2018**, *19*, 19. [CrossRef] [PubMed]
86. Huang, Y.; Wen, L.L.; Xie, J.D.; Ouyang, H.D.; Chen, D.T.; Zeng, W.A. Antinociceptive effectiveness of the inhibition of NCX reverse-mode action in rodent neuropathic pain model. *Mol. Pain* **2019**, *15*, 1744806919864511. [CrossRef] [PubMed]
87. Ballarini, E.; Malacrida, A.; Rodriguez-Menendez, V.; Pozzi, E.; Canta, A.; Chiorazzi, A.; Monza, L.; Semperboni, S.; Meregalli, C.; Carozzi, V.A.; et al. Sodium-Calcium Exchanger 2: A Pivotal Role in Oxaliplatin Induced Peripheral Neurotoxicity and Axonal Damage? *Int. J. Mol. Sci.* **2022**, *23*, 10063. [CrossRef]
88. Stys, P.K.; Waxman, S.G.; Ransom, B.R. Ionic mechanisms of anoxic injury in mammalian CNS white matter: Role of Na^{+} channels and Na^{+}-Ca^{2+} exchanger. *J. Neurosci.* **1992**, *12*, 430–439. [CrossRef] [PubMed]
89. Reyes, R.C.; Verkhratsky, A.; Parpura, V. Plasmalemmal Na^{+}/Ca^{2+} exchanger modulates Ca^{2+}-dependent exocytotic release of glutamate from rat cortical astrocytes. *ASN Neuro* **2012**, *4*, AN20110059. [CrossRef]
90. Boscia, F.; Begum, G.; Pignataro, G.; Sirabella, R.; Cuomo, O.; Casamassa, A.; Sun, D.; Annunziato, L. Glial Na^{+}-dependent ion transporters in pathophysiological conditions. *Glia* **2016**, *64*, 1677–1697. [CrossRef] [PubMed]
91. Persson, A.K.; Hoeijmakers, J.G.J.; Estacion, M.; Black, J.A.; Waxman, S.G. Sodium Channels, Mitochondria, and Axonal Degeneration in Peripheral Neuropathy. *Trends Mol. Med.* **2016**, *22*, 377–390. [CrossRef]
92. Quednau, B.D.; Nicoll, D.A.; Philipson, K.D. Tissue specificity and alternative splicing of the Na^{+}/Ca^{2+} exchanger isoforms NCX1, NCX2, and NCX3 in rat. *Am. J. Physiol.* **1997**, *272*, C1250–C1261. [CrossRef] [PubMed]
93. Thurneysen, T.; Nicoll, D.A.; Philipson, K.D.; Porzig, H. Sodium/calcium exchanger subtypes NCX1, NCX2 and NCX3 show cell-specific expression in rat hippocampus cultures. *Brain Res. Mol. Brain Res.* **2002**, *107*, 145–156. [CrossRef] [PubMed]
94. Canitano, A.; Papa, M.; Boscia, F.; Castaldo, P.; Sellitti, S.; Taglialatela, M.; Annunziato, L. Brain distribution of the Na^{+}/Ca^{2+} exchanger-encoding genes NCX1, NCX2, and NCX3 and their related proteins in the central nervous system. *Ann. N. Y. Acad. Sci.* **2002**, *976*, 394–404. [CrossRef]
95. Yu, L.; Colvin, R.A. Regional differences in expression of transcripts for Na^{+}/Ca^{2+} exchanger isoforms in rat brain. *Brain Res. Mol. Brain Res.* **1997**, *50*, 285–292. [CrossRef] [PubMed]
96. Jeon, D.; Yang, Y.M.; Jeong, M.J.; Philipson, K.D.; Rhim, H.; Shin, H.S. Enhanced learning and memory in mice lacking Na^{+}/Ca^{2+} exchanger 2. *Neuron* **2003**, *38*, 965–976. [CrossRef] [PubMed]
97. Persson, A.K.; Black, J.A.; Gasser, A.; Cheng, X.; Fischer, T.Z.; Waxman, S.G. Sodium-calcium exchanger and multiple sodium channel isoforms in intra-epidermal nerve terminals. *Mol. Pain* **2010**, *6*, 84. [CrossRef] [PubMed]
98. Engelhardt, M.; Vorwald, S.; Sobotzik, J.M.; Bennett, V.; Schultz, C. Ankyrin-B structurally defines terminal microdomains of peripheral somatosensory axons. *Brain Struct. Funct.* **2013**, *218*, 1005–1016. [CrossRef] [PubMed]
99. Persson, A.K.; Kim, I.; Zhao, P.; Estacion, M.; Black, J.A.; Waxman, S.G. Sodium channels contribute to degeneration of dorsal root ganglion neurites induced by mitochondrial dysfunction in an in vitro model of axonal injury. *J. Neurosci.* **2013**, *33*, 19250–19261. [CrossRef]
100. Molinaro, P.; Viggiano, D.; Nisticò, R.; Sirabella, R.; Secondo, A.; Boscia, F.; Pannaccione, A.; Scorziello, A.; Mehdawy, B.; Sokolow, S.; et al. Na^{+}-Ca^{2+} exchanger (NCX3) knock-out mice display an impairment in hippocampal long-term potentiation and spatial learning and memory. *J. Neurosci.* **2011**, *31*, 7312–7321. [CrossRef] [PubMed]
101. Gribkoff, V.K. The role of voltage-gated calcium channels in pain and nociception. *Semin. Cell Dev. Biol.* **2006**, *17*, 555–564. [CrossRef] [PubMed]
102. Lee, S. Pharmacological Inhibition of Voltage-gated Ca^{2+} Channels for Chronic Pain Relief. *Curr. Neuropharmacol.* **2013**, *11*, 606–620. [CrossRef] [PubMed]
103. Neher, E.; Sakaba, T. Multiple roles of calcium ions in the regulation of neurotransmitter release. *Neuron* **2008**, *59*, 861–872. [CrossRef] [PubMed]
104. Bagur, R.; Hajnóczky, G. Intracellular Ca^{2+} Sensing: Its Role in Calcium Homeostasis and Signaling. *Mol. Cell* **2017**, *66*, 780–788. [CrossRef] [PubMed]
105. Takahashi, M.; Seagar, M.J.; Jones, J.F.; Reber, B.F.; Catterall, W.A. Subunit structure of dihydropyridine-sensitive calcium channels from skeletal muscle. *Proc. Natl. Acad. Sci. USA* **1987**, *84*, 5478–5482. [CrossRef] [PubMed]
106. Dolphin, A.C. Calcium channel auxiliary $\alpha 2\delta$ and β subunits: Trafficking and one step beyond. *Nat. Rev. Neurosci.* **2012**, *13*, 542–555. [CrossRef] [PubMed]
107. Li, C.Y.; Zhang, X.L.; Matthews, E.A.; Li, K.W.; Kurwa, A.; Boroujerdi, A.; Gross, J.; Gold, M.S.; Dickenson, A.H.; Feng, G.; et al. Calcium channel alpha2delta1 subunit mediates spinal hyperexcitability in pain modulation. *Pain* **2006**, *125*, 20–34. [CrossRef] [PubMed]

108. Luan, C.; Ye, Y.; Singh, T.; Barghouth, M.; Eliasson, L.; Artner, I.; Zhang, E.; Renström, E. The calcium channel subunit gamma-4 is regulated by MafA and necessary for pancreatic beta-cell specification. *Commun. Biol.* **2019**, *2*, 106. [CrossRef] [PubMed]
109. Cain, S.M.; Snutch, T.P. Voltage-gated calcium channels and disease. *Biofactors* **2011**, *37*, 197–205. [CrossRef] [PubMed]
110. Ertel, E.A.; Campbell, K.P.; Harpold, M.M.; Hofmann, F.; Mori, Y.; Perez-Reyes, E.; Schwartz, A.; Snutch, T.P.; Tanabe, T.; Birnbaumer, L.; et al. Nomenclature of voltage-gated calcium channels. *Neuron* **2000**, *25*, 533–535. [CrossRef] [PubMed]
111. McGivern, J.G.; McDonough, S.I. Voltage-gated calcium channels as targets for the treatment of chronic pain. *Curr. Drug Targets CNS Neurol. Disord.* **2004**, *3*, 457–478. [CrossRef] [PubMed]
112. Cai, S.; Gomez, K.; Moutal, A.; Khanna, R. Targeting T-type/Cav3.2 channels for chronic pain. *Transl. Res.* **2021**, *234*, 20–30. [CrossRef] [PubMed]
113. Altier, C.; Zamponi, G.W. Targeting Ca^{2+} channels to treat pain: T-type versus N-type. *Trends Pharmacol. Sci.* **2004**, *25*, 465–470. [CrossRef] [PubMed]
114. Smith, M.T.; Cabot, P.J.; Ross, F.B.; Robertson, A.D.; Lewis, R.J. The novel N-type calcium channel blocker, AM336, produces potent dose-dependent antinociception after intrathecal dosing in rats and inhibits substance P release in rat spinal cord slices. *Pain* **2002**, *96*, 119–127. [CrossRef] [PubMed]
115. Saegusa, H.; Matsuda, Y.; Tanabe, T. Effects of ablation of N- and R-type Ca^{2+} channels on pain transmission. *Neurosci. Res.* **2002**, *43*, 1–7. [CrossRef] [PubMed]
116. Hoppanova, L.; Lacinova, L. Voltage-dependent Cav3.2 and Cav2.2 channels in nociceptive pathways. *Pflug. Arch.* **2022**, *474*, 421–434. [CrossRef]
117. Todorovic, S.M.; Jevtovic-Todorovic, V. Neuropathic pain: Role for presynaptic T-type channels in nociceptive signaling. *Pflug. Arch.* **2013**, *465*, 921–927. [CrossRef] [PubMed]
118. Jacus, M.O.; Uebele, V.N.; Renger, J.J.; Todorovic, S.M. Presynaptic Cav3.2 channels regulate excitatory neurotransmission in nociceptive dorsal horn neurons. *J. Neurosci.* **2012**, *32*, 9374–9382. [CrossRef] [PubMed]
119. Bourinet, E.; Alloui, A.; Monteil, A.; Barrère, C.; Couette, B.; Poirot, O.; Pages, A.; McRory, J.; Snutch, T.P.; Eschalier, A.; et al. Silencing of the Cav3.2 T-type calcium channel gene in sensory neurons demonstrates its major role in nociception. *EMBO J.* **2005**, *24*, 315–324. [CrossRef] [PubMed]
120. Chukyo, A.; Chiba, T.; Kambe, T.; Yamamoto, K.; Kawakami, K.; Taguchi, K.; Abe, K. Oxaliplatin-induced changes in expression of transient receptor potential channels in the dorsal root ganglion as a neuropathic mechanism for cold hypersensitivity. *Neuropeptides* **2018**, *67*, 95–101. [CrossRef] [PubMed]
121. Wu, L.J.; Sweet, T.B.; Clapham, D.E. International Union of Basic and Clinical Pharmacology. LXXVI. Current progress in the mammalian TRP ion channel family. *Pharmacol. Rev.* **2010**, *62*, 381–404. [CrossRef] [PubMed]
122. Feng, S. TRPC Channel Structure and Properties. *Adv. Exp. Med. Biol.* **2017**, *976*, 9–23. [CrossRef] [PubMed]
123. Seebohm, G.; Schreiber, J.A. Beyond Hot and Spicy: TRPV Channels and their Pharmacological Modulation. *Cell. Physiol. Biochem.* **2021**, *55*, 108–130. [CrossRef] [PubMed]
124. Chen, M.; Li, X. Role of TRPV4 channel in vasodilation and neovascularization. *Microcirculation* **2021**, *28*, e12703. [CrossRef] [PubMed]
125. Naert, R.; López-Requena, A.; Talavera, K. TRPA1 Expression and Pathophysiology in Immune Cells. *Int. J. Mol. Sci.* **2021**, *22*, 11460. [CrossRef] [PubMed]
126. Zholos, A.; Johnson, C.; Burdyga, T.; Melanaphy, D. TRPM channels in the vasculature. *Adv. Exp. Med. Biol.* **2011**, *704*, 707–729. [CrossRef] [PubMed]
127. Abuammar, H.; Bhattacharjee, A.; Simon-Vecsei, Z.; Blastyák, A.; Csordás, G.; Páli, T.; Juhász, G. Ion Channels and Pumps in Autophagy: A Reciprocal Relationship. *Cells* **2021**, *10*, 3537. [CrossRef] [PubMed]
128. Cavaletti, G.; Alberti, P.; Canta, A.; Carozzi, V.; Cherchi, L.; Chiorazzi, A.; Crippa, L.; Marmiroli, P.; Meregalli, C.; Pozzi, E.; et al. Translation of paclitaxel-induced peripheral neurotoxicity from mice to patients: The importance of model selection. *Pain* **2024**. [CrossRef] [PubMed]
129. Adelsberger, H.; Quasthoff, S.; Grosskreutz, J.; Lepier, A.; Eckel, F.; Lersch, C. The chemotherapeutic oxaliplatin alters voltage-gated Na^+ channel kinetics on rat sensory neurons. *Eur. J. Pharmacol.* **2000**, *406*, 25–32. [CrossRef] [PubMed]
130. Chang, W.; Berta, T.; Kim, Y.H.; Lee, S.; Lee, S.Y.; Ji, R.R. Expression and Role of Voltage-Gated Sodium Channels in Human Dorsal Root Ganglion Neurons with Special Focus on $Na_V1.7$, Species Differences, and Regulation by Paclitaxel. *Neurosci. Bull.* **2018**, *34*, 4–12. [CrossRef]
131. Lee, J.H.; Gang, J.; Yang, E.; Kim, W.; Jin, Y.H. Bee Venom Acupuncture Attenuates Oxaliplatin-Induced Neuropathic Pain by Modulating Action Potential Threshold in A-Fiber Dorsal Root Ganglia Neurons. *Toxins* **2020**, *12*, 737. [CrossRef] [PubMed]
132. Verma, P.; Eaton, M.; Kienle, A.; Flockerzi, D.; Yang, Y.; Ramkrishna, D. Examining Sodium and Potassium Channel Conductances Involved in Hyperexcitability of Chemotherapy-Induced Peripheral Neuropathy: A Mathematical and Cell Culture-Based Study. *Front. Comput. Neurosci.* **2020**, *14*, 564980. [CrossRef]
133. Tomaszewski, A.; Büsselberg, D. Cisplatin modulates voltage gated channel currents of dorsal root ganglion neurons of rats. *Neurotoxicology* **2007**, *28*, 49–58. [CrossRef] [PubMed]
134. Brenneman, D.E.; Kinney, W.A.; Ward, S.J. Knockdown siRNA Targeting the Mitochondrial Sodium-Calcium Exchanger-1 Inhibits the Protective Effects of Two Cannabinoids Against Acute Paclitaxel Toxicity. *J. Mol. Neurosci.* **2019**, *68*, 603–619. [CrossRef] [PubMed]

135. Li, Y.; Tatsui, C.E.; Rhines, L.D.; North, R.Y.; Harrison, D.S.; Cassidy, R.M.; Johansson, C.A.; Kosturakis, A.K.; Edwards, D.D.; Zhang, H.; et al. Dorsal root ganglion neurons become hyperexcitable and increase expression of voltage-gated T-type calcium channels (Cav3.2) in paclitaxel-induced peripheral neuropathy. *Pain* **2017**, *158*, 417–429. [CrossRef] [PubMed]
136. Leo, M.; Schmitt, L.I.; Erkel, M.; Melnikova, M.; Thomale, J.; Hagenacker, T. Cisplatin-induced neuropathic pain is mediated by upregulation of N-type voltage-gated calcium channels in dorsal root ganglion neurons. *Exp. Neurol.* **2017**, *288*, 62–74. [CrossRef]
137. Schmitt, L.I.; Leo, M.; Kleinschnitz, C.; Hagenacker, T. Oxaliplatin Modulates the Characteristics of Voltage-Gated Calcium Channels and Action Potentials in Small Dorsal Root Ganglion Neurons of Rats. *Mol. Neurobiol.* **2018**, *55*, 8842–8855. [CrossRef] [PubMed]
138. Tomita, S.; Sekiguchi, F.; Deguchi, T.; Miyazaki, T.; Ikeda, Y.; Tsubota, M.; Yoshida, S.; Nguyen, H.D.; Okada, T.; Toyooka, N.; et al. Critical role of Cav3.2 T-type calcium channels in the peripheral neuropathy induced by bortezomib, a proteasome-inhibiting chemotherapeutic agent, in mice. *Toxicology* **2019**, *413*, 33–39. [CrossRef] [PubMed]
139. Materazzi, S.; Fusi, C.; Benemei, S.; Pedretti, P.; Patacchini, R.; Nilius, B.; Prenen, J.; Creminon, C.; Geppetti, P.; Nassini, R. TRPA1 and TRPV4 mediate paclitaxel-induced peripheral neuropathy in mice via a glutathione-sensitive mechanism. *Pflug. Arch.* **2012**, *463*, 561–569. [CrossRef] [PubMed]
140. Nassini, R.; Gees, M.; Harrison, S.; De Siena, G.; Materazzi, S.; Moretto, N.; Failli, P.; Preti, D.; Marchetti, N.; Cavazzini, A.; et al. Oxaliplatin elicits mechanical and cold allodynia in rodents via TRPA1 receptor stimulation. *Pain* **2011**, *152*, 1621–1631. [CrossRef] [PubMed]
141. Sánchez, J.C.; Muñoz, L.V.; Ehrlich, B.E. Modulating TRPV4 channels with paclitaxel and lithium. *Cell Calcium* **2020**, *91*, 102266. [CrossRef] [PubMed]
142. Ta, L.E.; Bieber, A.J.; Carlton, S.M.; Loprinzi, C.L.; Low, P.A.; Windebank, A.J. Transient Receptor Potential Vanilloid 1 is essential for cisplatin-induced heat hyperalgesia in mice. *Mol. Pain* **2010**, *6*, 15. [CrossRef]
143. Trevisan, G.; Materazzi, S.; Fusi, C.; Altomare, A.; Aldini, G.; Lodovici, M.; Patacchini, R.; Geppetti, P.; Nassini, R. Novel therapeutic strategy to prevent chemotherapy-induced persistent sensory neuropathy by TRPA1 blockade. *Cancer Res.* **2013**, *73*, 3120–3131. [CrossRef] [PubMed]
144. Ertilav, K.; Nazıroğlu, M.; Ataizi, Z.S.; Yıldızhan, K. Melatonin and Selenium Suppress Docetaxel-Induced TRPV1 Activation, Neuropathic Pain and Oxidative Neurotoxicity in Mice. *Biol. Trace Elem. Res.* **2021**, *199*, 1469–1487. [CrossRef] [PubMed]
145. Anand, U.; Otto, W.R.; Anand, P. Sensitization of capsaicin and icilin responses in oxaliplatin treated adult rat DRG neurons. *Mol. Pain* **2010**, *6*, 82. [CrossRef] [PubMed]
146. Leo, M.; Schmitt, L.I.; Küsterarent, P.; Kutritz, A.; Rassaf, T.; Kleinschnitz, C.; Hendgen-Cotta, U.B.; Hagenacker, T. Platinum-Based Drugs Cause Mitochondrial Dysfunction in Cultured Dorsal Root Ganglion Neurons. *Int. J. Mol. Sci.* **2020**, *21*. [CrossRef] [PubMed]
147. Sánchez, J.C.; Muñoz, L.V.; Galindo-Márquez, M.L.; Valencia-Vásquez, A.; García, A.M. Paclitaxel Regulates TRPA1 Function and Expression Through PKA and PKC. *Neurochem. Res.* **2023**, *48*, 295–304. [CrossRef] [PubMed]
148. Caudle, R.M.; Neubert, J.K. Effects of Oxaliplatin on Facial Sensitivity to Cool Temperatures and TRPM8 Expressing Trigeminal Ganglion Neurons in Mice. *Front. Pain Res.* **2022**, *3*, 868547. [CrossRef] [PubMed]
149. Nieto, F.R.; Entrena, J.M.; Cendán, C.M.; Del Pozo, E.; Vela, J.M.; Baeyens, J.M. Tetrodotoxin inhibits the development and expression of neuropathic pain induced by paclitaxel in mice. *Pain* **2008**, *137*, 520–531. [CrossRef] [PubMed]
150. Makker, P.G.S.; White, D.; Lees, J.G.; Parmar, J.; Goldstein, D.; Park, S.B.; Howells, J.; Moalem-Taylor, G. Acute changes in nerve excitability following oxaliplatin treatment in mice. *J. Neurophysiol.* **2020**, *124*, 232–244. [CrossRef] [PubMed]
151. Alberti, P.; Canta, A.; Chiorazzi, A.; Fumagalli, G.; Meregalli, C.; Monza, L.; Pozzi, E.; Ballarini, E.; Rodriguez-Menendez, V.; Oggioni, N.; et al. Topiramate prevents oxaliplatin-related axonal hyperexcitability and oxaliplatin induced peripheral neurotoxicity. *Neuropharmacology* **2020**, *164*, 107905. [CrossRef]
152. Braden, K.; Stratton, H.J.; Salvemini, D.; Khanna, R. Small molecule targeting NaV1.7 via inhibition of the CRMP2-Ubc9 interaction reduces and prevents pain chronification in a mouse model of oxaliplatin-induced neuropathic pain. *Neurobiol. Pain* **2022**, *11*, 100082. [CrossRef] [PubMed]
153. Di Cesare Mannelli, L.; Lucarini, E.; Micheli, L.; Mosca, I.; Ambrosino, P.; Soldovieri, M.V.; Martelli, A.; Testai, L.; Taglialatela, M.; Calderone, V.; et al. Effects of natural and synthetic isothiocyanate-based H. *Neuropharmacology* **2017**, *121*, 49–59. [CrossRef] [PubMed]
154. Jia, S.; Wei, G.; Bono, J.; Pan, Z.; Zheng, B.; Wang, B.; Adaralegbe, A.; Tenorio, C.; Bekker, A.; Tao, Y.X. TET1 overexpression attenuates paclitaxel-induced neuropathic pain through rescuing K. *Life Sci.* **2022**, *297*, 120486. [CrossRef]
155. Kagiava, A.; Tsingotjidou, A.; Emmanouilides, C.; Theophilidis, G. The effects of oxaliplatin, an anticancer drug, on potassium channels of the peripheral myelinated nerve fibres of the adult rat. *Neurotoxicology* **2008**, *29*, 1100–1106. [CrossRef] [PubMed]
156. Kanbara, T.; Nakamura, A.; Shibasaki, M.; Mori, T.; Suzuki, T.; Sakaguchi, G.; Kanemasa, T. Morphine and oxycodone, but not fentanyl, exhibit antinociceptive effects mediated by G-protein inwardly rectifying potassium (GIRK) channels in an oxaliplatin-induced neuropathy rat model. *Neurosci. Lett.* **2014**, *580*, 119–124. [CrossRef] [PubMed]
157. Lucarini, E.; Micheli, L.; Trallori, E.; Citi, V.; Martelli, A.; Testai, L.; De Nicola, G.R.; Iori, R.; Calderone, V.; Ghelardini, C.; et al. Effect of glucoraphanin and sulforaphane against chemotherapy-induced neuropathic pain: K_v7 potassium channels modulation by H_2S release in vivo. *Phytother. Res.* **2018**, *32*, 2226–2234. [CrossRef] [PubMed]

158. Yilmaz, E.; Gold, M.S. Paclitaxel-induced increase in NCX activity in subpopulations of nociceptive afferents: A protective mechanism against chemotherapy-induced peripheral neuropathy? *Cell Calcium* **2016**, *60*, 25–31. [CrossRef] [PubMed]
159. Nodera, H.; Spieker, A.; Sung, M.; Rutkove, S. Neuroprotective effects of Kv7 channel agonist, retigabine, for cisplatin-induced peripheral neuropathy. *Neurosci. Lett.* **2011**, *505*, 223–227. [CrossRef] [PubMed]
160. Kawakami, K.; Chiba, T.; Katagiri, N.; Saduka, M.; Abe, K.; Utsunomiya, I.; Hama, T.; Taguchi, K. Paclitaxel increases high voltage-dependent calcium channel current in dorsal root ganglion neurons of the rat. *J. Pharmacol. Sci.* **2012**, *120*, 187–195. [CrossRef] [PubMed]
161. Matsumoto, M.; Inoue, M.; Hald, A.; Xie, W.; Ueda, H. Inhibition of paclitaxel-induced A-fiber hypersensitization by gabapentin. *J. Pharmacol. Exp. Ther.* **2006**, *318*, 735–740. [CrossRef] [PubMed]
162. Okubo, K.; Nakanishi, H.; Matsunami, M.; Shibayama, H.; Kawabata, A. Topical application of disodium isostearyl 2-O-L-ascorbyl phosphate, an amphiphilic ascorbic acid derivative, reduces neuropathic hyperalgesia in rats. *Br. J. Pharmacol.* **2012**, *166*, 1058–1068. [CrossRef] [PubMed]
163. Sekiguchi, F.; Kawara, Y.; Tsubota, M.; Kawakami, E.; Ozaki, T.; Kawaishi, Y.; Tomita, S.; Kanaoka, D.; Yoshida, S.; Ohkubo, T.; et al. Therapeutic potential of RQ-00311651, a novel T-type Ca^{2+} channel blocker, in distinct rodent models for neuropathic and visceral pain. *Pain* **2016**, *157*, 1655–1665. [CrossRef] [PubMed]
164. Meregalli, C.; Maricich, Y.; Cavaletti, G.; Canta, A.; Carozzi, V.A.; Chiorazzi, A.; Newbold, E.; Marmiroli, P.; Ceresa, C.; Diani, A.; et al. Reversal of Bortezomib-Induced Neurotoxicity by Suvecaltamide, a Selective T-Type Ca-Channel Modulator, in Preclinical Models. *Cancers* **2021**, *13*, 5013. [CrossRef] [PubMed]
165. Sharma, J.; Maslov, L.N.; Singh, N.; Jaggi, A.S. Pain attenuating actions of vincristinet-preconditioning in chemotherapeutic agent-induced neuropathic pain: Key involvement of T-type calcium channels. *Fundam. Clin. Pharmacol.* **2020**, *34*, 336–344. [CrossRef] [PubMed]
166. Chen, Y.; Yang, C.; Wang, Z.J. Proteinase-activated receptor 2 sensitizes transient receptor potential vanilloid 1, transient receptor potential vanilloid 4, and transient receptor potential ankyrin 1 in paclitaxel-induced neuropathic pain. *Neuroscience* **2011**, *193*, 440–451. [CrossRef] [PubMed]
167. Hori, K.; Ozaki, N.; Suzuki, S.; Sugiura, Y. Upregulations of $P2X_3$ and ASIC3 involve in hyperalgesia induced by cisplatin administration in rats. *Pain* **2010**, *149*, 393–405. [CrossRef] [PubMed]
168. Quartu, M.; Carozzi, V.A.; Dorsey, S.G.; Serra, M.P.; Poddighe, L.; Picci, C.; Boi, M.; Melis, T.; Del Fiacco, M.; Meregalli, C.; et al. Bortezomib treatment produces nocifensive behavior and changes in the expression of TRPV1, CGRP, and substance P in the rat DRG, spinal cord, and sciatic nerve. *Biomed Res. Int.* **2014**, *2014*, 180428. [CrossRef] [PubMed]
169. Mao, Q.; Wu, S.; Gu, X.; Du, S.; Mo, K.; Sun, L.; Cao, J.; Bekker, A.; Chen, L.; Tao, Y.X. DNMT3a-triggered downregulation of K. *Int. J. Cancer* **2019**, *145*, 2122–2134. [CrossRef] [PubMed]
170. Pereira, V.; Busserolles, J.; Christin, M.; Devilliers, M.; Poupon, L.; Legha, W.; Alloui, A.; Aissouni, Y.; Bourinet, E.; Lesage, F.; et al. Role of the TREK2 potassium channel in cold and warm thermosensation and in pain perception. *Pain* **2014**, *155*, 2534–2544. [CrossRef]
171. Rapacz, A.; Obniska, J.; Koczurkiewicz, P.; Wójcik-Pszczoła, K.; Siwek, A.; Gryboś, A.; Rybka, S.; Karcz, A.; Pękala, E.; Filipek, B. Antiallodynic and antihyperalgesic activity of new 3,3-diphenyl-propionamides with anticonvulsant activity in models of pain in mice. *Eur. J. Pharmacol.* **2018**, *821*, 39–48. [CrossRef]
172. Sałat, K.; Furgała-Wojas, A.; Sałat, R. The Microglial Activation Inhibitor Minocycline, Used Alone and in Combination with Duloxetine, Attenuates Pain Caused by Oxaliplatin in Mice. *Molecules* **2021**, *26*, 3577. [CrossRef] [PubMed]
173. Alberti, P. Platinum-drugs induced peripheral neurotoxicity: Clinical course and preclinical evidence. *Expert Opin. Drug Metab. Toxicol.* **2019**, *15*, 487–497. [CrossRef] [PubMed]
174. Lucchetta, M.; Lonardi, S.; Bergamo, F.; Alberti, P.; Velasco, R.; Argyriou, A.A.; Briani, C.; Bruna, J.; Cazzaniga, M.; Cortinovis, D.; et al. Incidence of atypical acute nerve hyperexcitability symptoms in oxaliplatin-treated patients with colorectal cancer. *Cancer Chemother. Pharmacol.* **2012**, *70*, 899–902. [CrossRef] [PubMed]
175. Alberti, P.; Salvalaggio, A.; Argyriou, A.A.; Bruna, J.; Visentin, A.; Cavaletti, G.; Briani, C. Neurological Complications of Conventional and Novel Anticancer Treatments. *Cancers* **2022**, *14*, 6088. [CrossRef] [PubMed]
176. Chiorazzi, A.; Semperboni, S.; Marmiroli, P. Current View in Platinum Drug Mechanisms of Peripheral Neurotoxicity. *Toxics* **2015**, *3*, 304–321. [CrossRef] [PubMed]
177. Monza, L.; Fumagalli, G.; Chiorazzi, A.; Alberti, P. Addressing the Need of a Translational Approach in Peripheral Neuropathy Research: Morphology Meets Function. *Brain Sci.* **2021**, *11*, 139. [CrossRef] [PubMed]
178. Monza, L.; Fumagalli, G.; Chiorazzi, A.; Alberti, P. Translating morphology from bench side to bed side via neurophysiology: 8-min protocol for peripheral neuropathy research. *J. Neurosci. Methods* **2021**, *363*, 109323. [CrossRef] [PubMed]
179. Bostock, H.; Cikurel, K.; Burke, D. Threshold tracking techniques in the study of human peripheral nerve. *Muscle Nerve* **1998**, *21*, 137–158. [CrossRef]
180. Boërio, D.; Greensmith, L.; Bostock, H. Excitability properties of motor axons in the maturing mouse. *J. Peripher. Nerv. Syst.* **2009**, *14*, 45–53. [CrossRef]
181. Arnold, R.; Moldovan, M.; Rosberg, M.R.; Krishnan, A.V.; Morris, R.; Krarup, C. Nerve excitability in the rat forelimb: A technique to improve translational utility. *J. Neurosci. Methods* **2017**, *275*, 19–24. [CrossRef] [PubMed]

182. Park, S.B.; Goldstein, D.; Lin, C.S.; Krishnan, A.V.; Friedlander, M.L.; Kiernan, M.C. Acute abnormalities of sensory nerve function associated with oxaliplatin-induced neurotoxicity. *J. Clin. Oncol.* **2009**, *27*, 1243–1249. [CrossRef] [PubMed]
183. McHugh, J.C.; Tryfonopoulos, D.; Fennelly, D.; Crown, J.; Connolly, S. Electroclinical biomarkers of early peripheral neurotoxicity from oxaliplatin. *Eur. J. Cancer Care* **2012**, *21*, 782–789. [CrossRef] [PubMed]
184. Eckel, F.; Schmelz, R.; Adelsberger, H.; Erdmann, J.; Quasthoff, S.; Lersch, C. Prevention of oxaliplatin-induced neuropathy by carbamazepine. A pilot study. *Dtsch. Med. Wochenschr.* **2002**, *127*, 78–82. [CrossRef] [PubMed]
185. Li, L.; Shao, J.; Wang, J.; Liu, Y.; Zhang, Y.; Zhang, M.; Zhang, J.; Ren, X.; Su, S.; Li, Y.; et al. MiR-30b-5p attenuates oxaliplatin-induced peripheral neuropathic pain through the voltage-gated sodium channel Na. *Neuropharmacology* **2019**, *153*, 111–120. [CrossRef] [PubMed]
186. Azad, I.; Khan, T.; Ahmad, N.; Khan, A.R.; Akhter, Y. Updates on drug designing approach through computational strategies: A review. *Future Sci. OA* **2023**, *9*, FSO862. [CrossRef] [PubMed]

International Journal of
Molecular Sciences

MDPI

Review

Glutamate-Mediated Excitotoxicity in the Pathogenesis and Treatment of Neurodevelopmental and Adult Mental Disorders

Noemi Nicosia [1,2,†], Mattia Giovenzana [1,2,†], Paulina Misztak [1], Jessica Mingardi [1,*] and Laura Musazzi [1,3,*]

1 School of Medicine and Surgery, University of Milano-Bicocca, 20900 Monza, Italy; n.nicosia@campus.unimib.it (N.N.); m.giovenzana6@campus.unimib.it (M.G.); paulina.misztak@unimib.it (P.M.)
2 PhD Program in Neuroscience, School of Medicine and Surgery, University of Milano-Bicocca, 20900 Monza, Italy
3 Fondazione IRCCS San Gerardo dei Tintori, 20900 Monza, Italy
* Correspondence: jessica.mingardi@unimib.it (J.M.); laura.musazzi@unimib.it (L.M.)
† These authors contributed equally to this work.

Abstract: Glutamate is the main excitatory neurotransmitter in the brain wherein it controls cognitive functional domains and mood. Indeed, brain areas involved in memory formation and consolidation as well as in fear and emotional processing, such as the hippocampus, prefrontal cortex, and amygdala, are predominantly glutamatergic. To ensure the physiological activity of the brain, glutamatergic transmission is finely tuned at synaptic sites. Disruption of the mechanisms responsible for glutamate homeostasis may result in the accumulation of excessive glutamate levels, which in turn leads to increased calcium levels, mitochondrial abnormalities, oxidative stress, and eventually cell atrophy and death. This condition is known as glutamate-induced excitotoxicity and is considered as a pathogenic mechanism in several diseases of the central nervous system, including neurodevelopmental, substance abuse, and psychiatric disorders. On the other hand, these disorders share neuroplasticity impairments in glutamatergic brain areas, which are accompanied by structural remodeling of glutamatergic neurons. In the current narrative review, we will summarize the role of glutamate-induced excitotoxicity in both the pathophysiology and therapeutic interventions of neurodevelopmental and adult mental diseases with a focus on autism spectrum disorders, substance abuse, and psychiatric disorders. Indeed, glutamatergic drugs are under preclinical and clinical development for the treatment of different mental diseases that share glutamatergic neuroplasticity dysfunctions. Although clinical evidence is still limited and more studies are required, the regulation of glutamate homeostasis is attracting attention as a potential crucial target for the control of brain diseases.

Keywords: glutamate; excitotoxicity; autism spectrum disorder; substance use disorders; schizophrenia; depression; therapeutics

1. Introduction

Glutamate (or glutamic acid) is a positively charged non-essential amino acid which, besides its metabolic roles [1], also represents the most important excitatory neurotransmitter in the central nervous system (CNS), where it drives approximately 70% of synapses [2]. The neurotransmitter glutamate regulates various brain processes ranging from cognitive functions, such as learning and memory, to mood, control of the sleep–wake cycle, pain perception, and motor function [3–5]. Glutamatergic transmission is mediated by specific ionotropic and metabotropic receptors located both at postsynaptic and presynaptic terminals. Ionotropic receptors, including AMPA (α-amino-3-hydroxy-5-methyl-a-isoxazolepropionate), NMDA (N-methyl-D-aspartate), and kainate receptors, are nonselective cation channels, allowing the passage of sodium and potassium ions and, in some cases, small amounts of calcium ions. They are primarily expressed at postsynaptic sites, and their activation produces excitatory currents. The interaction of glutamate with AMPA and kainate receptors induces the generation of a very rapid but weak excitatory postsynaptic current [6], which, however, is enough to allow the removal of the Mg^{2+}-dependent block of the NMDA receptor. In turn, the binding of glutamate and of the co-agonist glycine (or D-serine) promote the opening of NMDA receptors, which are permeable also to calcium ions, resulting in long-lasting postsynaptic currents [7]. In addition to ionotropic receptors, glutamate also acts through three groups of G protein-coupled metabotropic glutamate receptors (mGluRs), which modulate synaptic function more slowly than ionotropic receptors [8]. Group I includes mGlu1 and 5, Group II includes mGlu2 and 3, and Group III includes mGlu4, 6, 7, and 8. Group I is coupled to Gq/G11 and activates phospholipase Cβ, while group II and III are coupled predominantly to Gi/o proteins, thus leading to the inhibition of adenylyl cyclase and direct regulation of ion channels. Group I mGluRs are often localized postsynaptically, and their activation leads to cell depolarization and an increase in neuronal excitability. In contrast, group II and group III mGluRs are typically localized on presynaptic terminals, where they inhibit neurotransmitter release [9]. For a detailed description of ionotropic and metabotropic glutamatergic receptor signaling, activation mechanisms, and crosstalk, please refer to [10].

Glutamate dynamics play a critical role in maintaining glutamate homeostasis and are tightly regulated by neuron–astrocyte crosstalk [2]. Indeed, glutamatergic synapses are defined as "tripartite" as they are composed of a glutamate presynaptic terminal, a postsynaptic spine, and an astrocyte [11]. Since there are no enzymes able to metabolize glutamate in the extracellular space, excess glutamate is cleared from the extracellular space by neuronal and astrocytic excitatory amino acid transporters (EAAT1-5) [2]. In astrocytes, glutamate is converted by glutamine synthetase to glutamine, which is then released and uptaken by neurons, where it can be used to produce new glutamate. Glutamate is also released into extra-synaptic space by astrocytes via the cystine–glutamate antiporter, which thus contributes to the regulation of extracellular glutamate levels [12].

A dysfunctional alteration of glutamate homeostasis may lead to glutamate excitotoxicity, a condition in which the synaptic concentration of glutamate can reach up to 100 μM, leading to neurotoxicity and neuronal atrophy/death [13]. This condition may arise in the presence of (1) an impaired glutamate/glutamine recycle system, (2) dysfunctional reuptake and (3) altered expression and activity of glutamate receptors. An excessive synaptic glutamate concentration causes an overflow of intracellular calcium ions, which in turn leads to the activation of a transduction cascade [14]. In particular, excessive intracellular calcium levels lead to considerable excitotoxic damage by an alteration of the permeability of the mitochondrial membrane, resulting in the disruption of cell energy production, and by activation of calcium-dependent enzymes such as calpains, death-associated protein kinase 1 (DAPK1), and neuronal nitric oxide synthase (nNOS) with a consequent increase in production of nitric oxide [14,15]. This, in turn, leads to an increase in the production of reactive oxygen (ROS) species that, together with the hyperactivation of proteases and lipases, contribute to oxidative stress and cell death by apoptosis [14].

Glutamate excitotoxicity has received attention as a putative mechanism in the etiopathogenesis of several CNS disorders, including neurodevelopmental, neurological, neurodegenerative, and mental diseases [16–20].

The aim of this narrative review is to summarize evidence implying mechanisms of glutamate excitotoxicity in the pathophysiological mechanisms and therapy of neurodevelopmental disorders, with a focus on autism spectrum disorders (ASD), and adult mental disorders, with particular attention to substance abuse and psychiatric disorders. Among others, we selected these diseases because they have in common recent therapeutic approaches that include glutamatergic drugs, thus suggesting that direct modulation of glutamatergic transmission may exert therapeutic effects in these conditions.

Data for this review were collected using the PubMed database.

2. Involvement of Glutamate Excitotoxicity in Autism Spectrum Disorders

ASDs include a complex family of neurodevelopmental disorders characterized by repetitive behaviors, abnormalities in communication, and impaired social interaction skills, frequently together with other concurrent medical or psychiatric conditions [21,22].

DSM V defines ASD as a condition with "persistent deficits in social communication and social interaction across multiple contexts" and "restricted, repetitive patterns of behavior, interests, or activities" from early developmental phases and significantly impacting daily life [23].

Compelling evidence has highlighted a role for glutamate-dependent excitotoxicity in the pathophysiology of ASDs. Indeed, clinical studies consistently reported increased glutamate levels and receptor subunits in both the serum and brain areas of ASD patients, and a prolonged imbalance between excitatory and inhibitory transmissions was involved in etiopathogenetic processes [24–26]. Indeed, an overexcitation (or weak inhibition) of cortical function has been associated with a broad range of abnormalities in perception, memory, cognition, and motor control [27]. Intriguingly, genetic evidence has also clearly implicated glutamate receptors and transporter systems in ASD, since several polymorphisms of glutamate receptor subunits and glutamate transporter genes have been associated with the disease [28]. Moreover, rare mutations in structural proteins of the postsynaptic density regulating glutamatergic transmission, such as Shank proteins, have also been associated with ASD [29].

Accordingly, several animal models of ASD, including both pharmacological models and genetic models generated by mutations in genes found in patients with ASD, show glutamatergic alterations. Excitotoxicity, altered glutamate homeostasis, and receptor subunit expression and regulation were reported in the brain of a valproic acid model, one of the most widely used animal models of ASD [30–33]. Similarly, Shank mutant mice clearly exhibit glutamatergic impairments [34–37]. In both models, glutamatergic interventions can rescue ASD-like behavioral dysfunctions [35,37,38].

Glutamatergic Drugs for the Treatment of Autism Spectrum Disorders

Different glutamatergic approaches were considered for the management of ASDs.

Riluzole, an inhibitor of voltage-dependent sodium channels and NMDA/kainate receptor antagonist, and memantine, a non-selective NMDA receptor antagonist, were shown to improve ASD symptoms in both children and adult patients [39,40]. The therapeutic properties of low-dose intranasal ketamine, another NMDA receptor antagonist approved as an anesthetic and recently introduced at subanesthetic dose as a rapid-acting antidepressant (see Section 5.1), were also tested in adolescent and young adults with ASD, showing some therapeutic potential with limited and transient adverse effects [41,42]. Nevertheless, the clinical evidence is still very limited, and more studies are required to understand whether glutamatergic agents can be safe and effective therapeutic strategies for ASD.

mGluR5 antagonists have been the only class of glutamatergic modulators that has made significant advances in drug development for the treatment of ASD. However, despite the enthusiasm raised due to their success in preclinical phases, unfortunately mGluR5 antagonists have failed in phase III clinical trials on ASD patients for lack of efficacy [43,44].

Finally, another molecule with multiple actions, including antioxidant effects, reduction in cytokine activity, modulation of dopamine release, reversal of mitochondrial dysfunction, reduction in apoptosis, anti-inflammatory activity, increased neurogenesis, and regulation of glutamate homeostasis, has shown promising effects in ASD [45]. It is N-acetyl-L-cysteine (NAC), the acetylated precursor of L-cysteine, which has been shown to reduce hyperactivity and irritability and to enhance social awareness in ASD [46]. However, future trials with larger sample sizes, confounding effects controlled, and long-term follow-up are warranted.

3. Involvement of Glutamate Excitotoxicity in Substance Use Disorders

As defined by the DSM-5, substance-related disorders (SRDs) and addictive disorders encompass a spectrum of treatable mental health conditions attributed to the problematic consumption of various substances, such as alcohol, opioids, stimulants, cannabis, hallucinogens, inhalants, and others [23]. Glutamate is an important mediator in both the neurotoxicity induced by abused drugs and the development and maintenance of addiction. Indeed, the rewarding process implies the intricate and bidirectional interplay between glutamate and dopamine transmissions within the mesolimbic pathway [47]. In response to rewarding stimuli, glutamate modulates dopamine release by activating NMDA and AMPA receptors on dopaminergic neurons in the ventral tegmental area (VTA). As a result, dopaminergic inputs from VTA to glutamatergic neurons in the prefrontal cortex (PFC), amygdala, and hippocampus induce alterations of glutamate release and receptor expression [48]. Consequently, drug abuse induces dynamic alterations in plasticity mechanisms such as long-term potentiation (LTP) and long-term depression (LTD) in corticolimbic brain areas, in turn impacting processes related to mood, learning, and memory [49].

In this section, we will focus on selected classes of abused drugs which cause alterations of the glutamatergic system at different stages of exposure and withdrawal, focusing on works where glutamate-mediated excitotoxicity has been directly implicated.

3.1. Cocaine

Cocaine remains a prominent choice among drugs of abuse worldwide, renowned for its potent psychostimulant properties and its ability to induce euphoria. Similarly to other psychostimulants, cocaine induces its effects primarily through activation of the mesocorticolimbic dopaminergic system [50]. Nevertheless, a growing number of preclinical studies shed light on the critical role of glutamate homeostasis in neurotoxicity at different stages of cocaine exposure [51]. In fact, acute cocaine exposure of rats affected glutamate levels within VTA [52,53], whereas a challenge dose augmented its release in medial PFC (mPFC) after repeated treatment [54,55]. In line with these findings, further research employing microdialysis showed that glutamate levels increased within the rat striatum [56] and nucleus accumbens [57] after a challenge dose. Conversely, rats exposed to 20 days of cocaine self-administration exhibited a notable reduction in extracellular glutamate levels in the nucleus accumbens, then reversed upon the administration of a priming dose of the drug [58]. In addition, chronic cocaine exposure has been associated with changes in the expression levels of both ionotropic and metabotropic glutamate receptors within the mPFC in rats [59] and the caudate nucleus of non-human primates [60].

3.2. Amphetamines

Alongside cocaine, amphetamine (AMPH) and its derivatives including 3,4-methylenedioxymethamphetamine (MDMA) are widely recognized as some of the most misused psychostimulants, especially among adolescents. These substances trigger an excessive release of monoamines at the synaptic level by binding membrane-located transporters of dopamine [61], norepinephrine, and serotonin [62] but also lead to a sustained rise in glutamate, leading to neurotoxicity [63]. Studies on rats showed that repeated exposure to methamphetamine augmented glutamate levels within various brain regions including the striatum [64], hippocampus [65], VTA [66], nucleus accumbens, and PFC [67]. Accordingly, amphetamines also induced significant alterations in the expression of AMPA receptor subunits GluA1 and GluA2 in the nucleus accumbens of rats undergoing a repeated treatment [68]. Moreover, MK-801 prevented chronic methamphetamine-induced neurotoxicity by blocking NMDA receptors [69], while repeated exposure worsens neuronal damage in the striatum through mechanisms involving AMPA receptor-mediated excitotoxicity and calpain-specific spectrin proteolysis, a process implicated in cellular apoptosis [70]. In addition, further evidence from in vivo studies supports the involvement of the nitric oxide pathway [71] and the tumor necrosis factor (TNF)-α [72] in methamphetamine-induced excitotoxicity.

Similarly to other amphetamines, MDMA, commonly known as ecstasy, has been shown to promote a sustained increase in extracellular glutamate release within the hippocampus in rats, which was then attenuated by treatment with ketanserin/fluoxetine [73] and ketoprofen [74]. Furthermore, MDMA reduces the parvalbumin-positive gamma-aminobutyric acid (GABA)ergic neurons in the hippocampus, suggesting that MDMA-induced damage involves an excitatory/inhibitory imbalance [74]. Intriguingly, some findings point to the critical involvement of glial cell activation in the neuronal damage and neurotoxic effects induced by MDMA exposure. Indeed, MDMA exposure led to microglia activation in mouse striatum, resulting in neuronal death along with an increased expression of glial fibrillary acidic protein (GFAP) [75].

3.3. Ketamine

Since its introduction in the 1960s as a dissociative anesthetic, ketamine has drawn attention for its recreational use due to its dissociative effects. Ketamine acts as a non-competitive antagonist of NMDA receptors primarily targeting inhibitory neurons in corticolimbic brain areas, thereby amplifying glutamatergic transmission [76]. Accordingly, daily exposure to ketamine was reported to cause glutamate-induced neurotoxicity, oxidative stress, and apoptosis in both the cerebral cortex and hippocampus [77,78]. Moreover, prolonged ketamine exposure was reported to enhance NMDA receptor expression and ROS production in rat primary forebrain cultures, ultimately culminating in neuronal death [79,80].

It is also worth mentioning that early prenatal or postnatal exposure to ketamine as an anesthetic can induce neurodevelopmental alterations leading to long-term cognitive impairment and learning deficits [81]. This has been associated with an overexpression of NMDA receptors, mitochondrial dysfunction, oxidative stress, and defects in neurogenetic pathways [82].

3.4. Ethanol

Ethanol, the primary psychoactive compound found in alcoholic beverages, is responsible for intoxicating effects, and its overconsumption represents a serious societal challenge, significantly contributing to worldwide morbidity [83]. Ethanol passes through the blood–brain barrier (BBB) and potentiates GABAergic transmission, thus affecting excitatory/inhibitory balance [84]. Importantly, alcohol also directly modulates excitatory glutamatergic neurotransmission. Indeed, while acute ethanol intake induces the down-regulation of postsynaptic NMDA receptors, chronic exposure increases the expression of these receptors resulting in excitotoxic cascade events, including neuronal death [83]. Post-mortem analyses of brains from individuals with a history of alcohol abuse showed that ethanol further potentiates glutamatergic transmission in several brain regions by decreasing the expression of GLT-1 transporters [85,86] and damaging astrocytes [87], which are responsible for at least 90% of glutamate reuptake [83,88–90].

Evidence from preclinical studies have provided insights into the cellular neurobiological mechanisms implied in ethanol induced excitotoxicity. For instance, in vitro chronic ethanol exposure was reported to determine abnormal synaptic transmission and excitotoxic effects in hippocampal slices involving the dysregulation of ionotropic and metabotropic glutamate receptor subunit expression [91,92]. On the other hand, chronic alcohol exposure in rats has been shown to trigger calpain activity both in the cerebral cortex and cerebellum, confirming its role in cellular damage induced by alcohol consumption [93]. Furthermore, chronic alcohol exposure was also reported to increase the levels of the stress hormone glucocorticoid, which, in turn, overactivate NMDA receptors, particularly the NR2A and NR2B subunits, exacerbating neuronal cell death induced by alcohol excitotoxicity [83].

Importantly, ethanol exposure during early neurodevelopmental phases dramatically impacts synaptogenesis and triggers massive neuronal apoptosis, that in turn may contribute to mental disability syndromes including fetal alcohol spectrum disorders (reviewed in [94]). In addition, combined abuse of ethanol and ketamine or methamphetamine has been shown to synergically exacerbate glutamate-induced excitotoxicity, leading to neurotoxic consequences [95,96].

3.5. Therapeutic Approaches for Substance Use Disorders Based on the Rescue of Glutamate-Induced Neurotoxicity

Despite the valuable insight provided by existing research on excitotoxicity induced by substance-related disorders, further investigations are needed to fully clarify these mechanisms in order to identify potential targets for therapeutic interventions. Nevertheless, given the crucial role of glutamate in the pathophysiology of addiction, therapeutic strategies targeting glutamate-induced neurotoxicity hold promise for the treatment of substance-related disorders. In this regard, cation channel blockers (Lamotrigine) and NMDA receptor antagonists (Acamprosate, Amantadine, Memantine, and MK-801) have gained much consideration in the development of therapeutic interventions, since they have been reported to protect neurons from excitotoxicity [88,97–99]. Among these, memantine and MK-801 have shown efficacy in reducing cocaine and alcohol craving and relapse in preclinical models by blocking excessive calcium influx through NMDA receptors [100–104]. However, the clinical evidence is conflicting and inconclusive [105]. On the other hand, AMPA receptor antagonists such as topiramate have been more successful in showing neuroprotective properties against glutamate neurotoxicity [61,106] and have demonstrated efficacy in reducing alcohol intake and preventing relapse in human subjects [107] and increasing abstinence in cocaine consumers [108].

Moreover, preclinical studies have demonstrated that cannabidiol, an active compound found in cannabis, exerts anti-addictive properties (reviewed in [109,110]). Interestingly, cannabidiol attenuates excitotoxicity induced by various drugs of abuse by inhibiting the release of glutamate and reducing the activity of NMDA receptors. Therefore, through the modulation of glutamatergic neurotransmission, cannabinoids may help in preventing excessive calcium influx into neurons with protective effects against neuronal damage [111]. Cannabidiol is currently under investigation in clinical trials for the treatment of alcohol abuse [110].

Finally, serotonergic psychedelics, thanks to their neuroplastic action and capability to increase brain network connectivity, are under study as new treatment options for substance use disorders [112,113]. Although classic hallucinogens exert their pharmacological effects primarily through the serotonergic system, acting as agonists of the 5-HT_{2A} receptor, their action goes beyond this target and also involves the modulation of glutamatergic transmission (see Section 5.1).

4. Glutamatergic Alterations in Schizophrenia

Schizophrenia is a chronic psychiatric disorder with a prevalence estimated around 1% globally and is characterized by a broad symptomatology extending from positive (i.e., hallucinations and delusion) to negative symptoms (i.e., anhedonia and asociality) and cognitive deficits, thus severely impacting on the life of patients and their families [20]. Although the first theory of schizophrenia, derived from evidence of antipsychotic effects of antidopaminergic drugs, is centered on dopaminergic dysfunction, a role for the glutamate system has been speculated since the observation that NMDA receptor antagonists like ketamine or phencyclidine induced a schizophrenic-like state in healthy subjects and an aggravation of symptoms in schizophrenic patients [20,114,115]. Of note, the psychotic effects of NMDA receptor antagonists persist in the absence of dopamine activity or treatment with dopamine antagonists, further highlighting the importance of glutamatergic alterations in the induction of psychotic symptoms [116,117].

Although the NMDA receptor antagonists may be believed to cause a hypofunction of glutamatergic transmission, the fact that ketamine and phencyclidine have high affinity for NMDA receptors localized on GABAergic interneurons gives reason to the induction of increased glutamatergic transmission through disinhibition processes [118–120]. This leads to glutamate hyperexcitability, downstream stimulation of the mesolimbic pathway, and schizophrenic symptoms. At the same time, several lines of evidence have shown the importance of an excitatory/inhibitory imbalance in schizophrenia pathophysiology. Indeed, GABAergic disfunctions were reported in different brain areas of patients [117].

Preclinical models of schizophrenia have been based on the treatment of rodents with phencyclidine/phencyclidine-like drugs or on genetically modified animals with impaired glutamate receptor expression/function in specific brain areas (i.e., NMDA receptor subunit zeta1 knockdown, epsilon1 knockout, and zeta1 point mutant mice) [118,121]. Importantly, several studies using both mice and rats reported increased extracellular glutamate levels in cortical and subcortical regions following injection with phencyclidine or ketamine. Early studies reported neuronal vacuolization and necrosis induced by both phencyclidine and MK-801 [122]. Interestingly, low doses induced reversible vacuolar changes, while higher doses or a prolonged treatment with both phencyclidine or MK-801 produced irreversible and more widespread damages [123,124]. Furthermore, phencyclidine was shown to upregulate the hsp70 stress gene in the cortex, hippocampus, and basal nuclei of the amygdala of rats, contributing to increased intracellular calcium levels and subsequent apoptotic processes [125].

Interestingly, Schobel and colleagues demonstrated that excessive glutamate levels in specific subregions of the hippocampus in mice following repeated administration of ketamine were associated with neuronal hypermetabolism and could be responsible for consequent neuronal atrophy [126]. Importantly, in the same study, the authors showed that subjects at clinical high-risk for a psychotic disorder showed hypermetabolism in the hippocampus [126].

Moreover, imaging studies on schizophrenic patients consistently reported volumetric alterations of glutamatergic cortical and non-cortical brain regions together with white matter abnormalities [127] and changes in glutamate levels [118], thus confirming that glutamatergic dysfunction plays a role in the functional and morphological changes underlying schizophrenia. Importantly, clinical evidence strongly supports the hypothesis that the volumetric reductions and cortical thinning observed in schizophrenic patients may be related to neuroanatomical compromise through an excitotoxic process [128].

Further evidence linking psychosis with glutamatergic alterations comes from anti-NMDA receptor encephalitis, a now well-established autoimmune disorder presenting with schizophrenia-like symptoms caused by autoantibodies against the NMDA receptor [129]. This leads to isolated psychotic presentations which efficiently respond to immunotherapies.

Therapeutic Approaches for Schizophrenia Based on Glutamatergic Approaches

Low levels of the NMDA receptor co-agonist D-serine were found in the cerebrospinal fluid and post-mortem brains of schizophrenic patients, suggesting a possible functional contribution to NMDA receptor hypofunction. Moreover, schizophrenic patients were shown to have higher levels of D-amino acid oxidase, the peroxisomal flavoenzyme responsible for the metabolism of D-serine [130,131]. Early clinical trials employing D-serine as a supplementary treatment to antipsychotics showed significant improvements in positive, negative, and cognitive symptoms in schizophrenic patients. Although larger-scale studies have obtained conflicting results, more phase II and III trials are ongoing [132]. Moreover, both preclinical and clinical studies have considered the co-administration of D-serine analogs together with D-amino acid oxidase inhibitors, such as Luvadaxistat and sodium benzoate, as therapeutic approaches to enhance NMDA receptor function in schizophrenia [133]. In a randomized, double-blind, placebo-controlled study, Lin and colleagues observed a beneficial therapeutic effect of the addition of sodium benzoate to clozapine in the treatment of schizophrenic patients which positively correlated with antioxidant effects [134]. Yet, these results have not been replicated in larger studies. Moreover, preclinical studies have shown that the potential antioxidant properties of D-amino acid oxidase inhibitors might be dependent on the dosage because high doses are associated with increased oxidative stress [131,135,136].

Inhibitors of the glycine transporter-1 (GlyT1), which is responsible for glycine reuptake from the synaptic space, have been developed as well and tested for the treatment of schizophrenia. Sarcosine, an endogenous amino acid analog to D-serine acting both as a competitive inhibitor of GlyT1 and as a co-agonist of the NMDA receptor, has shown promising therapeutic benefits when combined with antipsychotic medication [137–139]. Iclepertin (BI 425809) is a selective GlyT1 inhibitor currently located in clinical phase III that showed pro-cognitive effects in patients with schizophrenia [140] and memory-enhancing effects in rodent cognition tasks, together with a decrease in the deficit in EEG parameters induced by MK-801 [141].

Another strategy adopted to compensate for the NMDA receptor hypofunction observed in schizophrenia is the positive modulation of the metabotropic glutamate receptors mGluR5 [142,143]. Indeed, post-mortem studies have reported reduced expression of mGluR5 in brain areas of schizophrenic patients compared to healthy controls, suggesting an involvement in pathophysiological processes [142–144].

Preclinical evidence has shown that mGluR5 positive allosteric modulators ameliorate cognitive impairment and negative symptoms in NMDA antagonist-induced models [145]. It is important to underline that, since in addition to directly increasing cellular excitability, mGluR5 physically interacts with NMDA receptor subunits to improve their activity, the use of positive allosteric modulators instead of full mGluR5 agonists seems to avoid the risk of excitotoxicity due to direct and prolonged activation of glutamate receptors [146]. However, the therapeutic potential of mGluR5 positive allosteric modulators has not yet been tested clinically.

Finally, mGluR2/3 agonists are also under evaluation for their possible antipsychotic effect. Preclinical studies have already provided exciting results, and pomaglumetad has entered clinical phase development. Pomaglumetad has been considered both in monotherapy and in combination with other treatments, although inconsistent results were observed in both [132]. A recent exploratory analysis suggested that pomaglumetad might be more effective in the early stages of the disease [147].

5. Glutamatergic Alterations in Major Depressive Disorder and Post-Traumatic Stress Disorder

Stress-related mental disorders, such as major depressive disorder (MDD) and post-traumatic stress disorder (PTSD), are common mental health conditions that severely impact patients' well-being and represent a global burden on society [148]. Patients suffering from MDD experience persistent low mood and anhedonia eventually associated with metabolic dysfunctions, sleep disturbances, cognitive impairments, and increased suicidal risk. On the other hand, PTSD is mostly characterized by dysfunctional processing of fear associated with recurrence of intrusive memories, hyperarousal and/or avoidance, and emotional distancing. Convergent post-mortem and brain imaging studies have indicated morphological and functional changes in the brain of both MDD and PTSD patients, particularly in areas where glutamatergic transmission is predominant [19,149–151]. Reduced volume and decreased activity of HPC and PFC were reported in MDD and were associated with disease severity and lack of therapeutic response [149]. Conversely, amygdala and nucleus accumbens volume and function have been found to be increased in MDD patients [152,153]. Similarly, human studies reported alterations of the excitatory/inhibitory balance in corticolimbic areas of PTSD patients [19], as well as HPC and PFC volume reductions and increased amygdala volume in some patients [151]. Overall, this evidence strongly supports the importance of glutamatergic transmission in the etiopathogenetic process of stress-related disorders.

Accordingly, preclinical models based on stress exposure strongly support that stress dramatically affects glutamatergic transmission in the same brain areas affected in patients [154]. As for HPC and PFC, acute stress has been consistently shown to rapidly enhance glutamatergic transmission by increasing glutamate release and inducing changes in glutamate receptor activation and trafficking (reviewed in [3,155]). Conversely, chronic exposure to stress has been mostly associated with deficits in glutamatergic transmission in PFC and HPC [154,156,157]. Indeed, chronic stress was shown to induce changes in the levels and function of both ionotropic and metabotropic glutamate receptors, leading to a reduction in synaptic strength and participation in dysregulated glutamate release [154,156]. Preclinical models based on both acute and chronic stress have also shown neuronal dendritic simplification and spine reductions in HPC and PFC, implying that glutamatergic dysfunctions may underlie the architectural abnormalities measured in patients [152,158]. Importantly, functional and morphological alterations induced by chronic stress were selectively observed in vulnerable animals and not resilient ones, confirming that the glutamatergic changes play a role in the maladaptive response to chronic stress [159–161]. Although the specific mechanisms causing functional impairment and dendritic simplification in corticolimbic areas under chronic stress exposure have not yet been identified, a main hypothesis implicates glutamatergic transmission [154,156]. Indeed, it has been speculated that the increase in glutamate release and transmission rapidly induced by stress may lead, in the long term, to adaptive excitotoxic processes, in turn causing a reduction in synaptic density and strength. Accordingly, rapid-acting antidepressants induce a transient surge in prefrontal glutamatergic neurotransmission, which is associated with rapid and sustained cortical connectivity (see Section 5.1).

Differently, studies focusing on the amygdala and nucleus accumbens mostly agree in reporting that both acute and chronic stress increase glutamatergic transmission, mainly in the basolateral component that receives connections from the PFC, leading to the consolidation of emotional memories [162,163]. Indeed, the amygdala is a central hub integrating and processing information related to fear and anxiety, with important implications in the regulation of memory, motivation, and autonomic responses [162].

Finally, it is worth noting that dysfunctional glutamatergic signaling was also registered in adult animals that were exposed to early life stress or corticosterone administration in adolescence, suggesting that stress-induced glutamatergic changes may be long-lasting and contribute to the pathogenesis of depressive disorders later in life [164,165].

5.1. Glutamatergic Rapid-Acting Antidepressant Drugs

Traditional antidepressant drugs, such as selective serotonin reuptake inhibitors (SSRIs), primarily acting by increasing the synaptic bioavailability of monoamines, were also shown to modulate glutamatergic transmission [166,167]. Indeed, monoaminergic antidepressants were demonstrated to regulate the expression and function of NMDA and AMPA receptors, to modulate glutamatergic synaptic transmission, and to rescue morphological changes in corticolimbic brain areas [168–170].

Accordingly, accumulated knowledge on the etiopathogenetic processes of MDD and PTSD converge in attributing a major role to neuroplasticity impairments and dysfunction of the glutamatergic system [171]. In this context, the evidence of rapid and long-lasting antidepressant properties of the non-competitive NMDA receptor antagonist ketamine administered at subanesthetic dose revolutionized research on antidepressants, which had not seen drugs with new mechanisms of action for decades [156]. Indeed, even though traditional antidepressants are effective in most patients, their efficacy is limited by a delay of therapeutic onset of several weeks and by a high percentage of non-response and disease recurrence [172]. Preclinical studies have revealed that the rapid antidepressant effect of ketamine is associated with the restoration of dysfunctional glutamatergic transmission and promotion of neuroplasticity in corticolimbic areas [150,173–175].

This amount of research led to the authorization of an intranasal formulation of esketamine, the (S)-enantiomer of ketamine, for the management of treatment-resistant

(TRD) patients in combination with a classic antidepressant in 2019 [176], and ketamine is also receiving attention for the treatment of PTSD and other psychiatric conditions [177,178].

Despite the enthusiasm raised by the rapid psychotropic action of ketamine, the abuse potential and other possible adverse effects represent significant limitations on wider use of the drug in therapy. Nevertheless, the success of ketamine has supported the investigation of antidepressant properties of other glutamatergic drugs.

Several glutamatergic drugs including selective antagonists of the GluN2B subunit (CP-101606; MK-0657), NMDA receptor partial agonists (D-Cycloserine), and glycine site modulators (GLYX-13) have been investigated in recent years [172,179]. However, all these drugs have failed in clinical trials due to a lack of antidepressant efficacy in patients and have been thus abandoned. Other glutamatergic drugs still under preclinical/early clinical development include the following: R-ketamine, hydroxynorketamine, modulators of metabotropic receptors (especially mGluR2/3 antagonists), dextromethorphan, dextromethadone, and nitrous oxide [177,179–181].

At the same time, since ketamine is a dissociative psychedelic drug, it has also reignited interest of research in unveiling the therapeutic potential of other psychedelic drugs. Indeed, hallucinogens like psilocybin, LSD, and MDMA are being studied not only as innovative, fast-acting antidepressants but also as drugs for the treatment of alcohol and substance use disorders, as well as for the management of PTSD [182–185]. Importantly, despite primarily targeting serotonergic transmission, most psychedelics were reported to downstream modulate glutamatergic transmission as a mechanism to promote neuroplasticity [186,187]. Indeed, psychedelics were shown to increase glutamate release and transmission predominantly in cortical brain regions, in turn activating neurotrophic cascades, inducing long-term structural plasticity and modifications of cortical functional connectivity [188].

6. Discussion and Future Perspectives

For decades, neurodevelopmental and adult mental disorders have been studied in parallel, with limited consideration of possible common etiopathogenetic factors and therapeutic strategies. The same happened with neuropsychiatric and substance use disorders. However, convergent recent evidence is highlighting the importance of glutamatergic regulation in all these clinical conditions. Impaired glutamatergic transmission may lead to excitation/inhibition imbalance, which possibly may activate excitotoxic processes, in turn affecting neuroplasticity and eventually neuronal survival [13]. Indeed, even though excitotoxicity is generally considered to be associated with neuronal death, overexcitation can also have a dramatic long-term impact on neuronal structure and networks, leading to glutamatergic hypofunction [149,189]. Accordingly, it is worth noticing that ASD, substance use disorders, and psychiatric disorders share neuroplasticity alterations in glutamatergic brain areas, thus impacting mood, motivation, and cognition. Since homeostatic levels of glutamate are required for the physiological functioning of the brain [16,190–193], it should not be surprising that pharmacological strategies potentiating glutamatergic transmission were shown to restore neuroplasticity and to exert therapeutic effects in both neurodevelopmental and adult mental diseases (Table 1).

Intriguingly, among different drugs, ketamine is of particular interest because, at high doses, it can precipitate neuropsychiatric disorders (specifically, psychotic symptoms) [20,114,115], and at low doses, it has been shown to open a window of neuroplasticity [194] and is associated with rapid antidepressant, anti-PTSD, and anti-ASD actions [177,195]. Similarly, classical hallucinogens are revolutionizing the neuropsychopharmacology field, opening new paths for the management of mental disorders [182,183]. Hallucinogens are showing promising therapeutic effects not only in the management of MDD and PTSD but also for substance use diseases [113]. Again, at first glance, this might seem contradictory but not when one considers that therapeutic effects pass through the rapid regulation of neuroplasticity and connectivity [186,187].

The possibility to regulate glutamatergic transmission through mGluR modulation is attracting attention as well [196,197]. Unfortunately, to date, this approach is not yet producing particularly encouraging results, and some molecules targeting mGluRs have recently failed in clinical phases (as in the case of mGluR5 antagonists for ASD). More studies are warranted to unveil the therapeutic potential of mGluR modulators which, compared to ionotropic glutamate receptor drugs, have the advantage to offer the possibility to fine tune glutamatergic transmission. Indeed, it must be recognized that glutamatergic transmission is distributed throughout the central nervous system and plays crucial roles in several brain functions, ranging from mood to memory and cognition [3–5]. Thus, if on one hand glutamatergic drugs have a high therapeutic potential in diseases associated with alterations of neuroplasticity, on the other hand, they might induce dangerous effects. First of all, we cannot fail to mention that glutamatergic drugs may induce dissociative effects and may have abuse potential, as in the case of ketamine. At the same time, an excessive direct glutamatergic activation may lead to psychotic symptoms, epileptogenic effects, and excitotoxicity. Milder modulation of glutamatergic transmission, as permitted by mGluR drugs (especially in the case of positive and negative modulators) or indirect glutamatergic agents, seems to be safer but also with lower therapeutic efficacy. More studies are required to fully exploit the therapeutic potential of glutamatergic drugs in mental disorders.

Table 1. Therapeutic strategies targeting glutamatergic transmission for the treatment of neurodevelopmental and adult mental disorders.

Drug	Target	Disorder	Refs.
Acamprosate	NMDA receptor antagonist	SUD	[88]
Amantadine	Non-competitive NMDA receptor antagonist	SUD	[99]
Cannabidiol	TRPV1 agonist; 5-HT1A agonist; indirect CB1 and CB2 agonist	SUD	[198] Reviewed in [109,110]
D-cycloserine	Partial NMDA receptor agonist	SUD	[99]
		Schizophrenia	[132]
Dextromethadone	NMDA receptor antagonist	MDD	[179]
Dextromethorphan	Non-competitive NMDA receptor antagonist; SERT/NET blocker; Sigma σ1 receptor agonist	MDD	[181]
Hydroxynorketamine	AMPA receptor activator	MDD	[177]
Iclepertin (BI 425809)	GlyT1 inhibitor	Schizophrenia	[140]
Ketamine/S-ketamine	Non-competitive NMDA receptor antagonist	ASD	[41,42]
		SUD	Reviewed in [177]
		MDD/TRD	[176] Reviewed in [150,173–175,177,181]
		PTSD	[178,199]
		Other psychiatric conditions	Reviewed in [177]
R-ketamine	Non-competitive NMDA receptor antagonist	MDD	[177]
Lamotrigine	Voltage-gated sodium and calcium channel blocker	SUD	[99]
LSD	5-HT2A receptor agonist; D2/D3 receptor agonist	MDD	Reviewed in [183]
		SUD	[200]
Luvadaxistat	D-amino acid oxidase inhibitor	Schizophrenia	[133]
MDMA	Releaser and/or reuptake inhibitor of presynaptic serotonin, dopamine, and norepinephrine	PTSD	[201] Reviewed in [183]

Table 1. *Cont.*

Drug	Target	Disorder	Refs.
Memantine	Non-selective NMDA receptor antagonist	ASD	[39,40]
		SUD	[100,102]
MK-801 (Dizocilpine)	NMDA receptor antagonist	SUD	[101,104]
N-acetyl-L-cysteine	mGluR2/3 agonist	ASD	[45,46]
Nitrous oxide	NMDA receptor antagonist	MDD	[181]
Pomaglumetad	mGluR2/3 agonist	Schizophrenia	[132,147]
Psilocybin	5-HT1A, 5-HT2A and 5-HT2C activator	MDD	[182] Reviewed in [183]
		SUD	[184]
Riluzole	Voltage-dependent sodium channel inhibitor; NMDA/kainate receptor antagonist	ASD	[39,40]
Sarcosine	GlyT1 inhibitor; NMDA receptor co-agonist	Schizophrenia	[137–139]
Sodium benzoate	D-amino acid oxidase inhibitor	Schizophrenia	[134]
Topiramate	Voltage-gated sodium channel blocker; AMPA/kainate receptor antagonist	SUD	[61,106]
Traditional antidepressants	SSRI; SNRI; TCA	MDD	[166,167]
CDPPB, ADX47273, DFB, CHPG, LSN2463359, LSN2814617	mGluR5 positive allosteric modulator	Schizophrenia	Reviewed in [145]
		SUD	[97]
TS-161, RO4995819	mGluR2/3 antagonists/allosteric modulators	MDD	Reviewed in: [177,181]
LY379268	mGluR2/3 agonist	SUD	Reviewed in [97,99]

NMDA: N-methyl-D-aspartate, TRPV1: transient receptor potential cation channel subfamily V member 1, CB1: cannabinoid receptor 1, CB2: cannabinoid receptor 2, 5-HT1A: 5-hydroxytryptamine receptor subtype 1A, ASD: autism spectrum disorder, MDD: major depressive disorder, SUD: substance use disorder, SERT: serotonin transporter, NET: norepinephrine transporter, AMPA: α-amino-3-hydroxy-5-methyl-a-isoxazolepropionate, PTSD: post-traumatic stress disorder, GlyT1: Glycine transporter type-1, TRD: treatment-resistant depression, LSD: lysergic acid diethylamide, D2: dopamine receptor 2, D3: dopamine receptor 3, MDMA: 3,4-Methylenedioxymethamphetamine, SSRI: selective serotonin reuptake inhibitor, SNRI: serotonin and norepinephrine reuptake inhibitor, TCA: tricyclic antidepressant.

7. Conclusions

Overall, etiopathogenetic and pharmacological studies converge in highlighting a crucial role of glutamate homeostasis and neuroplasticity in both neurodevelopmental and adult mental disorders (Figure 1). Glutamate-dependent excitotoxic processes can lead to glutamatergic dysfunctions, in turn impacting brain activity and plasticity. Neuroplastic glutamatergic drugs may have the potential to rescue pathological conditions and restore physiological neuroplasticity.

Although the clinical evidence is still limited and not always encouraging, glutamatergic modulation is representing a very interesting novel target for drug development, which is sparking research interest in different fields. Despite failures, which are an inherent part of the process, this is paving the way to novel neuroplasticity-related therapeutic strategies that could lay the foundation for next-generation pharmacotherapies to treat different brain disorders.

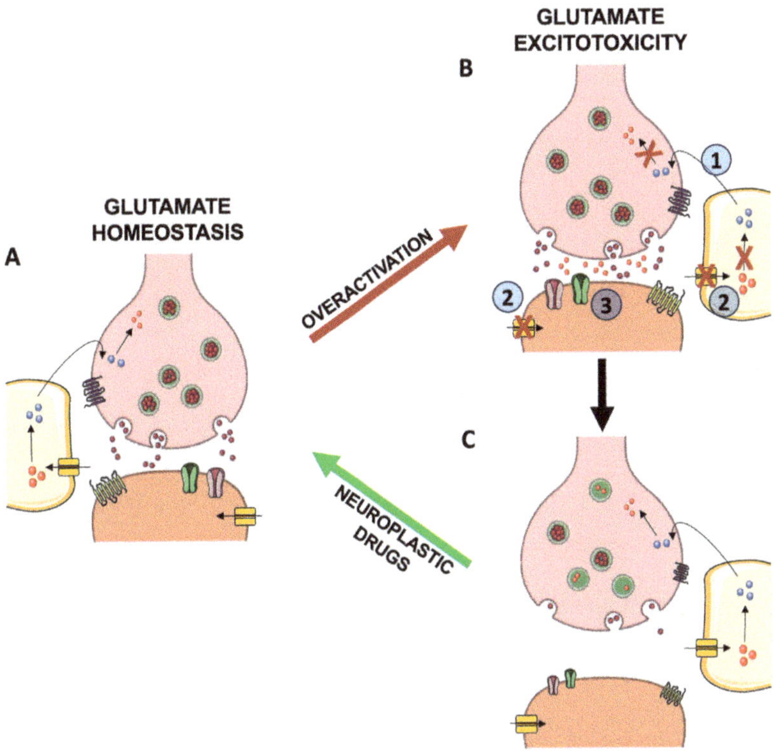

Figure 1. Neuroplastic drugs rescue glutamate homeostasis reverting glutamate-induced excitotoxicity and its consequences. (**A**) Physiological levels of glutamate at synapses are required for physiological brain activity, especially in those brain regions, such as the hippocampus and prefrontal cortex, implied in mood and cognitive functions and with a prevalence of glutamatergic neurons. Besides neurons, glia are also involved in the maintenance of glutamate homeostasis participating with neuronal synaptic terminals in the "tripartite synapse" to maintain balanced glutamate levels. Glutamate (red spheres) is synthesized from glutamine (blue spheres) supplied by glial cells and stored in vesicles until release throughout fusion with the presynaptic membrane. Once in the extracellular space, glutamate can bind to ionotropic (AMPA, NMDA; in pink and green) and metabotropic (mGluR; in yellow and purple) glutamate receptors at both presynaptic and postsynaptic terminals initiating several responses, including membrane depolarization, activation of intracellular messenger cascades, and modulation of local protein synthesis. Glutamate synaptic clearance is mediated by excitatory amino acid transporters (EAATs; in dark yellow) located both on neurons and astrocytes, where the reuptaken glutamate is converted into glutamine by glutamine synthetase, which is then released and ready to enter the cycle again. (**B**) The dysfunction of the mechanisms adopted to regulate synaptic glutamate levels may lead to glutamate excitotoxicity where the excessive accumulation of glutamate causes neurotoxicity and eventually cell atrophy/death. This condition may be caused by (1) impaired glutamate/glutamine recycle system, (2) dysfunctional reuptake, and (3) altered expression and activity of glutamate receptors and has been proposed as a pathogenic mechanism for neurodevelopmental and adult mental disorders. (**C**) In the long-term, the overactivation of the glutamate system may also induce impairments of neuronal structures and networks, leading to glutamatergic hypofunction. Accordingly, neurodevelopmental and adult mental disorders share neuroplasticity alterations in glutamatergic brain areas. Importantly, drugs restoring glutamatergic transmission have been demonstrated to exert therapeutic effects in these conditions together with restoring neuroplasticity. Created with smart.servier.com.

Author Contributions: Conceptualization, L.M.; writing—original draft preparation, N.N., M.G., P.M., J.M. and L.M.; writing—review and editing, N.N., M.G., P.M., J.M. and L.M.; supervision, J.M. and L.M.; funding acquisition, L.M. All authors have read and agreed to the published version of the manuscript.

Funding: This work was supported by the Cariplo Foundation (biomedical science Prog. 2019–3357).

Data Availability Statement: No new data were created or analyzed in this study. Data sharing is not applicable to this article.

Conflicts of Interest: The authors declare no conflicts of interest. The funders had no role in the design of the study; in the collection, analyses, or interpretation of data; in the writing of the manuscript; or in the decision to publish the results.

References

1. Brosnan, J.T.; Brosnan, M.E. Glutamate: A Truly Functional Amino Acid. *Amino Acids* **2013**, *45*, 413–418. [CrossRef] [PubMed]
2. Popoli, M.; Yan, Z.; McEwen, B.S.; Sanacora, G. The Stressed Synapse: The Impact of Stress and Glucocorticoids on Glutamate Transmission. *Nat. Rev. Neurosci.* **2012**, *13*, 22–37. [CrossRef] [PubMed]
3. Sanacora, G.; Treccani, G.; Popoli, M. Towards a Glutamate Hypothesis of Depression. *Neuropharmacology* **2012**, *62*, 63–77. [CrossRef] [PubMed]
4. Xie, R.-G.; Xu, G.-Y.; Wu, S.-X.; Luo, C. Presynaptic Glutamate Receptors in Nociception. *Pharmacol. Ther.* **2023**, *251*, 108539. [CrossRef] [PubMed]
5. Cox, M.F.; Hascup, E.R.; Bartke, A.; Hascup, K.N. Friend or Foe? Defining the Role of Glutamate in Aging and Alzheimer's Disease. *Front. Aging* **2022**, *3*, 929474. [CrossRef] [PubMed]
6. Lisman, J. Glutamatergic Synapses Are Structurally and Biochemically Complex Because of Multiple Plasticity Processes: Long-Term Potentiation, Long-Term Depression, Short-Term Potentiation and Scaling. *Philos. Trans. R Soc. B Biol. Sci.* **2017**, *372*, 20160260. [CrossRef] [PubMed]
7. Pérez-Alvarez, A.; Araque, A. Astrocyte-Neuron Interaction at Tripartite Synapses. *Curr. Drug Targets* **2013**, *14*, 1220–1224. [CrossRef] [PubMed]
8. Ribeiro, F.M.; Vieira, L.B.; Pires, R.G.W.; Olmo, R.P.; Ferguson, S.S.G. Metabotropic Glutamate Receptors and Neurodegenerative Diseases. *Pharmacol. Res.* **2017**, *115*, 179–191. [CrossRef]
9. Niswender, C.M.; Conn, P.J. Metabotropic Glutamate Receptors: Physiology, Pharmacology, and Disease. *Annu. Rev. Pharmacol. Toxicol.* **2010**, *50*, 295–322. [CrossRef]
10. Reiner, A.; Levitz, J. Glutamatergic Signaling in the Central Nervous System: Ionotropic and Metabotropic Receptors in Concert. *Neuron* **2018**, *98*, 1080–1098. [CrossRef]
11. Perea, G.; Navarrete, M.; Araque, A. Tripartite Synapses: Astrocytes Process and Control Synaptic Information. *Trends Neurosci.* **2009**, *32*, 421–431. [CrossRef] [PubMed]
12. Papouin, T.; Dunphy, J.; Tolman, M.; Foley, J.C.; Haydon, P.G. Astrocytic Control of Synaptic Function. *Philos. Trans. R. Soc. B Biol. Sci.* **2017**, *372*, 20160154. [CrossRef] [PubMed]
13. CHOI, D. Glutamate Neurotoxicity and Diseases of the Nervous System. *Neuron* **1988**, *1*, 623–634. [CrossRef] [PubMed]
14. Velasco, M.; Quintero, J.R.; Castillo, M.C.; Rojas, M.; Bautista, J.; Martinez, M.S.; Salazar, J.; Mendoza, R.; Bermudez, V. Excitotoxicity: An Organized Crime at The Cellular Level. *J. Neurol. Neurosci.* **2017**, *8*, 193. [CrossRef]
15. Verma, M.; Lizama, B.N.; Chu, C.T. Excitotoxicity, Calcium and Mitochondria: A Triad in Synaptic Neurodegeneration. *Transl. Neurodegener.* **2022**, *11*, 3. [CrossRef] [PubMed]
16. Olloquequi, J.; Cornejo-Córdova, E.; Verdaguer, E.; Soriano, F.X.; Binvignat, O.; Auladell, C.; Camins, A. Excitotoxicity in the Pathogenesis of Neurological and Psychiatric Disorders: Therapeutic Implications. *J. Psychopharmacol.* **2018**, *32*, 265–275. [CrossRef]
17. Bossy-Wetzel, E.; Schwarzenbacher, R.; Lipton, S.A. Molecular Pathways to Neurodegeneration. *Nat. Med.* **2004**, *10*, S2–S9. [CrossRef] [PubMed]
18. Dantzer, R.; Walker, A.K. Is There a Role for Glutamate-Mediated Excitotoxicity in Inflammation-Induced Depression? *J. Neural Transm.* **2014**, *121*, 925–932. [CrossRef] [PubMed]
19. Averill, L.A.; Purohit, P.; Averill, C.L.; Boesl, M.A.; Krystal, J.H.; Abdallah, C.G. Glutamate Dysregulation and Glutamatergic Therapeutics for PTSD: Evidence from Human Studies. *Neurosci. Lett.* **2017**, *649*, 147–155. [CrossRef]
20. Uno, Y.; Coyle, J.T. Glutamate Hypothesis in Schizophrenia. *Psychiatry Clin. Neurosci.* **2019**, *73*, 204–215. [CrossRef]
21. Essa, M.M.; Braidy, N.; Vijayan, K.R.; Subash, S.; Guillemin, G.J. Excitotoxicity in the Pathogenesis of Autism. *Neurotox. Res.* **2013**, *23*, 393–400. [CrossRef] [PubMed]
22. Lai, M.-C.; Lombardo, M.V.; Baron-Cohen, S. Autism. *Lancet* **2014**, *383*, 896–910. [CrossRef] [PubMed]
23. American Psychiatric Association. *Diagnostic and Statistical Manual of Mental Disorders*; American Psychiatric Association: Washington, DC, USA, 2013; ISBN 0-89042-555-8.

24. Rojas, D.C. The Role of Glutamate and Its Receptors in Autism and the Use of Glutamate Receptor Antagonists in Treatment. *J. Neural Transm.* **2014**, *121*, 891–905. [CrossRef] [PubMed]
25. Montanari, M.; Martella, G.; Bonsi, P.; Meringolo, M. Autism Spectrum Disorder: Focus on Glutamatergic Neurotransmission. *Int. J. Mol. Sci.* **2022**, *23*, 3861. [CrossRef]
26. Hollestein, V.; Poelmans, G.; Forde, N.J.; Beckmann, C.F.; Ecker, C.; Mann, C.; Schäfer, T.; Moessnang, C.; Baumeister, S.; Banaschewski, T.; et al. Excitatory/Inhibitory Imbalance in Autism: The Role of Glutamate and GABA Gene-Sets in Symptoms and Cortical Brain Structure. *Transl. Psychiatry* **2023**, *13*, 18. [CrossRef]
27. Rubenstein, J.L.R.; Merzenich, M.M. Model of Autism: Increased Ratio of Excitation/Inhibition in Key Neural Systems. *Genes. Brain Behav.* **2003**, *2*, 255–267. [CrossRef]
28. Nisar, S.; Bhat, A.A.; Masoodi, T.; Hashem, S.; Akhtar, S.; Ali, T.A.; Amjad, S.; Chawla, S.; Bagga, P.; Frenneaux, M.P.; et al. Genetics of Glutamate and Its Receptors in Autism Spectrum Disorder. *Mol. Psychiatry* **2022**, *27*, 2380–2392. [CrossRef] [PubMed]
29. Monteiro, P.; Feng, G. SHANK Proteins: Roles at the Synapse and in Autism Spectrum Disorder. *Nat. Rev. Neurosci.* **2017**, *18*, 147–157. [CrossRef] [PubMed]
30. Zohny, S.M.; Habib, M.Z.; Mohamad, M.I.; Elayat, W.M.; Elhossiny, R.M.; El-Salam, M.F.A.; Hassan, G.A.M.; Aboul-Fotouh, S. Memantine/Aripiprazole Combination Alleviates Cognitive Dysfunction in Valproic Acid Rat Model of Autism: Hippocampal CREB/BDNF Signaling and Glutamate Homeostasis. *Neurotherapeutics* **2023**, *20*, 464–483. [CrossRef]
31. Habib, M.Z.; Elnahas, E.M.; Aboul-Ela, Y.M.; Ebeid, M.A.; Tarek, M.; Sadek, D.R.; Negm, E.A.; Abdelhakam, D.A.; Aboul-Fotouh, S. Risperidone Impedes Glutamate Excitotoxicity in a Valproic Acid Rat Model of Autism: Role of ADAR2 in AMPA GluA2 RNA Editing. *Eur. J. Pharmacol.* **2023**, *955*, 175916. [CrossRef]
32. Maisterrena, A.; Matas, E.; Mirfendereski, H.; Balbous, A.; Marchand, S.; Jaber, M. The State of the Dopaminergic and Glutamatergic Systems in the Valproic Acid Mouse Model of Autism Spectrum Disorder. *Biomolecules* **2022**, *12*, 1691. [CrossRef] [PubMed]
33. D'Antoni, S.; Schiavi, S.; Buzzelli, V.; Giuffrida, S.; Feo, A.; Ascone, F.; Busceti, C.L.; Nicoletti, F.; Trezza, V.; Catania, M.V. Group I and Group II Metabotropic Glutamate Receptors Are Upregulated in the Synapses of Infant Rats Prenatally Exposed to Valproic Acid. *Psychopharmacology* **2023**, *240*, 2617–2629. [CrossRef] [PubMed]
34. Guo, B.; Chen, J.; Chen, Q.; Ren, K.; Feng, D.; Mao, H.; Yao, H.; Yang, J.; Liu, H.; Liu, Y.; et al. Anterior Cingulate Cortex Dysfunction Underlies Social Deficits in Shank3 Mutant Mice. *Nat. Neurosci.* **2019**, *22*, 1223–1234. [CrossRef] [PubMed]
35. Moutin, E.; Sakkaki, S.; Compan, V.; Bouquier, N.; Giona, F.; Areias, J.; Goyet, E.; Hemonnot-Girard, A.-L.; Seube, V.; Glasson, B.; et al. Restoring Glutamate Receptosome Dynamics at Synapses Rescues Autism-like Deficits in Shank3-Deficient Mice. *Mol. Psychiatry* **2021**, *26*, 7596–7609. [CrossRef]
36. Ma, B.; Shan, X.; Yu, J.; Zhu, T.; Li, R.; Lv, H.; Cheng, H.; Zhang, T.; Wang, L.; Wei, F.; et al. Social Deficits via Dysregulated Rac1-Dependent Excitability Control of Prefrontal Cortical Neurons and Increased GABA/Glutamate Ratios. *Cell Rep.* **2022**, *41*, 111722. [CrossRef] [PubMed]
37. Vicidomini, C.; Ponzoni, L.; Lim, D.; Schmeisser, M.J.; Reim, D.; Morello, N.; Orellana, D.; Tozzi, A.; Durante, V.; Scalmani, P.; et al. Pharmacological Enhancement of MGlu5 Receptors Rescues Behavioral Deficits in SHANK3 Knock-out Mice. *Mol. Psychiatry* **2017**, *22*, 689–702. [CrossRef] [PubMed]
38. Qin, L.; Ma, K.; Yan, Z. Chemogenetic Activation of Prefrontal Cortex in Shank3-Deficient Mice Ameliorates Social Deficits, NMDAR Hypofunction, and Sgk2 Downregulation. *iScience* **2019**, *17*, 24–35. [CrossRef] [PubMed]
39. Shamabadi, A.; Karimi, H.; Arabzadeh Bahri, R.; Motavaselian, M.; Akhondzadeh, S. Emerging Drugs for the Treatment of Irritability Associated with Autism Spectrum Disorder. *Expert Opin. Emerg. Drugs* **2024**, *29*, 45–56. [CrossRef] [PubMed]
40. Doyle, C.A.; McDougle, C.J. Pharmacotherapy to Control Behavioral Symptoms in Children with Autism. *Expert Opin. Pharmacother.* **2012**, *13*, 1615–1629. [CrossRef]
41. Wink, L.K.; Reisinger, D.L.; Horn, P.; Shaffer, R.C.; O'Brien, K.; Schmitt, L.; Dominick, K.R.; Pedapati, E.V.; Erickson, C.A. Brief Report: Intranasal Ketamine in Adolescents and Young Adults with Autism Spectrum Disorder—Initial Results of a Randomized, Controlled, Crossover, Pilot Study. *J. Autism Dev. Disord.* **2021**, *51*, 1392–1399. [CrossRef]
42. Kolevzon, A.; Levy, T.; Barkley, S.; Bedrosian-Sermone, S.; Davis, M.; Foss-Feig, J.; Halpern, D.; Keller, K.; Kostic, A.; Layton, C.; et al. An Open-Label Study Evaluating the Safety, Behavioral, and Electrophysiological Outcomes of Low-Dose Ketamine in Children with ADNP Syndrome. *Hum. Genet. Genom. Adv.* **2022**, *3*, 100138. [CrossRef] [PubMed]
43. Annamneedi, A.; Gora, C.; Dudas, A.; Leray, X.; Bozon, V.; Crépieux, P.; Pellissier, L.P. Towards the Convergent Therapeutic Potential of G Protein-coupled Receptors in Autism Spectrum Disorders. *Br. J. Pharmacol.* **2023**. *Early View*. [CrossRef] [PubMed]
44. Jacquemont, S.; Berry-Kravis, E.; Hagerman, R.; von Raison, F.; Gasparini, F.; Apostol, G.; Ufer, M.; Des Portes, V.; Gomez-Mancilla, B. The Challenges of Clinical Trials in Fragile X Syndrome. *Psychopharmacology* **2014**, *231*, 1237–1250. [CrossRef] [PubMed]
45. Bradlow, R.C.J.; Berk, M.; Kalivas, P.W.; Back, S.E.; Kanaan, R.A. The Potential of N-Acetyl-L-Cysteine (NAC) in the Treatment of Psychiatric Disorders. *CNS Drugs* **2022**, *36*, 451–482. [CrossRef]
46. Lee, T.-M.; Lee, K.-M.; Lee, C.-Y.; Lee, H.-C.; Tam, K.-W.; Loh, E.-W. Effectiveness of *N*-Acetylcysteine in Autism Spectrum Disorders: A Meta-Analysis of Randomized Controlled Trials. *Aust. N. Z. J. Psychiatry* **2021**, *55*, 196–206. [CrossRef] [PubMed]
47. Koob, G.F.; Volkow, N.D. Neurobiology of Addiction: A Neurocircuitry Analysis. *Lancet Psychiatry* **2016**, *3*, 760–773. [CrossRef] [PubMed]

48. D'Souza, M.S. Brain and Cognition for Addiction Medicine: From Prevention to Recovery Neural Substrates for Treatment of Psychostimulant-Induced Cognitive Deficits. *Front. Psychiatry* **2019**, *10*, 448029. [CrossRef] [PubMed]
49. Gonçalves, J.; Baptista, S.; Silva, A.P. Psychostimulants and Brain Dysfunction: A Review of the Relevant Neurotoxic Effects. *Neuropharmacology* **2014**, *87*, 135–149. [CrossRef] [PubMed]
50. Nestler, E. The Neurobiology of Cocaine Addiction. *Sci. Pract. Perspect.* **2005**, *3*, 4–10. [CrossRef]
51. Niedzielska-Andres, E.; Pomierny-Chamioło, L.; Andres, M.; Walczak, M.; Knackstedt, L.A.; Filip, M.; Przegaliński, E. Cocaine Use Disorder: A Look at Metabotropic Glutamate Receptors and Glutamate Transporters. *Pharmacol. Ther.* **2021**, *221*, 107797. [CrossRef]
52. Kalivas, P.; Duffy, P. D1 Receptors Modulate Glutamate Transmission in the Ventral Tegmental Area. *J. Neurosci.* **1995**, *15*, 5379–5388. [CrossRef] [PubMed]
53. Kalivas, P.W.; Duffy, P. Repeated Cocaine Administration Alters Extracellular Glutamate in the Ventral Tegmental Area. *J. Neurochem.* **1998**, *70*, 1497–1502. [CrossRef] [PubMed]
54. Williams, J.M.; Steketee, J.D. Cocaine Increases Medial Prefrontal Cortical Glutamate Overflow in Cocaine-sensitized Rats: A Time Course Study. *Eur. J. Neurosci.* **2004**, *20*, 1639–1646. [CrossRef] [PubMed]
55. Xie, X.; Steketee, J.D. Repeated Exposure to Cocaine Alters the Modulation of Mesocorticolimbic Glutamate Transmission by Medial Prefrontal Cortex Group II Metabotropic Glutamate Receptors. *J. Neurochem.* **2008**, *107*, 186–196. [CrossRef] [PubMed]
56. Gabriele, A.; Pacchioni, A.M.; See, R.E. Dopamine and Glutamate Release in the Dorsolateral Caudate Putamen Following Withdrawal from Cocaine Self-Administration in Rats. *Pharmacol. Biochem. Behav.* **2012**, *103*, 373–379. [CrossRef] [PubMed]
57. McFarland, K.; Lapish, C.C.; Kalivas, P.W. Prefrontal Glutamate Release into the Core of the Nucleus Accumbens Mediates Cocaine-Induced Reinstatement of Drug-Seeking Behavior. *J. Neurosci.* **2003**, *23*, 3531–3537. [CrossRef] [PubMed]
58. Miguéns, M.; Del Olmo, N.; Higuera-Matas, A.; Torres, I.; García-Lecumberri, C.; Ambrosio, E. Glutamate and Aspartate Levels in the Nucleus Accumbens during Cocaine Self-Administration and Extinction: A Time Course Microdialysis Study. *Psychopharmacology* **2008**, *196*, 303–313. [CrossRef] [PubMed]
59. Ben-Shahar, O.; Keeley, P.; Cook, M.; Brake, W.; Joyce, M.; Nyffeler, M.; Heston, R.; Ettenberg, A. Changes in Levels of D1, D2, or NMDA Receptors during Withdrawal from Brief or Extended Daily Access to IV Cocaine. *Brain Res.* **2007**, *1131*, 220–228. [CrossRef] [PubMed]
60. Beveridge, T.J.R.; Smith, H.R.; Nader, M.A.; Porrino, L.J. Group II Metabotropic Glutamate Receptors in the Striatum of Non-Human Primates: Dysregulation Following Chronic Cocaine Self-Administration. *Neurosci. Lett.* **2011**, *496*, 15–19. [CrossRef]
61. Shrestha, P.; Katila, N.; Lee, S.; Seo, J.H.; Jeong, J.-H.; Yook, S. Methamphetamine Induced Neurotoxic Diseases, Molecular Mechanism, and Current Treatment Strategies. *Biomed. Pharmacother.* **2022**, *154*, 113591. [CrossRef]
62. Yamamoto, B.K.; Raudensky, J. The Role of Oxidative Stress, Metabolic Compromise, and Inflammation in Neuronal Injury Produced by Amphetamine-Related Drugs of Abuse. *J. Neuroimmune Pharmacol.* **2008**, *3*, 203–217. [CrossRef] [PubMed]
63. Serra, M.; Simola, N.; Pollack, A.E.; Costa, G. Brain Dysfunctions and Neurotoxicity Induced by Psychostimulants in Experimental Models and Humans: An Overview of Recent Findings. *Neural Regen. Res.* **2024**, *19*, 1908–1918. [CrossRef]
64. Mark, K.A.; Quinton, M.S.; Russek, S.J.; Yamamoto, B.K. Dynamic Changes in Vesicular Glutamate Transporter 1 Function and Expression Related to Methamphetamine-Induced Glutamate Release. *J. Neurosci.* **2007**, *27*, 6823–6831. [CrossRef] [PubMed]
65. Rocher, C.; Gardier, A.M. Effects of repeated systemic administration of *d*-Fenfluramine on serotonin and glutamate release in rat ventral hippocampus: Comparison with methamphetamine using in vivo microdialysis. *Naunyn Schmiedebergs Arch. Pharmacol.* **2001**, *363*, 422–428. [CrossRef]
66. Zhang, Y.; Loonam, T.M.; Noailles, P.H.; Angulo, J.A. Comparison of Cocaine- and Methamphetamine-Evoked Dopamine and Glutamate Overflow in Somatodendritic and Terminal Field Regions of the Rat Brain during Acute, Chronic, and Early Withdrawal Conditions. *Ann. N. Y. Acad. Sci.* **2001**, *937*, 93–120. [CrossRef] [PubMed]
67. Reid, M.S.; Hsu, K.; Berger, S.P. Cocaine and Amphetamine Preferentially Stimulate Glutamate Release in the Limbic System: Studies on the Involvement of Dopamine. *Synapse* **1997**, *27*, 95–105. [CrossRef]
68. Lu, W.; Wolf, M.E. Repeated Amphetamine Administration Alters AMPA Receptor Subunit Expression in Rat Nucleus Accumbens and Medial Prefrontal Cortex. *Synapse* **1999**, *32*, 119–131. [CrossRef]
69. Baptista, S.; Bento, A.R.; Gonçalves, J.; Bernardino, L.; Summavielle, T.; Lobo, A.; Fontes-Ribeiro, C.; Malva, J.O.; Agasse, F.; Silva, A.P. Neuropeptide Y Promotes Neurogenesis and Protection against Methamphetamine-Induced Toxicity in Mouse Dentate Gyrus-Derived Neurosphere Cultures. *Neuropharmacology* **2012**, *62*, 2413–2423. [CrossRef]
70. Staszewski, R.D.; Yamamoto, B.K. Methamphetamine-induced Spectrin Proteolysis in the Rat Striatum. *J. Neurochem.* **2006**, *96*, 1267–1276. [CrossRef]
71. Itzhak, Y.; Gandia, C.; Huang, P.L.; Ali, S.F. Resistance of Neuronal Nitric Oxide Synthase-Deficient Mice to Methamphetamine-Induced Dopaminergic Neurotoxicity. *J. Pharmacol. Exp. Ther.* **1998**, *284*, 1040–1047.
72. Canedo, T.; Portugal, C.C.; Socodato, R.; Almeida, T.O.; Terceiro, A.F.; Bravo, J.; Silva, A.I.; Magalhães, J.D.; Guerra-Gomes, S.; Oliveira, J.F.; et al. Astrocyte-Derived TNF and Glutamate Critically Modulate Microglia Activation by Methamphetamine. *Neuropsychopharmacology* **2021**, *46*, 2358–2370. [CrossRef] [PubMed]
73. Anneken, J.H.; Gudelsky, G.A. MDMA Produces a Delayed and Sustained Increase in the Extracellular Concentration of Glutamate in the Rat Hippocampus. *Neuropharmacology* **2012**, *63*, 1022–1027. [CrossRef] [PubMed]

74. Anneken, J.H.; Cunningham, J.I.; Collins, S.A.; Yamamoto, B.K.; Gudelsky, G.A. MDMA Increases Glutamate Release and Reduces Parvalbumin-Positive GABAergic Cells in the Dorsal Hippocampus of the Rat: Role of Cyclooxygenase. *J. Neuroimmune Pharmacol.* **2013**, *8*, 58–65. [CrossRef] [PubMed]
75. Thomas, D.M.; Dowgiert, J.; Geddes, T.J.; Francescutti-Verbeem, D.; Liu, X.; Kuhn, D.M. Microglial Activation Is a Pharmacologically Specific Marker for the Neurotoxic Amphetamines. *Neurosci. Lett.* **2004**, *367*, 349–354. [CrossRef] [PubMed]
76. Zorumski, C.F.; Izumi, Y.; Mennerick, S. Ketamine: NMDA Receptors and Beyond. *J. Neurosci.* **2016**, *36*, 11158–11164. [CrossRef] [PubMed]
77. Onaolapo, A.Y.; Ayeni, O.J.; Ogundeji, M.O.; Ajao, A.; Onaolapo, O.J.; Owolabi, A.R. Subchronic Ketamine Alters Behaviour, Metabolic Indices and Brain Morphology in Adolescent Rats: Involvement of Oxidative Stress, Glutamate Toxicity and Caspase-3-Mediated Apoptosis. *J. Chem. Neuroanat.* **2019**, *96*, 22–33. [CrossRef]
78. Sampaio, T.B.; de Oliveira, L.F.; Constantino, L.C.; Costa, A.P.; Poluceno, G.G.; Martins, W.C.; Dal-Cim, T.; de Oliveira, K.A.; Ludka, F.K.; Prediger, R.D.; et al. Long-Term Neurobehavioral Consequences of a Single Ketamine Neonatal Exposure in Rats: Effects on Cellular Viability and Glutamate Transport in Frontal Cortex and Hippocampus. *Neurotox. Res.* **2018**, *34*, 649–659. [CrossRef] [PubMed]
79. Liu, F.; Patterson, T.A.; Sadovova, N.; Zhang, X.; Liu, S.; Zou, X.; Hanig, J.P.; Paule, M.G.; Slikker, W.; Wang, C. Ketamine-Induced Neuronal Damage and Altered N-Methyl-d-Aspartate Receptor Function in Rat Primary Forebrain Culture. *Toxicol. Sci.* **2013**, *131*, 548–557. [CrossRef]
80. Wang, C.; Sadovova, N.; Fu, X.; Schmued, L.; Scallet, A.; Hanig, J.; Slikker, W. The Role of the N-Methyl-d-Aspartate Receptor in Ketamine-Induced Apoptosis in Rat Forebrain Culture. *Neuroscience* **2005**, *132*, 967–977. [CrossRef]
81. McCann, M.E.; Soriano, S.G. Does General Anesthesia Affect Neurodevelopment in Infants and Children? *BMJ* **2019**, *367*, l6459. [CrossRef]
82. Kalopita, K.; Armakolas, A.; Philippou, A.; Zarros, A.; Angelogianni, P. Ketamine-induced neurotoxicity in neurodevelopment: A synopsis of main pathways based on recent in vivo experimental findings. *J. Anaesthesiol. Clin. Pharmacol.* **2021**, *37*, 37–42. [CrossRef] [PubMed]
83. Kamal, H.; Tan, G.C.; Ibrahim, S.F.; Shaikh, M.F.; Mohamed, I.N.; Mohamed, R.M.P.; Hamid, A.A.; Ugusman, A.; Kumar, J. Alcohol Use Disorder, Neurodegeneration, Alzheimer's and Parkinson's Disease: Interplay Between Oxidative Stress, Neuroimmune Response and Excitotoxicity. *Front. Cell. Neurosci.* **2020**, *14*, 282. [CrossRef] [PubMed]
84. Mira, R.G.; Tapia-Rojas, C.; Pérez, M.J.; Jara, C.; Vergara, E.H.; Quintanilla, R.A.; Cerpa, W. Alcohol Impairs Hippocampal Function: From NMDA Receptor Synaptic Transmission to Mitochondrial Function. *Drug Alcohol Depend.* **2019**, *205*, 107628. [CrossRef] [PubMed]
85. Kryger, R.; Wilce, P.A. The Effects of Alcoholism on the Human Basolateral Amygdala. *Neuroscience* **2010**, *167*, 361–371. [CrossRef] [PubMed]
86. Sari, Y.; Sreemantula, S.N. Neuroimmunophilin GPI-1046 Reduces Ethanol Consumption in Part through Activation of GLT1 in Alcohol-Preferring Rats. *Neuroscience* **2012**, *227*, 327–335. [CrossRef] [PubMed]
87. Ayers-Ringler, J.R.; Jia, Y.-F.; Qiu, Y.-Y.; Choi, D.-S. Role of Astrocytic Glutamate Transporter in Alcohol Use Disorder. *World J. Psychiatry* **2016**, *6*, 31. [CrossRef] [PubMed]
88. Rao, P.S.S.; Bell, R.L.; Engleman, E.A.; Sari, Y. Targeting Glutamate Uptake to Treat Alcohol Use Disorders. *Front. Neurosci.* **2015**, *9*, 104167. [CrossRef] [PubMed]
89. Sari, Y.; Sreemantula, S.N.; Lee, M.R.; Choi, D.-S. Ceftriaxone Treatment Affects the Levels of GLT1 and ENT1 As Well As Ethanol Intake in Alcohol-Preferring Rats. *J. Mol. Neurosci.* **2013**, *51*, 779–787. [CrossRef] [PubMed]
90. Flatscher-Bader, T.; Wilce, P.A. Impact of Alcohol Abuse on Protein Expression of Midkine and Excitatory Amino Acid Transporter 1 in the Human Prefrontal Cortex. *Alcohol. Clin. Exp. Res.* **2008**, *32*, 1849–1858. [CrossRef]
91. Gerace, E.; Landucci, E.; Bani, D.; Moroni, F.; Mannaioni, G.; Pellegrini-Giampietro, D.E. Glutamate Receptor-Mediated Neurotoxicity in a Model of Ethanol Dependence and Withdrawal in Rat Organotypic Hippocampal Slice Cultures. *Front. Neurosci.* **2019**, *12*, 1053. [CrossRef]
92. Reynolds, A.R.; Berry, J.N.; Sharrett-Field, L.; Prendergast, M.A. Ethanol Withdrawal Is Required to Produce Persisting N-Methyl-d-Aspartate Receptor-Dependent Hippocampal Cytotoxicity during Chronic Intermittent Ethanol Exposure. *Alcohol* **2015**, *49*, 219–227. [CrossRef] [PubMed]
93. Rajgopal, Y.; Vemuri, M.C. Calpain Activation and α-Spectrin Cleavage in Rat Brain by Ethanol. *Neurosci. Lett.* **2002**, *321*, 187–191. [CrossRef] [PubMed]
94. Olney, J.W.; Wozniak, D.F.; Jevtovic-Todorovic, V.; Ikonomidou, C. Glutamate Signaling and the Fetal Alcohol Syndrome. *Ment. Retard. Dev. Disabil. Res. Rev.* **2001**, *7*, 267–275. [CrossRef] [PubMed]
95. Zuo, D.; Liu, Y.; Liu, Z.; Cui, J.; Zhou, X.; Liu, Y.; Li, Z.; Wu, Y. Alcohol Aggravates Ketamine-Induced Behavioral, Morphological and Neurochemical Alterations in Adolescent Rats: The Involvement of CREB-Related Pathways. *Behav. Brain Res.* **2018**, *349*, 80–90. [CrossRef] [PubMed]
96. Blaker, A.L.; Moore, E.R.; Yamamoto, B.K. Serial Exposure to Ethanol Drinking and Methamphetamine Enhances Glutamate Excitotoxicity. *J. Neurochem.* **2019**, *151*, 749–763. [CrossRef] [PubMed]
97. Jensen, K.L.; Jensen, S.B.; Madsen, K.L. A Mechanistic Overview of Approaches for the Treatment of Psychostimulant Dependence. *Front. Pharmacol.* **2022**, *13*, 854176. [CrossRef] [PubMed]

98. Uys, J.; LaLumiere, R. Glutamate: The New Frontier in Pharmacotherapy for Cocaine Addiction. *CNS Neurol. Disord. Drug Targets* **2008**, *7*, 482–491. [CrossRef]
99. Schmidt, H.D.; Pierce, R.C. Cocaine-induced Neuroadaptations in Glutamate Transmission. *Ann. N. Y. Acad. Sci.* **2010**, *1187*, 35–75. [CrossRef] [PubMed]
100. Du, C.; Hua, Y.; Clare, K.; Park, K.; Allen, C.P.; Volkow, N.D.; Hu, X.-T.; Pan, Y. Memantine Attenuates Cocaine and NeuroHIV Neurotoxicity in the Medial Prefrontal Cortex. *Front. Pharmacol.* **2022**, *13*, 895006. [CrossRef]
101. Idrus, N.M.; McGough, N.N.H.; Riley, E.P.; Thomas, J.D. Administration of Memantine During Ethanol Withdrawal in Neonatal Rats: Effects on Long-Term Ethanol-Induced Motor Incoordination and Cerebellar Purkinje Cell Loss. *Alcohol. Clin. Exp. Res.* **2011**, *35*, 355–364. [CrossRef]
102. Idrus, N.M.; McGough, N.N.H.; Riley, E.P.; Thomas, J.D. Administration of Memantine During Withdrawal Mitigates Overactivity and Spatial Learning Impairments Associated with Neonatal Alcohol Exposure in Rats. *Alcohol. Clin. Exp. Res.* **2014**, *38*, 529–537. [CrossRef] [PubMed]
103. Idrus, N.M.; McGough, N.N.H.; Spinetta, M.J.; Thomas, J.D.; Riley, E.P. The Effects of a Single Memantine Treatment on Behavioral Alterations Associated with Binge Alcohol Exposure in Neonatal Rats. *Neurotoxicol. Teratol.* **2011**, *33*, 444–450. [CrossRef] [PubMed]
104. Young, B.W.; Sengelaub, D.R.; Steinmetz, J.E. MK-801 Administration during Neonatal Ethanol Withdrawal Attenuates Interpositus Cell Loss and Juvenile Eyeblink Conditioning Deficits. *Alcohol* **2010**, *44*, 359–369. [CrossRef] [PubMed]
105. Montemitro, C.; Angebrandt, A.; Wang, T.-Y.; Pettorruso, M.; Abulseoud, O.A. Mechanistic Insights into the Efficacy of Memantine in Treating Certain Drug Addictions. *Prog. Neuropsychopharmacol. Biol. Psychiatry* **2021**, *111*, 110409. [CrossRef] [PubMed]
106. Omidvari, S.; Azimzadeh, Z.; Rashnoo, F.; Tahmasebinia, F.; Keramatinia, A.; Roozbahany, N.A.; Abbaszadeh, H.A.; Darabi, S. Molecular Mechanisms and Treatment Strategies for Methamphetamine-induced Neurodegeneration, Inflammation and Neurotoxicity. *Acta Neurobiol. Exp.* **2023**, *83*, 414–431. [CrossRef]
107. Baltieri, D.A.; Daró, F.R.; Ribeiro, P.L.; De Andrade, A.G. Comparing Topiramate with Naltrexone in the Treatment of Alcohol Dependence. *Addiction* **2008**, *103*, 2035–2044. [CrossRef] [PubMed]
108. Kampman, K.M.; Pettinati, H.; Lynch, K.G.; Dackis, C.; Sparkman, T.; Weigley, C.; O'Brien, C.P. A Pilot Trial of Topiramate for the Treatment of Cocaine Dependence. *Drug Alcohol Depend.* **2004**, *75*, 233–240. [CrossRef]
109. Paulus, V.; Billieux, J.; Benyamina, A.; Karila, L. Cannabidiol in the Context of Substance Use Disorder Treatment: A Systematic Review. *Addict. Behav.* **2022**, *132*, 107360. [CrossRef] [PubMed]
110. Navarrete, F.; García-Gutiérrez, M.S.; Gasparyan, A.; Austrich-Olivares, A.; Manzanares, J. Role of Cannabidiol in the Therapeutic Intervention for Substance Use Disorders. *Front. Pharmacol.* **2021**, *12*, 626010. [CrossRef]
111. Eid, R.S.; Gobinath, A.R.; Galea, L.A.M. Sex Differences in Depression: Insights from Clinical and Preclinical Studies. *Prog. Neurobiol.* **2019**, *176*, 86–102. [CrossRef]
112. de Veen, B.T.H.; Schellekens, A.F.A.; Verheij, M.M.M.; Homberg, J.R. Psilocybin for Treating Substance Use Disorders? *Expert. Rev. Neurother.* **2017**, *17*, 203–212. [CrossRef]
113. De Gregorio, D.; Aguilar-Valles, A.; Preller, K.H.; Heifets, B.D.; Hibicke, M.; Mitchell, J.; Gobbi, G. Hallucinogens in Mental Health: Preclinical and Clinical Studies on LSD, Psilocybin, MDMA, and Ketamine. *J. Neurosci.* **2021**, *41*, 891–900. [CrossRef]
114. Pilowsky, L.S.; Bressan, R.A.; Stone, J.M.; Erlandsson, K.; Mulligan, R.S.; Krystal, J.H.; Ell, P.J. First in Vivo Evidence of an NMDA Receptor Deficit in Medication-Free Schizophrenic Patients. *Mol. Psychiatry* **2006**, *11*, 118–119. [CrossRef] [PubMed]
115. Stone, J.M.; Dietrich, C.; Edden, R.; Mehta, M.A.; De Simoni, S.; Reed, L.J.; Krystal, J.H.; Nutt, D.; Barker, G.J. Ketamine Effects on Brain GABA and Glutamate Levels with 1H-MRS: Relationship to Ketamine-Induced Psychopathology. *Mol. Psychiatry* **2012**, *17*, 664–665. [CrossRef] [PubMed]
116. Moghaddam, B.; Javitt, D. From Revolution to Evolution: The Glutamate Hypothesis of Schizophrenia and Its Implication for Treatment. *Neuropsychopharmacology* **2012**, *37*, 4–15. [CrossRef]
117. Egerton, A.; Modinos, G.; Ferrera, D.; McGuire, P. Neuroimaging Studies of GABA in Schizophrenia: A Systematic Review with Meta-Analysis. *Transl. Psychiatry* **2017**, *7*, e1147. [CrossRef] [PubMed]
118. Plitman, E.; Nakajima, S.; de la Fuente-Sandoval, C.; Gerretsen, P.; Chakravarty, M.M.; Kobylianskii, J.; Chung, J.K.; Caravaggio, F.; Iwata, Y.; Remington, G.; et al. Glutamate-Mediated Excitotoxicity in Schizophrenia: A Review. *Eur. Neuropsychopharmacol.* **2014**, *24*, 1591–1605. [CrossRef]
119. Nakazawa, K.; Zsiros, V.; Jiang, Z.; Nakao, K.; Kolata, S.; Zhang, S.; Belforte, J.E. GABAergic Interneuron Origin of Schizophrenia Pathophysiology. *Neuropharmacology* **2012**, *62*, 1574–1583. [CrossRef] [PubMed]
120. Farber, N.B.; Kim, S.H.; Dikranian, K.; Jiang, X.P.; Heinkel, C. Receptor Mechanisms and Circuitry Underlying NMDA Antagonist Neurotoxicity. *Mol. Psychiatry* **2002**, *7*, 32–43. [CrossRef]
121. Enomoto, T.; Noda, Y.; Nabeshima, T. Phencyclidine and Genetic Animal Models of Schizophrenia Developed in Relation to the Glutamate Hypothesis. *Methods Find. Exp. Clin. Pharmacol.* **2007**, *29*, 291. [CrossRef]
122. Olney, J.W.; Labruyere, J.; Price, M.T. Pathological Changes Induced in Cerebrocortical Neurons by Phencyclidine and Related Drugs. *Science* **1989**, *244*, 1360–1362. [CrossRef] [PubMed]
123. Corso, T.D.; Sesma, M.A.; Tenkova, T.I.; Der, T.C.; Wozniak, D.F.; Farber, N.B.; Olney, J.W. Multifocal Brain Damage Induced by Phencyclidine Is Augmented by Pilocarpine. *Brain Res.* **1997**, *752*, 1–14. [CrossRef] [PubMed]

124. Fix, A.S.; Horn, J.W.; Wightman, K.A.; Johnson, C.A.; Long, G.G.; Storts, R.W.; Farber, N.; Wozniak, D.F.; Olney, J.W. Neuronal Vacuolization and Necrosis Induced by the Noncompetitive N-Methyl-d-Aspartate (NMDA) Antagonist MK(+)801 (Dizocilpine Maleate): A Light and Electron Microscopic Evaluation of the Rat Retrosplenial Cortex. *Exp. Neurol.* **1993**, *123*, 204–215. [CrossRef] [PubMed]
125. Sharp, F.R.; Butman, M.; Koistinaho, J.; Aardalen, K.; Nakki, R.; Massa, S.M.; Swanson, R.A.; Sagar, S.M. Phencyclidine Induction of the Hsp70 Stress Gene in Injured Pyramidal Neurons Is Mediated via Multiple Receptors and Voltage Gated Calcium Channels. *Neuroscience* **1994**, *62*, 1079–1092. [CrossRef] [PubMed]
126. Schobel, S.A.; Chaudhury, N.H.; Khan, U.A.; Paniagua, B.; Styner, M.A.; Asllani, I.; Inbar, B.P.; Corcoran, C.M.; Lieberman, J.A.; Moore, H.; et al. Imaging Patients with Psychosis and a Mouse Model Establishes a Spreading Pattern of Hippocampal Dysfunction and Implicates Glutamate as a Driver. *Neuron* **2013**, *78*, 81–93. [CrossRef] [PubMed]
127. Colibazzi, T.; Wexler, B.E.; Bansal, R.; Hao, X.; Liu, J.; Sanchez-Peña, J.; Corcoran, C.; Lieberman, J.A.; Peterson, B.S. Anatomical Abnormalities in Gray and White Matter of the Cortical Surface in Persons with Schizophrenia. *PLoS ONE* **2013**, *8*, e55783. [CrossRef] [PubMed]
128. Shah, P.; Plitman, E.; Iwata, Y.; Kim, J.; Nakajima, S.; Chan, N.; Brown, E.E.; Caravaggio, F.; Torres, E.; Hahn, M.; et al. Glutamatergic Neurometabolites and Cortical Thickness in Treatment-Resistant Schizophrenia: Implications for Glutamate-Mediated Excitotoxicity. *J. Psychiatr. Res.* **2020**, *124*, 151–158. [CrossRef] [PubMed]
129. Pollak, T.A.; Lennox, B.R.; Müller, S.; Benros, M.E.; Prüss, H.; Tebartz van Elst, L.; Klein, H.; Steiner, J.; Frodl, T.; Bogerts, B.; et al. Autoimmune Psychosis: An International Consensus on an Approach to the Diagnosis and Management of Psychosis of Suspected Autoimmune Origin. *Lancet Psychiatry* **2020**, *7*, 93–108. [CrossRef]
130. Bendikov, I.; Nadri, C.; Amar, S.; Panizzutti, R.; Demiranda, J.; Wolosker, H.; Agam, G. A CSF and Postmortem Brain Study of D-Serine Metabolic Parameters in Schizophrenia. *Schizophr. Res.* **2007**, *90*, 41–51. [CrossRef]
131. Kuo, C.-Y.; Lin, C.-H.; Lane, H.-Y. Targeting D-Amino Acid Oxidase (DAAO) for the Treatment of Schizophrenia: Rationale and Current Status of Research. *CNS Drugs* **2022**, *36*, 1143–1153. [CrossRef]
132. Uliana, D.L.; Lisboa, J.R.F.; Gomes, F.V.; Grace, A.A. The Excitatory-Inhibitory Balance as a Target for the Development of Novel Drugs to Treat Schizophrenia. *Biochem. Pharmacol.* 2024, 116298, *in press*. [CrossRef] [PubMed]
133. Fradley, R.; Goetghebeur, P.; Miller, D.; Burley, R.; Almond, S.; Gruart i Massó, A.; Delgado García, J.M.; Zhu, B.; Howley, E.; Neill, J.C.; et al. Luvadaxistat: A Novel Potent and Selective d-Amino Acid Oxidase Inhibitor Improves Cognitive and Social Deficits in Rodent Models for Schizophrenia. *Neurochem. Res.* **2023**, *48*, 3027–3041. [CrossRef] [PubMed]
134. Lin, C.-H.; Lin, C.-H.; Chang, Y.-C.; Huang, Y.-J.; Chen, P.-W.; Yang, H.-T.; Lane, H.-Y. Sodium Benzoate, a D-Amino Acid Oxidase Inhibitor, Added to Clozapine for the Treatment of Schizophrenia: A Randomized, Double-Blind, Placebo-Controlled Trial. *Biol. Psychiatry* **2018**, *84*, 422–432. [CrossRef] [PubMed]
135. Olofinnade, A.T.; Onaolapo, A.Y.; Onaolapo, O.J.; Olowe, O.A. The Potential Toxicity of Food-Added Sodium Benzoate in Mice Is Concentration-Dependent. *Toxicol. Res.* **2021**, *10*, 561–569. [CrossRef] [PubMed]
136. Khan, I.S.; Dar, K.B.; Ganie, S.A.; Ali, M.N. Toxicological Impact of Sodium Benzoate on Inflammatory Cytokines, Oxidative Stress and Biochemical Markers in Male Wistar Rats. *Drug Chem. Toxicol.* **2022**, *45*, 1345–1354. [CrossRef] [PubMed]
137. Marchi, M.; Galli, G.; Magarini, F.M.; Mattei, G.; Galeazzi, G.M. Sarcosine as an Add-on Treatment to Antipsychotic Medication for People with Schizophrenia: A Systematic Review and Meta-Analysis of Randomized Controlled Trials. *Expert. Opin. Drug Metab. Toxicol.* **2021**, *17*, 483–493. [CrossRef] [PubMed]
138. Chang, C.-H.; Lin, C.-H.; Liu, C.-Y.; Chen, S.-J.; Lane, H.-Y. Efficacy and Cognitive Effect of Sarcosine (N-Methylglycine) in Patients with Schizophrenia: A Systematic Review and Meta-Analysis of Double-Blind Randomised Controlled Trials. *J. Psychopharmacol.* **2020**, *34*, 495–505. [CrossRef] [PubMed]
139. Fond, G.; Mallet, J.; Urbach, M.; Benros, M.E.; Berk, M.; Billeci, M.; Boyer, L.; Correll, C.U.; Fornaro, M.; Kulkarni, J.; et al. Adjunctive Agents to Antipsychotics in Schizophrenia: A Systematic Umbrella Review and Recommendations for Amino Acids, Hormonal Therapies and Anti-Inflammatory Drugs. *BMJ Ment. Health* **2023**, *26*, e300771. [CrossRef] [PubMed]
140. Fleischhacker, W.W.; Podhorna, J.; Gröschl, M.; Hake, S.; Zhao, Y.; Huang, S.; Keefe, R.S.E.; Desch, M.; Brenner, R.; Walling, D.P.; et al. Efficacy and Safety of the Novel Glycine Transporter Inhibitor BI 425809 Once Daily in Patients with Schizophrenia: A Double-Blind, Randomised, Placebo-Controlled Phase 2 Study. *Lancet Psychiatry* **2021**, *8*, 191–201. [CrossRef]
141. Rosenbrock, H.; Dorner-Ciossek, C.; Giovannini, R.; Schmid, B.; Schuelert, N. Effects of the Glycine Transporter-1 Inhibitor Iclepertin (BI 425809) on Sensory Processing, Neural Network Function, and Cognition in Animal Models Related to Schizophrenia. *J. Pharmacol. Exp. Ther.* **2022**, *382*, 223–232. [CrossRef]
142. Newell, K.A.; Matosin, N. Rethinking Metabotropic Glutamate Receptor 5 Pathological Findings in Psychiatric Disorders: Implications for the Future of Novel Therapeutics. *BMC Psychiatry* **2014**, *14*, 23. [CrossRef]
143. Wang, H.-Y.; MacDonald, M.L.; Borgmann-Winter, K.E.; Banerjee, A.; Sleiman, P.; Tom, A.; Khan, A.; Lee, K.-C.; Roussos, P.; Siegel, S.J.; et al. MGluR5 Hypofunction Is Integral to Glutamatergic Dysregulation in Schizophrenia. *Mol. Psychiatry* **2020**, *25*, 750–760. [CrossRef]
144. Fatemi, S.H.; Folsom, T.D.; Rooney, R.J.; Thuras, P.D. MRNA and Protein Expression for Novel GABAA Receptors θ and P2 Are Altered in Schizophrenia and Mood Disorders; Relevance to FMRP-MGluR5 Signaling Pathway. *Transl. Psychiatry* **2013**, *3*, e271. [CrossRef]

145. Matosin, N.; Newell, K.A. Metabotropic Glutamate Receptor 5 in the Pathology and Treatment of Schizophrenia. *Neurosci. Biobehav. Rev.* **2013**, *37*, 256–268. [CrossRef]
146. Rook, J.M.; Xiang, Z.; Lv, X.; Ghoshal, A.; Dickerson, J.W.; Bridges, T.M.; Johnson, K.A.; Foster, D.J.; Gregory, K.J.; Vinson, P.N.; et al. Biased $mGlu_5$-Positive Allosteric Modulators Provide In Vivo Efficacy without Potentiating $mGlu_5$ Modulation of NMDAR Currents. *Neuron* **2015**, *86*, 1029–1040. [CrossRef]
147. Kinon, B.J.; Millen, B.A.; Zhang, L.; McKinzie, D.L. Exploratory Analysis for a Targeted Patient Population Responsive to the Metabotropic Glutamate 2/3 Receptor Agonist Pomaglumetad Methionil in Schizophrenia. *Biol. Psychiatry* **2015**, *78*, 754–762. [CrossRef] [PubMed]
148. James, S.L.; Abate, D.; Abate, K.H.; Abay, S.M.; Abbafati, C.; Abbasi, N.; Abbastabar, H.; Abd-Allah, F.; Abdela, J.; Abdelalim, A.; et al. Global, Regional, and National Incidence, Prevalence, and Years Lived with Disability for 354 Diseases and Injuries for 195 Countries and Territories, 1990–2017: A Systematic Analysis for the Global Burden of Disease Study 2017. *Lancet* **2018**, *392*, 1789–1858. [CrossRef] [PubMed]
149. Duman, R.S.; Sanacora, G.; Krystal, J.H. Altered Connectivity in Depression: GABA and Glutamate Neurotransmitter Deficits and Reversal by Novel Treatments. *Neuron* **2019**, *102*, 75–90. [CrossRef]
150. Lener, M.S.; Niciu, M.J.; Ballard, E.D.; Park, M.; Park, L.T.; Nugent, A.C.; Zarate, C.A. Glutamate and Gamma-Aminobutyric Acid Systems in the Pathophysiology of Major Depression and Antidepressant Response to Ketamine. *Biol. Psychiatry* **2017**, *81*, 886–897. [CrossRef] [PubMed]
151. Abdallah, C.G.; Averill, L.A.; Akiki, T.J.; Raza, M.; Averill, C.L.; Gomaa, H.; Adikey, A.; Krystal, J.H. The Neurobiology and Pharmacotherapy of Posttraumatic Stress Disorder. *Annu. Rev. Pharmacol. Toxicol.* **2019**, *59*, 171–189. [CrossRef] [PubMed]
152. McEwen, B.S.; Nasca, C.; Gray, J.D. Stress Effects on Neuronal Structure: Hippocampus, Amygdala, and Prefrontal Cortex. *Neuropsychopharmacology* **2016**, *41*, 3–23. [CrossRef] [PubMed]
153. Abdallah, C.G.; Jackowski, A.; Salas, R.; Gupta, S.; Sato, J.R.; Mao, X.; Coplan, J.D.; Shungu, D.C.; Mathew, S.J. The Nucleus Accumbens and Ketamine Treatment in Major Depressive Disorder. *Neuropsychopharmacology* **2017**, *42*, 1739–1746. [CrossRef] [PubMed]
154. Sanacora, G.; Yan, Z.; Popoli, M. The Stressed Synapse 2.0: Pathophysiological Mechanisms in Stress-Related Neuropsychiatric Disorders. *Nat. Rev. Neurosci.* **2022**, *23*, 86–103. [CrossRef] [PubMed]
155. Musazzi, L.; Tornese, P.; Sala, N.; Popoli, M. What Acute Stress Protocols Can Tell Us About PTSD and Stress-Related Neuropsychiatric Disorders. *Front. Pharmacol.* **2018**, *9*, 358334. [CrossRef] [PubMed]
156. Abdallah, C.G.; Sanacora, G.; Duman, R.S.; Krystal, J.H. The Neurobiology of Depression, Ketamine and Rapid-Acting Antidepressants: Is It Glutamate Inhibition or Activation? *Pharmacol. Ther.* **2018**, *190*, 148–158. [CrossRef] [PubMed]
157. Musazzi, L.; Treccani, G.; Popoli, M. Functional and Structural Remodeling of Glutamate Synapses in Prefrontal and Frontal Cortex Induced by Behavioral Stress. *Front. Psychiatry* **2015**, *6*, 125121. [CrossRef] [PubMed]
158. Musazzi, L.; Tornese, P.; Sala, N.; Popoli, M. Acute or Chronic? A Stressful Question. *Trends Neurosci.* **2017**, *40*, 525–535. [CrossRef] [PubMed]
159. Mingardi, J.; Ndoj, E.; Bonifacino, T.; Misztak, P.; Bertoli, M.; La Via, L.; Torazza, C.; Russo, I.; Milanese, M.; Bonanno, G.; et al. Functional and Molecular Changes in the Prefrontal Cortex of the Chronic Mild Stress Rat Model of Depression and Modulation by Acute Ketamine. *Int. J. Mol. Sci.* **2023**, *24*, 10814. [CrossRef]
160. Tornese, P.; Sala, N.; Bonini, D.; Bonifacino, T.; La Via, L.; Milanese, M.; Treccani, G.; Seguini, M.; Ieraci, A.; Mingardi, J.; et al. Chronic Mild Stress Induces Anhedonic Behavior and Changes in Glutamate Release, BDNF Trafficking and Dendrite Morphology Only in Stress Vulnerable Rats. The Rapid Restorative Action of Ketamine. *Neurobiol. Stress* **2019**, *10*, 100160. [CrossRef] [PubMed]
161. Derosa, S.; Misztak, P.; Mingardi, J.; Mazzini, G.; Müller, H.K.; Musazzi, L. Changes in Neurotrophic Signaling Pathways in Brain Areas of the Chronic Mild Stress Rat Model of Depression as a Signature of Ketamine Fast Antidepressant Response/Non-Response. *Prog. Neuropsychopharmacol. Biol. Psychiatry* **2024**, *128*, 110871. [CrossRef]
162. Roozendaal, B.; McEwen, B.S.; Chattarji, S. Stress, Memory and the Amygdala. *Nat. Rev. Neurosci.* **2009**, *10*, 423–433. [CrossRef]
163. Zhang, W.-H.; Zhang, J.-Y.; Holmes, A.; Pan, B.-X. Amygdala Circuit Substrates for Stress Adaptation and Adversity. *Biol. Psychiatry* **2021**, *89*, 847–856. [CrossRef]
164. Qin, X.; Liu, X.-X.; Wang, Y.; Wang, D.; Song, Y.; Zou, J.-X.; Pan, H.-Q.; Zhai, X.-Z.; Zhang, Y.-M.; Zhang, Y.-B.; et al. Early Life Stress Induces Anxiety-like Behavior during Adulthood through Dysregulation of Neuronal Plasticity in the Basolateral Amygdala. *Life Sci.* **2021**, *285*, 119959. [CrossRef]
165. Bertholomey, M.L.; Nagarajan, V.; Smith, D.M.; Torregrossa, M.M. Sex- and Age-Dependent Effects of Chronic Corticosterone Exposure on Depressive-like, Anxiety-like, and Fear-Related Behavior: Role of Amygdala Glutamate Receptors in the Rat. *Front. Behav. Neurosci.* **2022**, *16*, 950000. [CrossRef]
166. Pehrson, A.L.; Sanchez, C. Serotonergic Modulation of Glutamate Neurotransmission as a Strategy for Treating Depression and Cognitive Dysfunction. *CNS Spectr.* **2014**, *19*, 121–133. [CrossRef]
167. Musazzi, L.; Treccani, G.; Mallei, A.; Popoli, M. The Action of Antidepressants on the Glutamate System: Regulation of Glutamate Release and Glutamate Receptors. *Biol. Psychiatry* **2013**, *73*, 1180–1188. [CrossRef]
168. Duman, R.S. Synaptic Plasticity and Mood Disorders. *Mol. Psychiatry* **2002**, *7*, S29–S34. [CrossRef]
169. Krishnan, V.; Nestler, E.J. The Molecular Neurobiology of Depression. *Nature* **2008**, *455*, 894–902. [CrossRef]

170. Harmer, C.J.; Duman, R.S.; Cowen, P.J. How Do Antidepressants Work? New Perspectives for Refining Future Treatment Approaches. *Lancet Psychiatry* **2017**, *4*, 409–418. [CrossRef]
171. Kadriu, B.; Musazzi, L.; Henter, I.D.; Graves, M.; Popoli, M.; Zarate, C.A. Glutamatergic Neurotransmission: Pathway to Developing Novel Rapid-Acting Antidepressant Treatments. *Int. J. Neuropsychopharmacol.* **2019**, *22*, 119–135. [CrossRef]
172. Lopez-Munoz, F.; Alamo, C. Monoaminergic Neurotransmission: The History of the Discovery of Antidepressants from 1950s Until Today. *Curr. Pharm. Des.* **2009**, *15*, 1563–1586. [CrossRef]
173. Krystal, J.H.; Kavalali, E.T.; Monteggia, L.M. Ketamine and Rapid Antidepressant Action: New Treatments and Novel Synaptic Signaling Mechanisms. *Neuropsychopharmacology* **2024**, *49*, 41–50. [CrossRef]
174. Zanos, P.; Gould, T.D. Mechanisms of Ketamine Action as an Antidepressant. *Mol. Psychiatry* **2018**, *23*, 801–811. [CrossRef]
175. Krystal, J.H.; Kaye, A.P.; Jefferson, S.; Girgenti, M.J.; Wilkinson, S.T.; Sanacora, G.; Esterlis, I. Ketamine and the Neurobiology of Depression: Toward next-Generation Rapid-Acting Antidepressant Treatments. *Proc. Natl. Acad. Sci. USA* **2023**, *120*, e2305772120. [CrossRef]
176. Kim, J.; Farchione, T.; Potter, A.; Chen, Q.; Temple, R. Esketamine for Treatment-Resistant Depression—First FDA-Approved Antidepressant in a New Class. *N. Engl. J. Med.* **2019**, *381*, 1–4. [CrossRef]
177. Johnston, J.N.; Kadriu, B.; Kraus, C.; Henter, I.D.; Zarate, C.A. Ketamine in Neuropsychiatric Disorders: An Update. *Neuropsychopharmacology* **2024**, *49*, 23–40. [CrossRef]
178. Feder, A.; Rutter, S.B.; Schiller, D.; Charney, D.S. The Emergence of Ketamine as a Novel Treatment for Posttraumatic Stress Disorder. In *Advances in Pharmacology*; Elsevier: Amsterdam, The Netherlands, 2020; pp. 261–286.
179. Gonda, X.; Dome, P.; Neill, J.C.; Tarazi, F.I. Novel Antidepressant Drugs: Beyond Monoamine Targets. *CNS Spectr.* **2023**, *28*, 6–15. [CrossRef]
180. Johnston, J.N.; Henter, I.D.; Zarate, C.A. The Antidepressant Actions of Ketamine and Its Enantiomers. *Pharmacol. Ther.* **2023**, *246*, 108431. [CrossRef]
181. Borbély, É.; Simon, M.; Fuchs, E.; Wiborg, O.; Czéh, B.; Helyes, Z. Novel Drug Developmental Strategies for Treatment-resistant Depression. *Br. J. Pharmacol.* **2022**, *179*, 1146–1186. [CrossRef]
182. Chi, T.; Gold, J.A. A Review of Emerging Therapeutic Potential of Psychedelic Drugs in the Treatment of Psychiatric Illnesses. *J. Neurol. Sci.* **2020**, *411*, 116715. [CrossRef]
183. Inserra, A.; De Gregorio, D.; Gobbi, G. Psychedelics in Psychiatry: Neuroplastic, Immunomodulatory, and Neurotransmitter Mechanisms. *Pharmacol. Rev.* **2021**, *73*, 202–277. [CrossRef]
184. Acero, V.P.; Cribas, E.S.; Browne, K.D.; Rivellini, O.; Burrell, J.C.; O'Donnell, J.C.; Das, S.; Cullen, D.K. Bedside to Bench: The Outlook for Psychedelic Research. *Front. Pharmacol.* **2023**, *14*, 1240295. [CrossRef]
185. dos Santos, R.G.; Bouso, J.C.; Alcázar-Córcoles, M.Á.; Hallak, J.E.C. Efficacy, Tolerability, and Safety of Serotonergic Psychedelics for the Management of Mood, Anxiety, and Substance-Use Disorders: A Systematic Review of Systematic Reviews. *Expert. Rev. Clin. Pharmacol.* **2018**, *11*, 889–902. [CrossRef]
186. Aleksandrova, L.R.; Phillips, A.G. Neuroplasticity as a Convergent Mechanism of Ketamine and Classical Psychedelics. *Trends Pharmacol. Sci.* **2021**, *42*, 929–942. [CrossRef]
187. Savalia, N.K.; Shao, L.-X.; Kwan, A.C. A Dendrite-Focused Framework for Understanding the Actions of Ketamine and Psychedelics. *Trends Neurosci.* **2021**, *44*, 260–275. [CrossRef]
188. Kwan, A.C.; Olson, D.E.; Preller, K.H.; Roth, B.L. The Neural Basis of Psychedelic Action. *Nat. Neurosci.* **2022**, *25*, 1407–1419. [CrossRef]
189. Duman, C.H.; Duman, R.S. Spine Synapse Remodeling in the Pathophysiology and Treatment of Depression. *Neurosci. Lett.* **2015**, *601*, 20–29. [CrossRef]
190. Kalivas, P.W. The Glutamate Homeostasis Hypothesis of Addiction. *Nat. Rev. Neurosci.* **2009**, *10*, 561–572. [CrossRef]
191. Bauminger, H.; Gaisler-Salomon, I. Beyond NMDA Receptors: Homeostasis at the Glutamate Tripartite Synapse and Its Contributions to Cognitive Dysfunction in Schizophrenia. *Int. J. Mol. Sci.* **2022**, *23*, 8617. [CrossRef]
192. Sarawagi, A.; Soni, N.D.; Patel, A.B. Glutamate and GABA Homeostasis and Neurometabolism in Major Depressive Disorder. *Front. Psychiatry* **2021**, *12*, 637863. [CrossRef]
193. Kelmendi, B.; Adams, T.G.; Yarnell, S.; Southwick, S.; Abdallah, C.G.; Krystal, J.H. PTSD: From Neurobiology to Pharmacological Treatments. *Eur. J. Psychotraumatol.* **2016**, *7*, 31858. [CrossRef] [PubMed]
194. Wu, H.; Savalia, N.K.; Kwan, A.C. Ketamine for a Boost of Neural Plasticity: How, but Also When? *Biol. Psychiatry* **2021**, *89*, 1030–1032. [CrossRef] [PubMed]
195. Andrade, C. Intranasal Drug Delivery in Neuropsychiatry: Focus on Intranasal Ketamine for Refractory Depression. *J. Clin. Psychiatry* **2015**, *76*, e628–e631. [CrossRef] [PubMed]
196. Pilc, A.; Chaki, S. Role of MGlu Receptors in Psychiatric Disorders—Recent Advances. *Pharmacol. Biochem. Behav.* **2023**, *232*, 173639. [CrossRef] [PubMed]
197. Musazzi, L. Targeting Metabotropic Glutamate Receptors for Rapid-Acting Antidepressant Drug Discovery. *Expert. Opin. Drug Discov.* **2021**, *16*, 147–157. [CrossRef] [PubMed]
198. Gobira, P.H.; Vilela, L.R.; Gonçalves, B.D.C.; Santos, R.P.M.; de Oliveira, A.C.; Vieira, L.B.; Aguiar, D.C.; Crippa, J.A.; Moreira, F.A. Cannabidiol, a Cannabis Sativa Constituent, Inhibits Cocaine-Induced Seizures in Mice: Possible Role of the MTOR Pathway and Reduction in Glutamate Release. *Neurotoxicology* **2015**, *50*, 116–121. [CrossRef] [PubMed]

199. Davis, M.T.; DellaGiogia, N.; Maruff, P.; Pietrzak, R.H.; Esterlis, I. Acute Cognitive Effects of Single-Dose Intravenous Ketamine in Major Depressive and Posttraumatic Stress Disorder. *Transl. Psychiatry* **2021**, *11*, 205. [CrossRef] [PubMed]
200. Krebs, T.S.; Johansen, P.-Ø. Lysergic Acid Diethylamide (LSD) for Alcoholism: Meta-Analysis of Randomized Controlled Trials. *J. Psychopharmacol.* **2012**, *26*, 994–1002. [CrossRef]
201. Mithoefer, M.C.; Wagner, M.T.; Mithoefer, A.T.; Jerome, L.; Martin, S.F.; Yazar-Klosinski, B.; Michel, Y.; Brewerton, T.D.; Doblin, R. Durability of Improvement in Post-Traumatic Stress Disorder Symptoms and Absence of Harmful Effects or Drug Dependency after 3,4-Methylenedioxymethamphetamine-Assisted Psychotherapy: A Prospective Long-Term Follow-up Study. *J. Psychopharmacol.* **2013**, *27*, 28–39. [CrossRef]

International Journal of
Molecular Sciences

MDPI

Review

HIV-Associated Neurocognitive Disorder: A Look into Cellular and Molecular Pathology

Landon John-Patrick Thompson [1], Jessica Genovese [1], Zhenzi Hong [1], Meera Vir Singh [2] and Vir Bahadur Singh [1,*]

1 Department of Life Sciences, Albany College of Pharmacy and Health Sciences, Albany, NY 12208, USA
2 Department of Neurology, University of Rochester, Rochester, NY 14642, USA
* Correspondence: vir.singh@acphs.edu

Abstract: Despite combined antiretroviral therapy (cART) limiting HIV replication to undetectable levels in the blood, people living with HIV continue to experience HIV-associated neurocognitive disorder (HAND). HAND is associated with neurocognitive impairment, including motor impairment, and memory loss. HIV has been detected in the brain within 8 days of estimated exposure and the mechanisms for this early entry are being actively studied. Once having entered into the central nervous system (CNS), HIV degrades the blood–brain barrier through the production of its gp120 and Tat proteins. These proteins are directly toxic to endothelial cells and neurons, and propagate inflammatory cytokines by the activation of immune cells and dysregulation of tight junction proteins. The BBB breakdown is associated with the progression of neurocognitive disease. One of the main hurdles for treatment for HAND is the latent pool of cells, which are insensitive to cART and prolong inflammation by harboring the provirus in long-lived cells that can reactivate, causing damage. Multiple strategies are being studied to combat the latent pool and HAND; however, clinically, these approaches have been insufficient and require further revisions. The goal of this paper is to aggregate the known mechanisms and challenges associated with HAND.

Keywords: HIV; HAND; neuroinflammation; microglia; latency; platelet

1. Introduction

Although combined antiretroviral therapy (cART) has made HIV a manageable and treatable disease, resulting in similar life expectancy for people living with HIV (PLWH) compared to their non-infected counterparts, HIV-associated neurocognitive disorder (HAND) remains a significant challenge [1,2]. For PLWH, the estimated percentage of those experiencing HAND is around 50%, with marginal differences observed across the United States, Europe, Africa, and Asia [3]. Clinical manifestations of HAND have been divided into three severities, asymptomatic neurocognitive impairment (ANI), mild neurocognitive impairment (MNI), and HIV-associated dementia (HAD). ANI is characterized by motor, memory, and executive functioning loss not impacting the daily life of those living with it. MNI is slightly more severe and is defined by impairments beginning to impact daily life. Lastly, HAD is the most severe and is associated with severe motor and memory loss, which usually leads to death within one year [4]. At a cellular level, the interactions that lead to the development and progression of HAND are complex and include both host and viral factors that are at the forefront of HIV research today. From early invasion into the brain to chronic inflammatory states lasting decades, many groups have elucidated various mechanisms associated with HAND. The goal of this paper is to describe the mechanisms associated with the onset and progression of HAND and discuss current management strategies.

2. Early Infiltration of HIV to the Central Nervous System

A 1992 case involving a blood transfusion contaminated with HIV led to the discovery that HIV enters the CNS within 15 days of exposure [5]. At this time, it was unknown how HIV entered the immune-privileged brain, especially early into infection. One of the prevailing theories was the "Trojan Horse" model. In this model, infected monocytes would be able to cross the blood–brain barrier (BBB) to gain access early in infection. By 2006, this theory garnered strong evidence showing that HIV-infected monocytes can cross the BBB through the expression of C-C motif receptor 2 (CCR2) on infected monocytes, which was attracted to secreted C-C motif ligand 2 (CCL2) [6]. This was further supported in 2010 by showing that HIV-infected monocytes expressed elevated levels of CCR2 and were more responsive to CCL2 [7]. In 2012, a study in Thailand demonstrated that HIV mRNA is detectable in the CNS within 8 days of suspected exposure [8]. This study sparked the interest of many in the field and has led to the development of several models of HIV to bypass the BBB to gain entry to the CNS. Recently, it has been shown that HIV can infect pericytes in the BBB [9]. The infection of pericytes can allow HIV to be secreted into the CNS where it can infect the resident microglia. This would provide a direct route for HIV to enter the brain without needing to hijack immune cells in order to bypass the BBB for entry.

Of interest to our group is the role of platelet activation and platelet leukocyte complexes (PLCs) in the infiltration of HIV into the brain. We showed that during HIV infection, there is an activation of platelets to secrete CD40 Ligand (sCD40L) [10]. Subsequently, we demonstrated that the sCD40L promotes the development of platelet monocyte complexes (PMCs) during HIV infection through the interaction of P-selectin on platelets and the P-selectin glycoprotein ligand-1 (PSGL-1) on monocytes [11]. Interestingly, the most increased phenotype of monocytes in the PMCs we found was CD16+ monocytes [10]. CD16+ monocytes display a more inflammatory phenotype than CD16− monocytes, prompting the notion that these circulating complexes are the most significant PLC for promoting chronic inflammation and it had been previously shown that during HIV infection, the vast majority of monocyte-derived immune cells that enter the brain are CD16+ [12,13]. These previous studies paired with our work show that PMCs, particularly CD16+, are significant in promoting the transmigration of inflammatory monocytes into the brain. We are currently analyzing different strategies to antagonize these complexes to reduce further transmigration to the brain, as well as reduce persistent inflammation associated with these cell complexes.

3. Loss of Blood–Brain Barrier Integrity

The deterioration of the BBB has been shown to progress to neurocognitive dysfunctions and dementia [14,15]. By 2002, it had been shown that breakdown of the BBB was one of the key markers of HAD [16]. The mechanisms behind HIV-associated BBB breakdown represent an interesting interplay between host factors and viral proteins. One of the first pathways characterized by the HIV-related dysregulation of the BBB was trans-activator of transcription (Tat) toxicity on brain microvascular endothelial cells (BMECs) [17]. Tat's primary function for HIV is to promote the transcription of long viral transcripts for the integrated provirus. It achieves this by binding to host factors in the P-TEFb complex and TAR region in short viral transcripts to cause the continual activation of CDK9 in the P-TEFb complex. This results in the expression of viral transcripts long enough to encode key proteins [18–20]. In Tat-mediated BMEC toxicity, Tat is able to induce the activation of the antioxidant nuclear factor-κB (NF-κB) and activator protein-1 (AP-1), which damage the BMEC and promote monocyte chemoattractant protein, which further propagates inflammation by recruiting circulating monocytes [17]. Tat-mediated toxicity has been one of our group's interests for many years and we have demonstrated an interesting interplay between Tat and sCD40L. We have shown that Tat is able to upregulate CD40 in monocytes and microglia through the activation of NF-κB. We also have shown that Tat, in combination with sCD40L, exacerbates the activation of NF-κB signaling and the release of tumor necrosis factor-alpha (TNF-α) [21]. Our results show that Tat alone is able to induce host factors that can further amplify the inflammation [21].

In addition to attracting and activating immune cells that propagate an inflammatory response, Tat also damages the BBB by weakening tight junctions and deregulating cellular migration through the BMEC layer. Specifically, Tat acts by downregulating the transcription of the tight junction proteins occludin and zonula occluden 1 and 2 (ZO-1 and ZO-2), while also promoting the accumulation of reactive oxygen species (ROS), advanced glycation end products (AGEs), and amyloid beta (Aβ) in the brain [22–24]. The downregulation of the tight junction proteins is associated with interference of Ras signaling through the stimulation of the Ras homolog gene family A (RhoA)/Rho-associated kinase (ROCK) pathway. The accumulation of ROS is associated with the activation of the Ras pathways stimulated, while the buildup of Aβ and AGE has been shown to be caused by the Tat-mediated upregulation of the Aβ transfer receptor lipoprotein receptor protein 1 (LRP-1), and the receptor of AGE (RAGE) [23,24]. Another pathway our group elucidated for the Tat-mediated disruption of tight junction proteins is through the inhibition of Sonic hedgehog (Shh) signaling. Using a transgenic Tat-expressing mouse model, we demonstrated that Tat expression decreased protein levels of Shh, Gli 1 (a key transcription factor for tight junction proteins), and the tight junction protein Claudin5 [25]. In addition, we also used a humanized mouse model to show that during HIV infection, a mimetic of Shh, the smoothened agonist (SAG), can prevent the downregulation of Gli1, occludin, and Claudin5 while also protecting from leukocyte invasion in acute and chronic phases of infection [26,27]. Together, our work and others have shown the different ways that Tat is involved in disrupting tight junctions and the overall disruption of the BBB during HIV infection.

Another key HIV protein that damages the BBB is gp120. gp120 had been shown to be secreted at low levels in the blood, although due to technical limitations, it was not shown until 1999 that gp120 is secreted at low levels in the brain as well [28]. This led to the question of whether gp120 alone is significant enough to cause neuronal cell death and/or the breakdown of the BBB. gp120 was first shown to be important to HAND by showing its toxicity to neurons by inducing apoptosis by the activation of the CXCR4 receptor on neurons [29]. After the direct pathway for neuronal cell death was characterized, further investigation with mouse models and macrophage co-culture systems demonstrated that neuronal cell death is predominately associated with the activation of macrophages and cell-to-cell signaling and the p38 MPAK cell death pathway [30]. Shortly after these findings, it was shown that gp160 and gp120 could induce apoptosis in endothelial cell lines as well [31]. Lastly, gp120 alone was shown to weaken BBB tight junctions and allow for elevated levels of monocyte transmigration [32]. This suggests that not only does gp120 have a role in the long-lasting BBB weakening, but it is also involved in early damage and immune cell invasion.

4. Activation of Immune Cells Exacerbates Neuroinflammation

4.1. Microglial Targeting of Synapses

The activation of microglia significantly contributes to HAND progression. Microglia of the CNS are essential to maintain the overall homeostasis of the brain. They are involved in critical functions that regulate brain development, neuroplasticity, and neuron injury repair [33]. These specialized cells become activated when small changes in the CNS are detected [34]. The dysfunction and overactivation of microglial cells are implicated in several neurodegenerative diseases.

A necessary receptor that is crucial for the survival and function of microglia is the colony-stimulating factor receptor (CSF1R) [35]. Previous studies using mouse models have shown that the inhibition of this receptor results in approximately 99% of brain-wide microglia depletion [36]. When activated, this receptor can regulate two microglial phenotypes: the M1 phenotype is pro-inflammatory and associated with neurotoxicity, while the M2 phenotype is anti-inflammatory and associated with neuroprotection [36,37]. Microglia's role within the CNS is tightly regulated by "on" and "off" signals that create a reactive/phagocytic or a quiescent state, respectively. These signals are released by neurons and astrocytes through the TGFβ2 signaling pathway [38–40].

Upon stimulus or injury, pro-inflammatory signals from microglia are transduced throughout the cellular network of resident CNS immune cells. As a result, inflammation occurs in localized areas of the brain, resulting in the upregulation of phagocytosis, cytokine secretion, inflammasome activity, and immune cell proliferation [41]. During HIV infection, Tat can induce a shift to the M1 phenotype [42]. While this is thought to be one of the reasons for prolonged damage of the CNS seen in HAND, M1 microglia stimulated with TNF-α and IFN-γ have been shown to contain HIV replication pre- and post-integration by the upregulation of APOBEC 3A [43]. This creates a challenging paradigm for preventing inflammation, as M1 activation is inherently leading to inflammatory cytokines, but also prevents inflammation by restricting HIV replication. This restriction has also been linked to inducing latency in microglia, which further complicates the picture for developing treatment strategies [44].

In part, infected microglia exert their effects by directly targeting the neuronal synapses. In a developing brain, these cells are involved in the establishment of neural circuits through synaptic pruning. However, dysregulated microglia can cause abnormal pruning that results in the loss of synaptic connections and degeneracy [45]. Several studies have provided mechanisms that suggest an interaction between virally induced/activated microglia and synaptic proteins that cause phagocytosis of the synapse.

Since the 1990s, neuron loss and cortical brain damage have been reported in PLWH. In such cases, correlations were observed between dysregulated fractalkine microglial receptors (CX3CR1/FKN) and synapse loss. A recent study investigating HIV-associated pain revealed a possible connection between synaptic degradation and the gp120-induced upregulation of both CX3CR1 and CX3CL1, a chemoattractant typically secreted by neurons during toxic insults. When treated with gp120 alone, an increase in CX3CR1 activity in microglia was seen along with a significant decrease in three markers of presynaptic health in both neuron–glial co-cultures and mouse models: Syn I, Syt-1, and PSD-95 [46].

The coupled response between increased glial receptors and increased neuronal cytokine expression perfectly positions microglia to target the synapse for phagocytosis. With regards to the CNS and PLWH, the ramifications of induced microglia could include the extensive elimination of neuronal circuits. This is especially concerning as the continuous activation of microglia due to HIV latency would only promote further neurodegeneration.

4.2. Recruitment of Circulating Monocytes

Circulating monocytes are recruited to the CNS by secreted CCL2. As with synaptic degradation, HIV viral protein gp120 is implicated in altering the activity of resident CNS immune cells [6]. This includes both microglia and astrocytes through an increased CCL2 expression upon gp120 treatment. As a result, monocytes are recruited to where they are needed, expand in population, and provide an immune response. The exposure to CCL2 upregulates the monocyte expression of adhesion molecules. Because of this, gp120 might induce monocyte interaction with endothelial cells and platelets [47].

The transudative mechanism of peripheral monocytes through endothelial cells (ECs) into subendothelial spaces allows monocytes to reach targeted locations. During an immune response, the activation of pro-inflammatory transcription factors like NF-κB has been shown to upregulate cell surface receptors collectively known as cellular adhesion molecules (CAMs/ICAMSs) [47,48]. It has been recognized for some time that HIV infection leads to the increased activation of NF-κB (subsequent increase in CAMs/ICAMS). More recent studies have shown that in addition to gp120, Tat also induces the expression of these adhesion molecules [49].

The upregulation of these cell surface receptors allows for increased interactions between peripheral monocytes and endothelial cells of the BBB. Because of this, monocytes are encouraged to migrate across the BBB and invade the CNS. As previously mentioned, activated monocytes form a complex with activated platelets, known as a platelet–monocyte complex (PMC). The formation of PMCs in addition to increased CAM/ICAM expression on both endothelial and monocytic cells further enhances the migration of monocytes across the BBB. This persistent cellular movement sustains the immune response and propagates pro-inflammatory signals within the CNS as invading immune cells are activated. Ultimately, in an attempt to localize and combat viral infection, the early activation of microglial and then peripheral monocytes exposes the double-edged sword of the host's immune response.

4.3. Other Long-Lasting Cytokine Pathways

There are several cytokine and chemokine cues from the environment that are implicated in the persistent neuroinflammation of neurodegenerative diseases. Direct evidence for this was produced in 2005 when Cartier et al. showed elevated levels of pro-inflammatory and apoptotic molecules (IL-1α, CXCR2, CCR3, CCR5, and TGF-β) in post-mortem brain tissue in subjects with Alzheimer's disease (AD) compared to age-matched controls [50]. Like other neurodegenerative diseases, the prognosis of HIV, a life-long viral infection, is intimately mediated by long-lasting dysregulated cytokine pathways.

Although the cytotoxic effects of Tat have been known for decades, in 2019, a strong connection was made between the downregulation of nod-like receptor CARD domain 5 (NLRC5) and NF-κB. Nod-like receptors are located on several glial cells and are dedicated to responding to pathogen-associated molecular patterns. In vitro exposure to Tat revealed a significant decrease in NLRC5 expression in the prefrontal cortex of mouse microglial cells [51]. NLRC5 is thought to downregulate the pro-inflammatory effects of NF-κB by blocking the phosphorylation of IKKα and IKKβ [52]. This is an important immunological break in the inflammatory cascade as the activation of NF-κB leads to inflammasome activity and inflammatory molecules. However, with Tat's ability to decrease NLRC5, this break is diminished and the host immune response is kept on.

The well-known immunological signaling pathway JAK-STAT is critical for the coordination of immune cells and is implicated in HIV neuropathology. These interactions are bolstered as the downstream cytokines of JAK-STAT travel through the CNS in cerebrospinal fluid to interact with immune cells deep within cervical lymph nodes [53]. In 2018, it was demonstrated that HIV-1 viral protein Vif associates with STAT1 and STAT3, leading to their degradation through ubiquitination, as opposed to activation through phosphorylation. Without STAT1 and STAT3, T cells fail to receive Type 1 interferon (IFN) survival signals, resulting in their depletion [54].

Over the past decade, however, there has been growing evidence that disruption within this pathway may affect HIV latency. Due to insufficient activation signals from JAK-STAT, there is a possibility that memory T cells may be blocked from becoming a latency reservoir. Studies have shown a marked reduction in the number of CD4+ T cells harboring integrated HIV DNA in both in vitro and in vivo experiments using FDA-approved Jak inhibitors tofacitinib and ruxolitinib [55]. The lack of viral integration within these cells further implicates JAK-STAT's involvement with HIV latency.

The mechanism behind both of these pathways creates a situation for the host in which early viral proteins elicit an initial host response that persists and becomes chaotic, distorted, and harmful to the individual. Further research into identifying the roles of the NLRC5-NF-κB signaling axis and the JAK-STAT pathway during an HIV infection is needed. Nevertheless, preliminary data provide a possible insight into the sustained neuroinflammation seen in PLWH. This information may be used in the future to develop novel treatments for HAND [51,55].

4.4. Biomarkers of HAND and Neurocognitive Impairment

There are two categories of biomarkers being investigated in the context of HAND, advanced neuroimaging markers and plasma/serum and CSF-based soluble markers. Among the neuroimaging markers, cerebral blood flow (CBF) and cerebrovascular reactivity (CVR) are markers for BBB dysfunction. Ances et al. have shown reduced resting CBF in the basal ganglia and visual cortex of PWLH compared to HIV-negative healthy controls (HCs) [56]. CBF abnormalities have also been documented in PWLH with cognitive impairment [56–58]. Callen et al. used changes in CBF induced by intravenous acetazolamide and found decreased CVR in the frontal lobe and basal ganglia, while another study used transcranial Doppler ultrasound and breath-holding to assess mean flow velocity changes to infer CVR [59,60]. Higher CVR was associated with a higher score on the Montreal Cognitive Assessment. In corroboration with these reports, our group found that CVR was a more sensitive measure of pre-cART neurovascular damage as compared to CBF [61]. In addition to these, axonal and myelin changes, extracellular free water (an index of inflammation), quantitative susceptibility mapping (to measure iron dysmetabolism), and white matter hyperintensities are other biomarkers that are being explored by our group and others [61–71].

Many different soluble markers in plasma, serum, and CSF have been investigated for their association with HAND such as markers of neuronal injury, endothelial dysfunction, and monocyte activation. A neurofilament light (NFL) chain is an essential scaffolding protein of the neuronal cytoskeleton, which is released into CSF and peripheral circulation upon axonal injury. Phosphorylated Tau is considered a marker for neuronal and synaptic loss. Glial fibrillary acidic protein (GFAP) is associated with glial activation. Multiple studies have shown that PLWH exhibit increased levels of NFL in blood plasma and CSF, especially in cART-naïve individuals [72–76]. A few reports have also found pTau and GFAP levels to be increased in PLWH; however, these findings have not been consistent throughout the different studies [77–80]. Lastly, endothelial activation has been measured via levels of the intracellular adhesion molecule (ICAM) and vascular cellular adhesion molecule (VCAM), and monocyte activation has been measured using levels of circulating monocytes and sCD14, sCD163, neopterin, etc. ICAM and VCAM are found to be elevated even in well-controlled HIV infection [61,81–84]. A study by Shikuma et al. showed that a higher antiretroviral monocyte efficacy score was linked with better cognitive performance [85]. HIV DNA levels in monocytes and soluble CD14 were found to be associated with VCI in PLWH [86,87]. In addition, our group and others have shown that CD16+ monocyte levels are also associated with a worse cognitive outcome in PLWH [88,89]. In resource-limited settings where MRI facilities are not readily available, these blood biomarkers could be more easily implemented. However, well-designed, prospective, longitudinal studies are needed to overcome the variations found across different studies.

5. Propagation of HIV-Associated Neuroinflammation

5.1. FKN Signaling

CX3CL1, or fractalkine (FKN), exists as an EC membrane-bound protein that serves cell-to-cell adhesion. However, when bound to a monocyte receptor (CX3CR1), it doubles as an immune cell attractant, promoting the recruitment of WBCs. Additionally, CX3CL1-CX3CR signaling is required for immune cells to communicate. This interaction suggests that the activation of the FKN pathway during an HIV CNS invasion may encourage further immune cell invasion and inflammation.

A study in 1997 showed that CX3CL1 is more highly expressed in the CNS compared to the surrounding periphery [90]. Additionally, this expression is increased during viral infections, such as HIV. As previously mentioned, the interaction between HIV Tat and sCD40L is able to activate the NF-κB pathway [21]. A downstream effect of this transcription factor is the upregulation of CX3CL1 expression. In doing so, further recruitment of immune cells such as CD8 T cells occurs [91]. The implications of CD8 T cell involvement include the release of potent and possibly toxic granules, along with dysregulated interactions with CD4+ T cells. Furthermore, the secondary activation of CD8 T cells during latent HIV infection has been shown to result in poor viral control [91].

The effects of altered CX3CL1 expression have been documented in other inflammatory diseases like coronary artery disease (CAD) and kidney disease. When comparing CX3CL1 mRNAs within the T cells of atherosclerotic plaques and peripheral blood, it was found that plaques contained a higher concentration [92]. In another study with similar findings, CX3CL1 expression in renal biopsies of patients with acute glomerulonephritis was elevated compared to negative controls [93]. In addition to an HIV infection of the CNS, a variety of neurodegenerative diseases involve pro-inflammatory signals and immune cell communication. Current evidence supports the involvement of the CX3CL1-CX3CR signaling axis and shows its possible association with other pro-inflammatory pathologies. The long-term dysregulated CX3CL1-CX3CR signaling response from the host during an HIV invasion of the CNS implicates immune cell migration, persistent inflammation, and possibly latency. However, there are other immunological pathways that link these disorders.

5.2. TNF-Alpha Signaling

Another important and well-studied immune molecule is TNF-α. Upon infection, TNF-α is released by macrophages and monocytes to stimulate inflammatory pathways and oxidative stress. Our previous work has shown that Tat in combination with sCD40L can increase TNF-α secretion through the activation of NF-κB [21]. Because of this, the early Tat gene of HIV is linked to the exacerbation of pro-inflammatory pathways. Two additional and important signaling effects of TNF-α include apoptosis and T cell differentiation [94]. Across the literature, the dysregulation of inflammatory pathways has emerged as a common cause of neurodegenerative diseases.

Although vastly different diseases, the neurodegeneration seen in Parkinson's Disease (PD) and amyotrophic lateral sclerosis (ALS) show similar molecular patterns to those with HIV. For patients with PD, elevated levels of TNF-α were found in serum samples. Perhaps more importantly though, elevated TNF-α was positively correlated with the severity of disease [95]. In patients with ALS, elevated levels of TNF-α were found, in association with gene mutations SOD1 and C9ORF72 [96].

Since the 21st century, there has also been evidence showing elevated TNF-α in subjects with a variety of psychiatric illnesses linked to neuroinflammation [97]. One such illness is schizophrenia, a complex psychiatric disorder marked by hallucinations and/or delusions. Not only was an association between elevated TNF-α serum levels and schizophrenia demonstrated, but a positive correlation between schizophrenic symptoms and TNF-α was found [38,98].

With regards to HIV in the CNS, the implications of dysregulated TNF-α involve a widespread inflammatory response that leads to tissue damage. A recent study investigating the mechanism behind HIV-induced TNF-α signaling in monocytes showed that HIV-1 gp120 produced increased levels of TNF-α in a dose-dependent manner [39]. Interestingly, this report also showed that even when viral entry was blocked, increased levels of TNF-α were still observed. This implies that viral entry is not a requirement to alter TNF-α expression.

The relationship between pro-inflammatory molecules and different neurodegenerative diseases represents immunological pathways that have gone awry. This dysfunction can be seen in their overlap from laboratory results such as serum biomarkers to clinical presentations like premature cognitive decline. The causes of PD, ALS, schizophrenia, and HAND arise from vastly different methods, but one of the underlying mechanisms heavily involved appears to include TNF-α. Further investigation into the pathogenic inflammatory effects of this molecule may provide insight into treatment and possibly a cure for these neurodegenerative diseases.

5.3. Cofactors of HAND Progression

A variety of cofactors have been identified to worsen HAND development. One major cofactor of interest is substance abuse. A 2008 study found that 71% of PLWH reported substance abuse while just 24% of PLWH seek treatment for substance abuse [99]. The interplay between HAND progression and substance abuse remains challenging to study; however, some reports have shed insight into this relationship. Studies using simian immunodeficiency virus models in macaques showed an increased influx of macrophages and viral load in the CNS with exposure to cocaine and methamphetamine [100,101]. Other studies have shown that HIV replication in microglia can be upregulated during substance abuse as a response to increased dopamine from drug use [102,103]. For opiates, reports have shown that morphine and Tat synergistically promote neuronal cell death [104,105]. Clinically, it is unclear how substance abuse can shape HAND progression; however, one study showed that a history of substance abuse did not increase the rate of neurocognitive impairments [106]. Noteworthy for this study is that the majority of substance abusers had over a year of abstinence prior to the start, so it is unclear how active drug use may contribute to HAND progression.

A second cofactor to HAND progression is co-infection with both chronic and acute pathogens. In particular, hepatitis C virus (HCV) has garnered attention due to exacerbating neurodegeneration in PLWH. For HIV+HCV+ individuals, the rate of neurocognitive impairments is estimated to be around 63% compared to just 10–49% for HCV+ and 50% for HIV+ [3,107,108]. HCV has been shown to infect and persist in microglia, which leads to a chronic stimulus and cytokine release [109]. Two of these cytokines include TNF-α and IL-1ß, which are key contributors to HAND [94,110,111]. Further investigations are needed to provide a more comprehensive understanding of the interaction of HCV and HIV co-infection, particularly during active HCV replication, and how it can lead to neurodegeneration [107]. Two other pathogens that have been documented to exacerbate HAND include the intracellular parasite *Leishmania* and the *Mycobacterium* bacteria. *L. donovani* infects dendritic cells and macrophages in the blood where it upregulates IL-4 and IL-6, and during the HIV co-infection of macrophages, this upregulation is enhanced [112]. The increase in IL-4 and IL-6 during *L. donovani* likely can damage the BBB and CNS, although more evidence is needed to confirm this. *M. tuberculosis*, like HIV, can infect macrophages for years and produce a chronic inflammatory state. The co-infection of *M. tuberculosis* with HIV in the context of AIDS progression has been well established; however, latent infection with both pathogens in the context of HAND remains less studied [113,114]. During active TB, an increased level of HIV is detectable in the blood and CSF [115,116]. In the context of HIV impacting *M. tuberculosis*, HIV targets CD4+ T cells that have been shown to neutralize *M. tuberculosis* by IFN-γ and IL-2 induction, which can worsen *M. tuberculosis* outcomes in patients [117]. For latent infections of HIV and *M. tuberculosis*, there has not been a detectable chronic increase in cytokines compared to HIV alone; however, there is an increase in T cell activation markers on both CD4+ and CD8+ T cells, namely CD38+ and HLA-DR expression [118]. HLA-DR and CD38+ expression on CD8+ T cells has been shown to be a marker for HAND, but it is not clear if it is directly causative of HAND [119]. Another possible way *M. tuberculosis* could contribute to HAND during more latent states could be intermittent reactivation, but further investigation is needed. Overall, more research is needed to determine the role of these cofactors and others in the progression of HAND.

6. Impact of Latency on Neuroinflammation

6.1. Latency in the Central Nervous System

One of the most challenging aspects of HAND is the latent pool of infected cells that persist during cART. The development of cellular, three-dimensional organoid, and small animal models of HIV-1 infection in microglia has greatly improved our molecular understanding of how intermittent emergence of HIV-1 from latency in microglial cells primarily contributes to both neuroinflammation in the CNS and the progression of HAND [40,120–124].

With an average age of 4.2 years, microglial cells' unique characteristic of slow cell division allows the persistence of HIV-1 in the brain throughout the life of the patients compared to other potentially infected cells in the brain [125,126]. Microglia also serve as the main producer of pro-inflammatory cytokines and partake in phagocytosis when purines released in the forms of ATP and ADP from damaged neurons bind to microglia purinergic receptors P2RY12 [127]. During an active infection, neurodegeneration can result from the persistency of pro-inflammatory markers, such as NO, ROS, IL-1ß, TNF-α, and IL-6 via AP-1 (FosB), NF-κB (RelA/NF-KBI), and Stat1 signaling in microglial cells [128]. A direct association has also been drawn between the persistent upregulation of pro-inflammatory markers, including IL-8, IL-1ß, and TNF-α, and reactivation of latent HIV [129–131].

In a case–control comparison study, direct evidence of HIV-1 latency associated with neurodegenerative disease from 10 autopsies out of 32 HIV-seropositive (HIV+) patients showed non-detectable p24 in the brain with a high amount of viral DNA [132]. Under normal circumstances, the physiological homeostasis of the CNS causes microglia to return to their "resting" state by signaling factors such as FKN, CX3CL1, CD200, TGFbeta1/2, and Wnt [133,134]. However, when there is an imbalance of physiological homeostasis with the prolonged overproduction of pro-inflammatory cytokines such as TNF-α, IL-1ß, IL-6, IL-8, CCL2, CCL5, ROS, and reactive nitrogen species (RNS) triggered by persistent invading viruses, there is a consequence of the exacerbation of neuroinflammation and a further neurodegenerative effect [133,135].

The periodic emergence of HIV-1 from latency within microglial cells may further contribute to neurological complications via the reactivation of more microglial latent reservoirs and proviruses from the recognition of inflammatory cytokines, such as TNF-α and IL-1ß that are associated with HAND [133,136]. Furthermore, the abnormal activation of microglial cells occurs in response to both (1) neuronal damage that results from chronic neuroinflammation and (2) inflammatory cytokines that are produced by reactive astrocyte neurodegeneration [137]. TNF-α triggers the activation of microglial cells, leading to the excessive activation of microglia, which leads to the accumulation of pro-inflammatory cytokines. Meanwhile, TNF-α induces the translocation of the transcription factor NF-κB to the nucleus and upregulates many inflammatory cytokines in primary astrocytes, resulting in neuronal death [110]. Moreover, TNF-α induces the expression of FKN, which enhances the adhesion, chemoattraction, recruitment, and activation of other inflammatory cells [138–140]. Similarly, the expression of IL-1β from microglia is tightly regulated by caspase-1, whose activation is triggered by cleavages of procaspase-1. Once IL-1 is activated, IL-1β activates astrocytes, which subsequently produce TNF-α [139]. Studies have shown that gp120 upregulates both the expression of IL-1β and TNF-α, which indirectly activates latently infected microglial cells [141–143]. The cleavage of IL-1 to its active form, IL-1β, has been shown to be a result of Tat and gp120-mediated activation of the NLRP3 inflammasome [144,145]. This interplay between neuronal damage and microglia activation in the context of latency in the CNS and the progression of HAND continues to be an important area of in-depth research.

Astrocytes have also been shown to support latent HIV reservoirs in the CNS [146,147]. In the viral production in the CNS, astrocytes are typically less studied as they have been shown to support limited viral production. This limited production can be overcome by a stimulus from IFN-γ and TNF-α [148,149]. In the context of harboring a latent pool, it has been shown that class I histone deacetylases and histone methyltransferase, SU(VAR)3-9, repress viral production and promote latency in astrocytes [149]. It is unclear how significantly the astrocytic latent pool contributes to the progression of HAND and further research is needed to fully understand its impact.

6.2. The Circulating Latent Pool

Microglial cells are considered the most challenging HIV reservoir in CNS. Likewise, circulating T cells are characterized as the major HIV reservoir in the peripheral blood. Most studies have found that memory T cells are the majority of latently infected cells. During HIV entry into different T cell subsets, the support of proviral transcription, proviral repression, and latency would be determined by the different transcriptional and intrinsic cellular factors that the virus encountered. Some of the essential host transcription factors, such as NF-κB, AP-1, NFAT, Sp1, and processive RNAP II, can assist efficient proviral transcription once the viral genome is integrated [150–152]. On the other hand, studies have shown that three main biochemical mechanisms limit the proviral transcription: (i) absence of the positive transcription factors like NF-κB, (ii) epigenetic changes on chromatin, and (iii) the presence of repressive factors [153,154]. Latent HIV infection appears to be mostly contained within the central CD4+ T cell compartment with a minority contained in naïve T cells [155,156]. Two cellular mechanisms for the generation of latent infection in CD4+ T cells could be categorized as follows: (1) during the transition from activated CD4+ T cells to a resting state and (2) the direct infection of resting CD4+ T cells [157,158].

Both naïve and activated CD4+ T cells can be infected and support latency. Resting CD4+ T cells can be infected through cell–cell contact by taking viral particles from dendritic cells that present with Siglec-1/CD169 and DC-SIGN receptors. In HIV-1-infected resting CD4+ T cells, no infectious particles were produced. Instead, these cells were observed to have provirus-integrated near transcriptionally active chromatin, which allows the expression of some spliced HIV RNA producing viral proteins, such as Gag and Nef but no to negligible production of Tat, Rev, and Env [159–161]. On the other hand, activated CD4+ T cells contain crucial metabolites such as nucleotides and amino acids that are required for HIV replication and viral transcription. Activated CD4+ T cells will then establish and maintain latency by returning back into quiescent resting memory CD4+ T cells or harbor latent HIV provirus. The subsequent reactivation of quiescent resting memory CD4+ T cells will be triggered by the activation of TCR with the same antigen, leading to the TCR-induced transcription factors and translocation into the nucleus [111]. Studies have shown that HIV-1 protein nef is capable of modulating the TCR activation signals to promote HIV replication and pathogenesis by increasing the expression of the TCR-induced transcription factors NF-kB, NFAT, and interleukin 2 (IL-2) in both quiescent and metabolically active CD4+ T cells [162–164]. This pool of latent cells in the blood is relevant to HAND as it could allow re-entry to the CNS even if the virus is eliminated in the brain.

7. How to Address Neuropathogenesis and CNS Latent Reservoir

7.1. Shock and Kill Approach

The main hurdle to preventing HAND is the latent reservoir in microglial cells and perivascular macrophages [165]. One of the most popular strategies being studied currently to eliminate the latent reservoir, not just in the brain, is the "shock and kill" approach. The "shock and kill" approach entails reactivating the latent reservoir by the use of latency-reactivating agents (LRAs). Using LRAs, the provirus in latent cells would be expressed, which would either trigger the cells to die by a cytopathic effect or recognition by other immune cells such as cytotoxic T lymphocytes [166]. Unfortunately, this strategy has several obstacles and limitations when applied to the CNS. Two of the major challenges are—(i) addressing the heterogenous nature of the latent reservoir and (ii) controlling the damage caused by reactivation. For now, addressing the diverse latent pool is more important for researchers than determining how to prevent damage during reactivation. To address this, a variety of broad-acting drug classes are being studied [167]. One drug class developed includes those acting on epigenetic markers around the provirus, which include histone methyltransferase inhibitors (HMTis) and histone deacetyl transferase inhibitors (HDACis). This mechanism of reactivation would work by relaxing the chromatin to promote transcription factors to enable proviral replication [166]. Other drugs aim to target intracellular and paracellular signaling pathways. Some of these work by enhancing protein kinases in the JAK-STAT pathway, inhibiting the degradation of NF-κB or directly stimulating cytokine receptors [168,169]. Unfortunately, these drugs, even in synergistic combinations, have been insufficient to cause enough reactivation to make the "shock and kill" a viable strategy for the time being [170]. A further challenge to the "shock and kill" approach is minimizing damage in the CNS during reactivation. The cell death in the CNS during "shock and kill" would likely be severe, as the latently infected cells would succumb to death, but possibly uninfected cells could perish by a bystander effect. It is unknown how this cell death could alter the brain and what the long-term consequences could be to the central nervous system.

7.2. Block and Lock Approach

On the flip side, another strategy termed the "block and lock" approach has gained traction in recent years to target the latent reservoir. This strategy aims to lock the latent reservoir into a "deep latency" wherein there is no production of the virus and very minimal to no transcription of the provirus. This would not eliminate the viral reservoir and confer sterilizing a cure but would lower intermittent reactivation and could potentially reduce neuroinflammation associated with viral shedding [171]. There have been a variety of compounds analyzed in the "block and lock" approach, but three classes that could see strong efficacy with regards to HAND are Tat inhibitors, LEDGINs, and NF-κB inhibitors.

As previously discussed, Tat plays a crucial role in the transcription of later viral transcripts as well as propagating neuroinflammation [19,21,22,24,25]. One particular Tat inhibitor of interest is didehydro-cortistatin A (dCA). dCA was shown to reduce the transcription of HIV and reactivation of latent cell lines and primary cells in vitro, as well as reducing reactivation in a mouse latency model [172–174]. dCA was also shown to create a restrictive epigenetic landscape around the HIV promoter by recruiting BAF and increasing the presence of deacetylated histone 3 at Nuc-1 [175].

LEDGINs are inhibitors of HIV integrase and have been shown to be effective in reducing HIV replication in the early and late stages of the cellular cycle [176,177]. Specifically, LEDGINs restrict early replication by allosterically inhibiting integrase. LEDGINs can also affect later stages of replication by promoting the formation of oligomers of integrase that result in the production of defective virions [176,178]. The early administration of LEDGINs could help to prevent the establishment of CNS latent reservoirs, and later administration could serve the "block and lock" approach to achieve a functional cure.

NF-κB inhibition is another path being investigated for the "block and lock" strategy. NF-κB was first shown to increase HIV replication around 50-fold back in 1987 [179]. More recently, there has been attention on inhibiting HSP90, which has been shown to regulate NF-κB, STAT5, and NFAT in a Tat-mediated manner [180,181]. One HSP90 inhibitor, AUY922, in combination with a reverse transcriptase inhibitor, EFdA, was even shown to prevent the reactivation of latent cells 11 weeks after administration in a mouse model [182]. Our group has been interested in sulforaphane (SFN), an isothyonate found in cruciferous vegetables, which has also been shown as an anticarcinogen [183,184]. We recently showed that SFN treatment can reduce the TNF-α-mediated reactivation of latently infected cells by attenuating NF-κB signaling by promoting the antioxidant NRF2 [185]. SFN remains of interest for the CNS as its antioxidant properties can help prevent disease progression associated with oxidative stress in HAND [186]. SFN has also been shown to disrupt the activation of the NLRP3 inflammasome, which could majorly reduce IL-1ß and TNF-α signaling mediated by microglia during HIV infection [187]. Our group is continuing to investigate how the dual anti-HIV and antioxidant effects of SFN could be used to combat HAND (Figure 1). With this said, NF-κB signaling remains one of the top pathways of interest to inhibit in order to promote the "block and lock approach".

7.3. Alternate Treatment and Concluding Remarks

Currently, the best approach to preventing HAND is through cART. cART helps to prevent the progression of ANI to MNI or HAD. Improving the CD4+ T cell count and suppressing viral load through cART has been shown to alleviate neurocognitive impairments [188]. Whether addressing cART, "shock and kill", or "block and lock", one large hurdle to overcome is getting the drugs to cross the BBB to travel to the microglia while avoiding neurotoxicity. Currently, antiretrovirals with a greater ability to permeate the BBB have been shown to reduce viral load but have also been shown to have a greater degree of neurotoxic effects and patients who cease taking them have even been shown to have an increase in neurocognitive functioning [189,190]. This creates a very challenging double-edged sword for treatment and management with no clear solution currently. For most patients on cART, the main symptoms experienced are mood disorders. Treatment for these symptoms has been successful with standard antidepressant drug classes like serotonin reuptake inhibitors [191,192]. A 2018 study found that a CCR2 and CCR5 inhibitor, cenicriviroc, was able to improve neurocognitive impairments in patients with cART [193]. In conclusion, while researchers have improved the treatment of HAND over the past 35 years, until the latent reservoir can be fully purged or driven to a highly non-productive state, we will continue to need cART and develop new drug regimens to improve and prevent neurocognitive impairments.

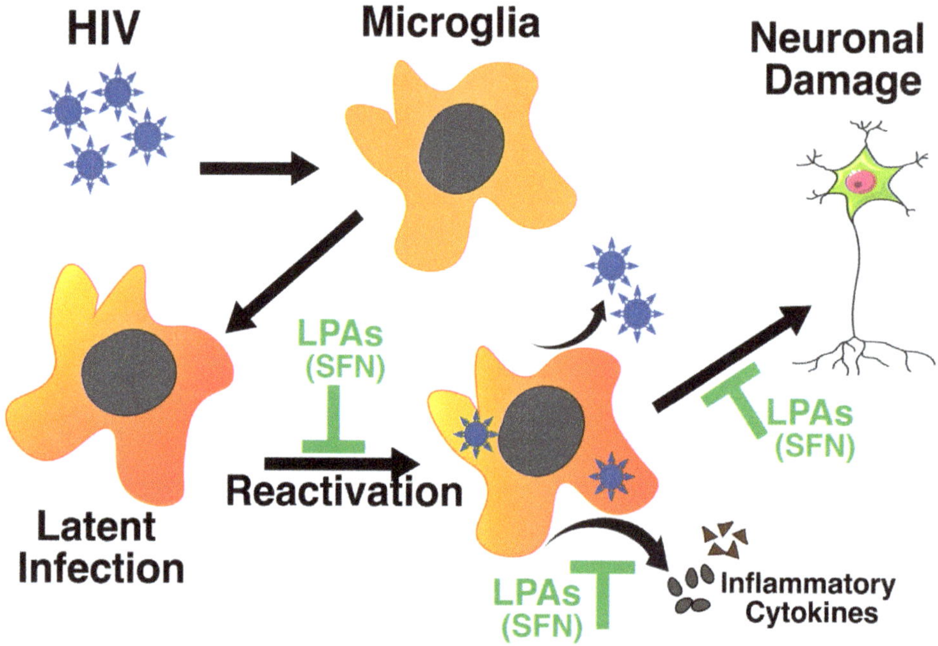

Figure 1. Schematic diagram depicting that microglial cells can be infected by HIV and establish a latent reservoir. However, this reservoir is subject to intermittent reactivation that produces virus particles, viral proteins, and inflammatory cytokines. Neurotoxic viral proteins and cytokines cause neuronal damage and contribute to progressive neurodegenerative consequences. The identification of novel latency promoting agents such as SFN is highly desirable as it can prevent the reactivation of HIV as well as inhibit cytokine release and thus may prevent HIV-associated neuropathology.

Funding: This work was supported by the National Institutes of Health (Grant# R15AI165295-01A1 from NIAID to V.B.S., Grant# R21AI162076 from NIAID to M.V.S., and Grant# R01HL160229 from NHLBI to M.V.S.).

Acknowledgments: We would like to thank Manjot Deol and Dhatri Sangasani for their valuable comments and proofreading the revised manuscript.

Conflicts of Interest: The authors have no conflict of interest.

References

1. Deeks, S.G.; Tracy, R.; Douek, D.C. Systemic Effects of Inflammation on Health during Chronic HIV Infection. *Immunity* **2013**, *39*, 633–645. [CrossRef]
2. Deeks, S.G.; Lewin, S.R.; Havlir, D.V. The End of AIDS: HIV Infection as a Chronic Disease. *Lancet* **2013**, *382*, 1525–1533. [CrossRef]
3. Zenebe, Y.; Necho, M.; Yimam, W.; Akele, B. Worldwide Occurrence of HIV-Associated Neurocognitive Disorders and Its Associated Factors: A Systematic Review and Meta-Analysis. *Front. Psychiatry* **2022**, *13*, 814362. [CrossRef]
4. Rojas-Celis, V.; Valiente-Echeverría, F.; Toro-Ascuy, D.; Soto-Rifo, R. New Challenges of HIV-1 Infection: How HIV-1 Attacks and Resides in the Central Nervous System. *Cells* **2019**, *8*, 1245. [CrossRef]

5. Davis, L.E.; Hjelle, B.L.; Miller, V.E.; Palmer, D.L.; Llewellyn, A.L.; Merlin, T.L.; Young, S.A.; Mills, R.G.; Wachsman, W.; Wiley, C.A. Early Viral Brain Invasion in Iatrogenic Human Immunodeficiency Virus Infection. *Neurology* **1992**, *42*, 1736. [CrossRef]
6. Eugenin, E.A.; Osiecki, K.; Lopez, L.; Goldstein, H.; Calderon, T.M.; Berman, J.W. CCL2/Monocyte Chemoattractant Protein-1 Mediates Enhanced Transmigration of Human Immunodeficiency Virus (HIV)-Infected Leukocytes across the Blood-Brain Barrier: A Potential Mechanism of HIV-CNS Invasion and NeuroAIDS. *J. Neurosci.* **2006**, *26*, 1098–1106. [CrossRef]
7. Hazleton, J.E.; Berman, J.W.; Eugenin, E.A. Novel mechanisms of central nervous system damage in HIV infection. *HIV/AIDS-Res. Palliat. Care* **2010**, *2*, 39–49.
8. Valcour, V.; Chalermchai, T.; Sailasuta, N.; Marovich, M.; Lerdlum, S.; Suttichom, D.; Suwanwela, N.C.; Jagodzinski, L.; Michael, N.; Spudich, S.; et al. Central Nervous System Viral Invasion and Inflammation during Acute HIV Infection. *J. Infect. Dis.* **2012**, *206*, 275–282. [CrossRef]
9. Bertrand, L.; Cho, H.J.; Toborek, M. Blood-Brain Barrier Pericytes as a Target for HIV-1 Infection. *Brain* **2019**, *142*, 502–511. [CrossRef]
10. Singh, M.V.; Davidson, D.C.; Kiebala, M.; Maggirwar, S.B. Detection of Circulating Platelet-Monocyte Complexes in Persons Infected with Human Immunodeficiency Virus Type-1. *J. Virol. Methods* **2012**, *181*, 170–176. [CrossRef] [PubMed]
11. Singh, M.V.; Davidson, D.C.; Jackson, J.W.; Singh, V.B.; Silva, J.; Ramirez, S.H.; Maggirwar, S.B. Characterization of Platelet–Monocyte Complexes in HIV-1–Infected Individuals: Possible Role in HIV-Associated Neuroinflammation. *J. Immunol.* **2014**, *192*, 4674–4684. [CrossRef]
12. Strauss-Ayali, D.; Conrad, S.M.; Mosser, D.M. Monocyte Subpopulations and Their Differentiation Patterns during Infection. *J. Leukoc. Biol.* **2007**, *82*, 244–252. [CrossRef]
13. Fischer-Smith, T.; Bell, C.; Croul, S.; Lewis, M.; Rappaport, J. Monocyte/Macrophage Trafficking in Acquired Immunodeficiency Syndrome Encephalitis: Lessons from Human and Nonhuman Primate Studies. *J. Neurovirol.* **2008**, *14*, 318–326. [CrossRef]
14. Kaul, M.; Garden, G.A.; Lipton, S.A. Pathways to Neuronal Injury and Apoptosis in HIV-Associated Dementia. *Nature* **2001**, *410*, 988–994. [CrossRef]
15. Cecchelli, R.; Berezowski, V.; Lundquist, S.; Culot, M.; Renftel, M.; Dehouck, M.-P.; Fenart, L. Modelling of the Blood–Brain Barrier in Drug Discovery and Development. *Nat. Rev. Drug Discov.* **2007**, *6*, 650–661. [CrossRef]
16. Avison, M.J.; Nath, A.; Greene-Avison, R.; Schmitt, F.A.; Bales, R.A.; Ethisham, A.; Greenberg, R.N.; Berger, J.R. Inflammatory Changes and Breakdown of Microvascular Integrity in Early Human Immunodeficiency Virus Dementia. *J. Neurovirol.* **2004**, *10*, 223–232. [CrossRef]
17. Toborek, M.; Lee, Y.W.; Pu, H.; Malecki, A.; Flora, G.; Garrido, R.; Hennig, B.; Bauer, H.; Nath, A. HIV-Tat Protein Induces Oxidative and Inflammatory Pathways in Brain Endothelium. *J. Neurochem.* **2003**, *84*, 169–179. [CrossRef]
18. Wada, T.; Orphanides, G.; Hasegawa, J.; Kim, D.-K.; Shima, D.; Yamaguchi, Y.; Fukuda, A.; Hisatake, K.; Oh, S.; Reinberg, D.; et al. FACT Relieves DSIF/NELF-Mediated Inhibition of Transcriptional Elongation and Reveals Functional Differences between P-TEFb and TFIIH. *Mol. Cell* **2000**, *5*, 1067–1072. [CrossRef]
19. Isel, C.; Karn, J. Direct Evidence That HIV-1 Tat Stimulates RNA Polymerase II Carboxyl-Terminal Domain Hyperphosphorylation during Transcriptional Elongation. *J. Mol. Biol.* **1999**, *290*, 929–941. [CrossRef]
20. Wada, T.; Takagi, T.; Yamaguchi, Y.; Ferdous, A.; Imai, T.; Hirose, S.; Sugimoto, S.; Yano, K.; Hartzog, G.A.; Winston, F.; et al. DSIF, a Novel Transcription Elongation Factor That Regulates RNA Polymerase II Processivity, Is Composed of Human Spt4 and Spt5 Homologs. *Genes. Dev.* **1998**, *12*, 343–356. [CrossRef]
21. Sui, Z.; Sniderhan, L.F.; Schifitto, G.; Phipps, R.P.; Gelbard, H.A.; Dewhurst, S.; Maggirwar, S.B. Functional Synergy between CD40 Ligand and HIV-1 Tat Contributes to Inflammation: Implications in HIV Type 1 Dementia. *J. Immunol.* **2007**, *178*, 3226–3236. [CrossRef]
22. Zhong, Y.; Smart, E.J.; Weksler, B.; Couraud, P.-O.; Hennig, B.; Toborek, M. Caveolin-1 Regulates Human Immunodeficiency Virus-1 Tat-Induced Alterations of Tight Junction Protein Expression via Modulation of the Ras Signaling. *J. Neurosci.* **2008**, *28*, 7788–7796. [CrossRef]
23. Jiang, W.; Huang, W.; Chen, Y.; Zou, M.; Peng, D.; Chen, D. HIV-1 Transactivator Protein Induces ZO-1 and Neprilysin Dysfunction in Brain Endothelial Cells via the Ras Signaling Pathway. *Oxid. Med. Cell Longev.* **2017**, *2017*, 3160360. [CrossRef]
24. Chen, Y.; Huang, W.; Jiang, W.; Wu, X.; Ye, B.; Zhou, X. HIV-1 Tat Regulates Occludin and A β Transfer Receptor Expression in Brain Endothelial Cells via Rho/ROCK Signaling Pathway. *Oxid. Med. Cell Longev.* **2016**, *2016*, 4196572. [CrossRef]
25. Khan, I.A.; Worrad, A.H.; Singh, M.V.; Maggirwar, S.B.; Singh, V.B. Human Immunodeficiency Virus-1 Tat Exerts Its Neurotoxic Effects by Downregulating Sonic Hedgehog Signaling. *J. Neurovirol.* **2022**, *28*, 305–311. [CrossRef]
26. Singh, V.B.; Singh, M.V.; Piekna-Przybylska, D.; Gorantla, S.; Poluektova, L.Y.; Maggirwar, S.B. Sonic Hedgehog Mimetic Prevents Leukocyte Infiltration into the CNS during Acute HIV Infection. *Sci. Rep.* **2017**, *7*, 9578. [CrossRef]
27. Singh, V.B.; Singh, M.V.; Gorantla, S.; Poluektova, L.Y.; Maggirwar, S.B. Smoothened Agonist Reduces Human Immunodeficiency Virus Type-1-Induced Blood-Brain Barrier Breakdown in Humanized Mice. *Sci. Rep.* **2016**, *6*, 26876. [CrossRef]
28. Altmeyer, R.; Mordelet, E.; Girard, M.; Vidal, C. Expression and Detection of Macrophage-Tropic HIV-1 Gp120 in the Brain Using Conformation-Dependent Antibodies. *Virology* **1999**, *259*, 314–323. [CrossRef]
29. Hesselgesser, J.; Taub, D.; Baskar, P.; Greenberg, M.; Hoxie, J.; Kolson, D.L.; Horuk, R. Neuronal Apoptosis Induced by HIV-1 Gp120 and the Chemokine SDF-1α Is Mediated by the Chemokine Receptor CXCR4. *Curr. Biol.* **1998**, *8*, 595–598. [CrossRef]

30. Kaul, M.; Lipton, S.A. Chemokines and Activated Macrophages in HIV Gp120-Induced Neuronal Apoptosis. *Proc. Natl. Acad. Sci. USA* **1999**, *96*, 8212–8216. [CrossRef]
31. Ullrich, C.K.; Groopman, J.E.; Ganju, R.K. HIV-1 Gp120- and Gp160-Induced Apoptosis in Cultured Endothelial Cells Is Mediated by Caspases. *Blood* **2000**, *96*, 1438–1442. [CrossRef]
32. Kanmogne, G.D.; Schall, K.; Leibhart, J.; Knipe, B.; Gendelman, H.E.; Persidsky, Y. HIV-1 Gp120 Compromises Blood-Brain Barrier Integrity and Enhance Monocyte Migration across Blood-Brain Barrier: Implication for Viral Neuropathogenesis. *J. Cereb. Blood Flow Metab.* **2007**, *27*, 123–134. [CrossRef]
33. Colonna, M.; Butovsky, O. Microglia Function in the Central Nervous System during Health and Neurodegeneration. *Annu. Rev. Immunol.* **2017**, *35*, 441–468. [CrossRef]
34. Ginhoux, F.; Prinz, M. Origin of Microglia: Current Concepts and Past Controversies. *Cold Spring Harb. Perspect. Biol.* **2015**, *7*, a020537. [CrossRef]
35. Green, K.N.; Crapser, J.D.; Hohsfield, L.A. To Kill a Microglia: A Case for CSF1R Inhibitors. *Trends Immunol.* **2020**, *41*, 771–784. [CrossRef]
36. Elmore, M.R.P.; Najafi, A.R.; Koike, M.A.; Dagher, N.N.; Spangenberg, E.E.; Rice, R.A.; Kitazawa, M.; Matusow, B.; Nguyen, H.; West, B.L.; et al. Colony-Stimulating Factor 1 Receptor Signaling Is Necessary for Microglia Viability, Unmasking a Microglia Progenitor Cell in the Adult Brain. *Neuron* **2014**, *82*, 380–397. [CrossRef]
37. Kwon, H.S.; Koh, S.-H. Neuroinflammation in Neurodegenerative Disorders: The Roles of Microglia and Astrocytes. *Transl. Neurodegener.* **2020**, *9*, 42. [CrossRef]
38. Karanikas, E.; Manganaris, S.; Ntouros, E.; Floros, G.; Antoniadis, D.; Garyfallos, G. Cytokines, Cortisol and IGF-1 in First Episode Psychosis and Ultra High Risk Males. Evidence for TNF-α, IFN-γ, TNF-β, IL-4 Deviation. *Asian J. Psychiatr.* **2017**, *26*, 99–103. [CrossRef]
39. Planès, R.; Serrero, M.; Leghmari, K.; BenMohamed, L.; Bahraoui, E. HIV-1 Envelope Glycoproteins Induce the Production of TNF-α and IL-10 in Human Monocytes by Activating Calcium Pathway. *Sci. Rep.* **2018**, *8*, 17215. [CrossRef]
40. Nosi, D.; Lana, D.; Giovannini, M.G.; Delfino, G.; Zecchi-Orlandini, S. Neuroinflammation: Integrated Nervous Tissue Response through Intercellular Interactions at the "Whole System" Scale. *Cells* **2021**, *10*, 1195. [CrossRef]
41. Augusto-Oliveira, M.; Arrifano, G.P.; Lopes-Araújo, A.; Santos-Sacramento, L.; Takeda, P.Y.; Anthony, D.C.; Malva, J.O.; Crespo-Lopez, M.E. What Do Microglia Really Do in Healthy Adult Brain? *Cells* **2019**, *8*, 1293. [CrossRef]
42. Silveira, D.B.; Américo, M.F.; Flores, N.P.; Terenzi, H.; Pinto, A.R. Pharmacological Inhibition of UPR Sensor PERK Attenuates HIV Tat-induced Inflammatory M1 Phenotype in Microglial Cells. *Cell Biochem. Funct.* **2022**, *40*, 177–188. [CrossRef] [PubMed]
43. Cassetta, L.; Kajaste-Rudnitski, A.; Coradin, T.; Saba, E.; Della Chiara, G.; Barbagallo, M.; Graziano, F.; Alfano, M.; Cassol, E.; Vicenzi, E.; et al. M1 Polarization of Human Monocyte-Derived Macrophages Restricts Pre and Postintegration Steps of HIV-1 Replication. *AIDS* **2013**, *27*, 1847–1856. [CrossRef]
44. Graziano, F.; Aimola, G.; Forlani, G.; Turrini, F.; Accolla, R.S.; Vicenzi, E.; Poli, G. Reversible Human Immunodeficiency Virus Type-1 Latency in Primary Human Monocyte-Derived Macrophages Induced by Sustained M1 Polarization. *Sci. Rep.* **2018**, *8*, 14249. [CrossRef]
45. Watson, Z.; Tang, S.-J. Aberrant Synaptic Pruning in CNS Diseases: A Critical Player in HIV-Associated Neurological Dysfunction? *Cells* **2022**, *11*, 1943. [CrossRef]
46. Ru, W.; Liu, X.; Bae, C.; Shi, Y.; Walikonis, R.; Mo Chung, J.; Tang, S.-J. Microglia Mediate HIV-1 Gp120-Induced Synaptic Degeneration in Spinal Pain Neural Circuits. *J. Neurosci.* **2019**, *39*, 8408–8421. [CrossRef]
47. Martins, P.A.d.C.; van Gils, J.M.; Mol, A.; Hordijk, P.L.; Zwaginga, J.J. Platelet Binding to Monocytes Increases the Adhesive Properties of Monocytes by Up-Regulating the Expression and Functionality of B1 and B2 Integrins. *J. Leukoc. Biol.* **2006**, *79*, 499–507. [CrossRef]
48. Regal-McDonald, K.; Patel, R.P. Selective Recruitment of Monocyte Subsets by Endothelial N-Glycans. *Am. J. Pathol.* **2020**, *190*, 947–957. [CrossRef]
49. Dhawan, S.; Puri, R.K.; Kumar, A.; Duplan, H.; Masson, J.M.; Aggarwal, B.B. Human Immunodeficiency Virus-1-Tat Protein Induces the Cell Surface Expression of Endothelial Leukocyte Adhesion Molecule-1, Vascular Cell Adhesion Molecule-1, and Intercellular Adhesion Molecule-1 in Human Endothelial Cells. *Blood* **1997**, *90*, 1535–1544.
50. Cartier, L.; Hartley, O.; Dubois-Dauphin, M.; Krause, K.-H. Chemokine Receptors in the Central Nervous System: Role in Brain Inflammation and Neurodegenerative Diseases. *Brain Res. Rev.* **2005**, *48*, 16–42. [CrossRef]
51. Periyasamy, P.; Thangaraj, A.; Bendi, V.S.; Buch, S. HIV-1 Tat-Mediated Microglial Inflammation Involves a Novel MiRNA-34a-NLRC5-NFκB Signaling Axis. *Brain Behav. Immun.* **2019**, *80*, 227–237. [CrossRef] [PubMed]
52. Meng, Q.; Cai, C.; Sun, T.; Wang, Q.; Xie, W.; Wang, R.; Cui, J. Reversible Ubiquitination Shapes NLRC5 Function and Modulates NF-KB Activation Switch. *J. Cell Biol.* **2015**, *211*, 1025–1040. [CrossRef] [PubMed]
53. Zhao, Z.; Nelson, A.R.; Betsholtz, C.; Zlokovic, B.V. Establishment and Dysfunction of the Blood-Brain Barrier. *Cell* **2015**, *163*, 1064–1078. [CrossRef] [PubMed]
54. Gargan, S.; Ahmed, S.; Mahony, R.; Bannan, C.; Napoletano, S.; O'Farrelly, C.; Borrow, P.; Bergin, C.; Stevenson, N.J. HIV-1 Promotes the Degradation of Components of the Type 1 IFN JAK/STAT Pathway and Blocks Anti-Viral ISG Induction. *EBioMedicine* **2018**, *30*, 203–216. [CrossRef] [PubMed]

55. Gavegnano, C.; Brehm, J.H.; Dupuy, F.P.; Talla, A.; Ribeiro, S.P.; Kulpa, D.A.; Cameron, C.; Santos, S.; Hurwitz, S.J.; Marconi, V.C.; et al. Novel Mechanisms to Inhibit HIV Reservoir Seeding Using Jak Inhibitors. *PLoS Pathog.* **2017**, *13*, e1006740. [CrossRef] [PubMed]
56. Ances, B.M.; Sisti, D.; Vaida, F.; Liang, C.L.; Leontiev, O.; Perthen, J.E.; Buxton, R.B.; Benson, D.; Smith, D.M.; Little, S.J.; et al. Resting Cerebral Blood Flow. *Neurology* **2009**, *73*, 702–708. [CrossRef] [PubMed]
57. Chang, L.; Ernst, T.; Leonido-Yee, M.; Speck, O. Perfusion MRI Detects RCBF Abnormalities in Early Stages of HIV–Cognitive Motor Complex. *Neurology* **2000**, *54*, 389. [CrossRef] [PubMed]
58. Ances, B.M.; Roc, A.C.; Wang, J.; Korczykowski, M.; Okawa, J.; Stern, J.; Kim, J.; Wolf, R.; Lawler, K.; Kolson, D.L.; et al. Caudate Blood Flow and Volume Are Reduced in HIV^+ Neurocognitively Impaired Patients. *Neurology* **2006**, *66*, 862–866. [CrossRef] [PubMed]
59. Callen, A.L.; Dupont, S.M.; Pyne, J.; Talbott, J.; Tien, P.; Calabrese, E.; Saloner, D.; Chow, F.C.; Narvid, J. The Regional Pattern of Abnormal Cerebrovascular Reactivity in HIV-Infected, Virally Suppressed Women. *J. Neurovirol.* **2020**, *26*, 734–742. [CrossRef]
60. Chow, F.C.; Wang, H.; Li, Y.; Mehta, N.; Hu, Y.; Han, Y.; Xie, J.; Lu, W.; Xu, W.; Li, T. Cerebral Vasoreactivity Evaluated by the Breath-Holding Challenge Correlates with Performance on a Cognitive Screening Test in Persons Living with Treated HIV Infection in China. *JAIDS J. Acquir. Immune Defic. Syndr.* **2018**, *79*, e101–e104. [CrossRef]
61. Singh, M.V.; Uddin, M.N.; Singh, V.B.; Peterson, A.N.; Murray, K.D.; Zhuang, Y.; Tyrell, A.; Wang, L.; Tivarus, M.E.; Zhong, J.; et al. Initiation of Combined Antiretroviral Therapy Confers Suboptimal Beneficial Effects on Neurovascular Function in People with HIV. *Front. Neurol.* **2023**, *14*, 1240300. [CrossRef] [PubMed]
62. Murray, K.D.; Singh, M.V.; Zhuang, Y.; Uddin, M.N.; Qiu, X.; Weber, M.T.; Tivarus, M.E.; Wang, H.Z.; Sahin, B.; Zhong, J.; et al. Pathomechanisms of HIV-Associated Cerebral Small Vessel Disease: A Comprehensive Clinical and Neuroimaging Protocol and Analysis Pipeline. *Front. Neurol.* **2020**, *11*, 595463. [CrossRef] [PubMed]
63. Murray, K.D.; Uddin, M.N.; Tivarus, M.E.; Sahin, B.; Wang, H.Z.; Singh, M.V.; Qiu, X.; Wang, L.; Spincemaille, P.; Wang, Y.; et al. Increased Risk for Cerebral Small Vessel Disease Is Associated with Quantitative Susceptibility Mapping in HIV Infected and Uninfected Individuals. *Neuroimage Clin.* **2021**, *32*, 102786. [CrossRef] [PubMed]
64. McMurtray, A.; Nakamoto, B.; Shikuma, C.; Valcour, V. Cortical Atrophy and White Matter Hyperintensities in HIV: The Hawaii Aging with HIV Cohort Study. *J. Stroke Cerebrovasc. Dis.* **2008**, *17*, 212–217. [CrossRef] [PubMed]
65. Su, T.; Wit, F.W.N.M.; Caan, M.W.A.; Schouten, J.; Prins, M.; Geurtsen, G.J.; Cole, J.H.; Sharp, D.J.; Richard, E.; Reneman, L.; et al. White Matter Hyperintensities in Relation to Cognition in HIV-Infected Men with Sustained Suppressed Viral Load on Combination Antiretroviral Therapy. *AIDS* **2016**, *30*, 2329–2339. [CrossRef] [PubMed]
66. Filippi, C.G.; Ulug, A.M.; Ryan, E.; Ferrando, S.J.; van Gorp, W. Diffusion Tensor Imaging of Patients with HIV and Normal-Appearing White Matter on MR Images of the Brain. *AJNR Am. J. Neuroradiol.* **2001**, *22*, 277–283.
67. Wu, Y.; Storey, P.; Cohen, B.A.; Epstein, L.G.; Edelman, R.R.; Ragin, A.B. Diffusion Alterations in Corpus Callosum of Patients with HIV. *AJNR Am. J. Neuroradiol.* **2006**, *27*, 656–660. [PubMed]
68. Chang, L.; Wong, V.; Nakama, H.; Watters, M.; Ramones, D.; Miller, E.N.; Cloak, C.; Ernst, T. Greater Than Age-Related Changes in Brain Diffusion of HIV Patients after 1 Year. *J. Neuroimmune Pharmacol.* **2008**, *3*, 265–274. [CrossRef] [PubMed]
69. Pfefferbaum, A.; Rosenbloom, M.J.; Rohlfing, T.; Kemper, C.A.; Deresinski, S.; Sullivan, E. V Frontostriatal Fiber Bundle Compromise in HIV Infection without Dementia. *AIDS* **2009**, *23*, 1977–1985. [CrossRef]
70. Tate, D.F.; Conley, J.; Paul, R.H.; Coop, K.; Zhang, S.; Zhou, W.; Laidlaw, D.H.; Taylor, L.E.; Flanigan, T.; Navia, B.; et al. Quantitative Diffusion Tensor Imaging Tractography Metrics Are Associated with Cognitive Performance among HIV-Infected Patients. *Brain Imaging Behav.* **2010**, *4*, 68–79. [CrossRef]
71. Zhu, T.; Zhong, J.; Hu, R.; Tivarus, M.; Ekholm, S.; Harezlak, J.; Ombao, H.; Navia, B.; Cohen, R.; Schifitto, G. Patterns of White Matter Injury in HIV Infection after Partial Immune Reconstitution: A DTI Tract-Based Spatial Statistics Study. *J. Neurovirol.* **2013**, *19*, 10–23. [CrossRef] [PubMed]
72. Peterson, J.; Gisslen, M.; Zetterberg, H.; Fuchs, D.; Shacklett, B.L.; Hagberg, L.; Yiannoutsos, C.T.; Spudich, S.S.; Price, R.W. Cerebrospinal Fluid (CSF) Neuronal Biomarkers across the Spectrum of HIV Infection: Hierarchy of Injury and Detection. *PLoS ONE* **2014**, *9*, e116081. [CrossRef]
73. Gisslén, M.; Price, R.W.; Andreasson, U.; Norgren, N.; Nilsson, S.; Hagberg, L.; Fuchs, D.; Spudich, S.; Blennow, K.; Zetterberg, H. Plasma Concentration of the Neurofilament Light Protein (NFL) Is a Biomarker of CNS Injury in HIV Infection: A Cross-Sectional Study. *EBioMedicine* **2016**, *3*, 135–140. [CrossRef]
74. Alagaratnam, J.; De Francesco, D.; Zetterberg, H.; Heslegrave, A.; Toombs, J.; Kootstra, N.A.; Underwood, J.; Gisslen, M.; Reiss, P.; Fidler, S.; et al. Correlation between Cerebrospinal Fluid and Plasma Neurofilament Light Protein in Treated HIV Infection: Results from the COBRA Study. *J. Neurovirol.* **2022**, *28*, 54–63. [CrossRef]
75. Guha, D.; Mukerji, S.S.; Chettimada, S.; Misra, V.; Lorenz, D.R.; Morgello, S.; Gabuzda, D. Cerebrospinal Fluid Extracellular Vesicles and Neurofilament Light Protein as Biomarkers of Central Nervous System Injury in HIV-Infected Patients on Antiretroviral Therapy. *AIDS* **2019**, *33*, 615–625. [CrossRef]
76. Ripamonti, E.; Edén, A.; Nilsson, S.; Sönnerborg, A.; Zetterberg, H.; Gisslén, M. Longitudinal Decline of Plasma Neurofilament Light Levels after Antiretroviral Initiation in People Living with HIV. *J. Intern. Med.* **2023**, *293*, 445–456. [CrossRef]

77. de Menezes, E.G.M.; Liu, J.S.; Bowler, S.A.; Giron, L.B.; D'Antoni, M.L.; Shikuma, C.M.; Abdel-Mohsen, M.; Ndhlovu, L.C.; Norris, P.J. Circulating Brain-Derived Extracellular Vesicles Expressing Neuroinflammatory Markers Are Associated with HIV-Related Neurocognitive Impairment. *Front. Immunol.* **2022**, *13*, 1033712. [CrossRef] [PubMed]
78. Ellis, R.J.; Chenna, A.; Petropoulos, C.J.; Lie, Y.; Curanovic, D.; Crescini, M.; Winslow, J.; Sundermann, E.; Tang, B.; Letendre, S.L. Higher Cerebrospinal Fluid Biomarkers of Neuronal Injury in HIV-Associated Neurocognitive Impairment. *J. Neurovirol.* **2022**, *28*, 438–445. [CrossRef] [PubMed]
79. Ozturk, T.; Kollhoff, A.; Anderson, A.M.; Christina Howell, J.; Loring, D.W.; Waldrop-Valverde, D.; Franklin, D.; Letendre, S.; Tyor, W.R.; Hu, W.T. Linked CSF Reduction of Phosphorylated Tau and IL-8 in HIV Associated Neurocognitive Disorder. *Sci. Rep.* **2019**, *9*, 8733. [CrossRef]
80. Sathler, M.F.; Doolittle, M.J.; Cockrell, J.A.; Nadalin, I.R.; Hofmann, F.; VandeWoude, S.; Kim, S. HIV and FIV Glycoproteins Increase Cellular Tau Pathology via CGMP-Dependent Kinase II Activation. *J. Cell Sci.* **2022**, *135*, jcs259764. [CrossRef]
81. Rönsholt, F.F.; Ullum, H.; Katzenstein, T.L.; Gerstoft, J.; Ostrowski, S.R. Persistent Inflammation and Endothelial Activation in HIV-1 Infected Patients after 12 Years of Antiretroviral Therapy. *PLoS ONE* **2013**, *8*, e65182. [CrossRef]
82. Ross, A.C.; O'Riordan, M.A.; Storer, N.; Dogra, V.; McComsey, G.A. Heightened Inflammation Is Linked to Carotid Intima-Media Thickness and Endothelial Activation in HIV-Infected Children. *Atherosclerosis* **2010**, *211*, 492–498. [CrossRef]
83. Guha, D.; Misra, V.; Yin, J.; Horiguchi, M.; Uno, H.; Gabuzda, D. Vascular Injury Markers Associated with Cognitive Impairment in People with HIV on Suppressive Antiretroviral Therapy. *AIDS* **2023**, *37*, 2137–2147. [CrossRef]
84. Saloner, R.; Sun-Suslow, N.; Morgan, E.E.; Lobo, J.; Cherner, M.; Ellis, R.J.; Heaton, R.K.; Grant, I.; Letendre, S.L.; Iudicello, J.E. Plasma Biomarkers of Vascular Dysfunction Uniquely Relate to a Vascular-Risk Profile of Neurocognitive Deficits in Virally-Suppressed Adults with HIV. *Brain Behav. Immun. Health* **2022**, *26*, 100560. [CrossRef]
85. Shikuma, C.M.; Nakamoto, B.; Shiramizu, B.; Liang, C.-Y.; DeGruttola, V.; Bennett, K.; Paul, R.; Kallianpur, K.; Chow, D.; Gavegnano, C.; et al. Antiretroviral Monocyte Efficacy Score Linked to Cognitive Impairment in HIV. *Antivir. Ther.* **2012**, *17*, 1233–1242. [CrossRef]
86. Kusao, I.; Shiramizu, B.; Liang, C.-Y.; Grove, J.; Agsalda, M.; Troelstrup, D.; Velasco, V.-N.; Marshall, A.; Whitenack, N.; Shikuma, C.; et al. Cognitive Performance Related to HIV-1-Infected Monocytes. *J. Neuropsychiatry Clin. Neurosci.* **2012**, *24*, 71–80. [CrossRef] [PubMed]
87. Muñoz-Nevárez, L.A.; Imp, B.M.; Eller, M.A.; Kiweewa, F.; Maswai, J.; Polyak, C.; Olwenyi, O.A.; Allen, I.E.; Rono, E.; Milanini, B.; et al. Monocyte Activation, HIV, and Cognitive Performance in East Africa. *J. Neurovirol.* **2020**, *26*, 52–59. [CrossRef] [PubMed]
88. Veenhuis, R.T.; Williams, D.W.; Shirk, E.N.; Monteiro Abreu, C.; Ferreira, E.A.; Coughlin, J.M.; Brown, T.T.; Maki, P.M.; Anastos, K.; Berman, J.W.; et al. Higher Circulating Intermediate Monocytes Are Associated with Cognitive Function in Women with HIV. *JCI Insight* **2021**, *6*, 146215. [CrossRef]
89. Singh, M.V.; Uddin, N.; Covacevich Vidalle, M.; Sutton, K.R.; Boodoo, Z.D.; Peterson, A.N.; Tyrell, A.; Brenner, R.; Madalina Tivarus, E.; Wang, H.Z.; et al. Role of Non-Classical Monocytes in HIV-Associated Vascular Cognitive Impairment. *medRxiv* **2023**. [CrossRef] [PubMed]
90. Pawelec, P.; Ziemka-Nalecz, M.; Sypecka, J.; Zalewska, T. The Impact of the CX3CL1/CX3CR1 Axis in Neurological Disorders. *Cells* **2020**, *9*, 2277. [CrossRef]
91. Wang, Z.-C.; Li, L.-H.; Bian, C.; Yang, L.; Lv, N.; Zhang, Y.-Q. Involvement of NF-KB and the CX3CR1 Signaling Network in Mechanical Allodynia Induced by Tetanic Sciatic Stimulation. *Neurosci. Bull.* **2018**, *34*, 64–73. [CrossRef]
92. Komissarov, A.; Potashnikova, D.; Freeman, M.L.; Gontarenko, V.; Maytesyan, D.; Lederman, M.M.; Vasilieva, E.; Margolis, L. Driving T Cells to Human Atherosclerotic Plaques: CCL3/CCR5 and CX3CL1/CX3CR1 Migration Axes. *Eur. J. Immunol.* **2021**, *51*, 1857–1859. [CrossRef]
93. Cormican, S.; Griffin, M.D. Fractalkine (CX3CL1) and Its Receptor CX3CR1: A Promising Therapeutic Target in Chronic Kidney Disease? *Front. Immunol.* **2021**, *12*, 664202. [CrossRef]
94. Aggarwal, B.B. Signalling Pathways of the TNF Superfamily: A Double-Edged Sword. *Nat. Rev. Immunol.* **2003**, *3*, 745–756. [CrossRef]
95. Williams-Gray, C.H.; Wijeyekoon, R.; Yarnall, A.J.; Lawson, R.A.; Breen, D.P.; Evans, J.R.; Cummins, G.A.; Duncan, G.W.; Khoo, T.K.; Burn, D.J.; et al. Serum Immune Markers and Disease Progression in an Incident Parkinson's Disease Cohort. *Mov. Disord.* **2016**, *31*, 995–1003. [CrossRef]
96. Mifflin, L.; Hu, Z.; Dufort, C.; Hession, C.C.; Walker, A.J.; Niu, K.; Zhu, H.; Liu, N.; Liu, J.S.; Levin, J.Z.; et al. A RIPK1-Regulated Inflammatory Microglial State in Amyotrophic Lateral Sclerosis. *Proc. Natl. Acad. Sci. USA* **2021**, *118*, e2025102118. [CrossRef]
97. Dowlati, Y.; Herrmann, N.; Swardfager, W.; Liu, H.; Sham, L.; Reim, E.K.; Lanctôt, K.L. A Meta-Analysis of Cytokines in Major Depression. *Biol. Psychiatry* **2010**, *67*, 446–457. [CrossRef]
98. Garcia-Alvarez, L.; Garcia-Portilla, M.P.; Gonzalez-Blanco, L.; Saiz Martinez, P.A.; de la Fuente-Tomas, L.; Menendez-Miranda, I.; Iglesias, C.; Bobes, J. Biomarcadores Sanguíneos Diferenciales de Las Dimensiones Psicopatológicas de La Esquizofrenia. *Rev. Psiquiatr. Salud Ment.* **2016**, *9*, 219–227. [CrossRef]
99. Korthuis, P.T.; Josephs, J.S.; Fleishman, J.A.; Hellinger, J.; Himelhoch, S.; Chander, G.; Morse, E.B.; Gebo, K.A. Substance Abuse Treatment in Human Immunodeficiency Virus: The Role of Patient–Provider Discussions. *J. Subst. Abus. Treat.* **2008**, *35*, 294–303. [CrossRef]

100. Bokhari, S.M.; Hegde, R.; Callen, S.; Yao, H.; Adany, I.; Li, Q.; Li, Z.; Pinson, D.; Yeh, H.-W.; Cheney, P.D.; et al. Morphine Potentiates Neuropathogenesis of SIV Infection in Rhesus Macaques. *J. Neuroimmune Pharmacol.* **2011**, *6*, 626–639. [CrossRef] [PubMed]
101. Marcondes, M.C.G.; Flynn, C.; Watry, D.D.; Zandonatti, M.; Fox, H.S. Methamphetamine Increases Brain Viral Load and Activates Natural Killer Cells in Simian Immunodeficiency Virus-Infected Monkeys. *Am. J. Pathol.* **2010**, *177*, 355–361. [CrossRef]
102. Gaskill, P.J.; Calderon, T.M.; Luers, A.J.; Eugenin, E.A.; Javitch, J.A.; Berman, J.W. Human Immunodeficiency Virus (HIV) Infection of Human Macrophages Is Increased by Dopamine. *Am. J. Pathol.* **2009**, *175*, 1148–1159. [CrossRef]
103. Gaskill, P.J.; Calderon, T.M.; Coley, J.S.; Berman, J.W. Drug Induced Increases in CNS Dopamine Alter Monocyte, Macrophage and T Cell Functions: Implications for HAND. *J. Neuroimmune Pharmacol.* **2013**, *8*, 621–642. [CrossRef]
104. Malik, S.; Khalique, H.; Buch, S.; Seth, P. A Growth Factor Attenuates HIV-1 Tat and Morphine Induced Damage to Human Neurons: Implication in HIV/AIDS-Drug Abuse Cases. *PLoS ONE* **2011**, *6*, e18116. [CrossRef]
105. Fitting, S.; Xu, R.; Bull, C.; Buch, S.K.; El-Hage, N.; Nath, A.; Knapp, P.E.; Hauser, K.F. Interactive Comorbidity between Opioid Drug Abuse and HIV-1 Tat. *Am. J. Pathol.* **2010**, *177*, 1397–1410. [CrossRef]
106. Byrd, D.A.; Fellows, R.P.; Morgello, S.; Franklin, D.; Heaton, R.K.; Deutsch, R.; Atkinson, J.H.; Clifford, D.B.; Collier, A.C.; Marra, C.M.; et al. Neurocognitive Impact of Substance Use in HIV Infection. *JAIDS J. Acquir. Immune Defic. Syndr.* **2011**, *58*, 154–162. [CrossRef]
107. Gill, A.J.; Kolson, D.L. Chronic Inflammation and the Role for Cofactors (Hepatitis C, Drug Abuse, Antiretroviral Drug Toxicity, Aging) in HAND Persistence. *Curr. HIV/AIDS Rep.* **2014**, *11*, 325–335. [CrossRef]
108. Chak, E.; Talal, A.H.; Sherman, K.E.; Schiff, E.R.; Saab, S. Hepatitis C Virus Infection in USA: An Estimate of True Prevalence. *Liver Int.* **2011**, *31*, 1090–1101. [CrossRef]
109. Wilkinson, J.; Radkowski, M.; Eschbacher, J.M.; Laskus, T. Activation of Brain Macrophages/Microglia Cells in Hepatitis C Infection. *Gut* **2010**, *59*, 1394–1400. [CrossRef] [PubMed]
110. Han, Y.; He, T.; Huang, D.; Pardo, C.A.; Ransohoff, R.M. TNF-α Mediates SDF-1α–Induced NF-KB Activation and Cytotoxic Effects in Primary Astrocytes. *J. Clin. Investig.* **2001**, *108*, 425–435. [CrossRef] [PubMed]
111. Hokello, J.; Sharma, A.L.; Dimri, M.; Tyagi, M. Insights into the HIV Latency and the Role of Cytokines. *Pathogens* **2019**, *8*, 137. [CrossRef]
112. Maksoud, S.; El Hokayem, J. The Cytokine/Chemokine Response in Leishmania/HIV Infection and Co-Infection. *Heliyon* **2023**, *9*, e15055. [CrossRef]
113. Whalen, C.; Horsburgh, C.R.; Hom, D.; Lahart, C.; Simberkoff, M.; Ellner, J. Accelerated Course of Human Immunodeficiency Virus Infection after Tuberculosis. *Am. J. Respir. Crit. Care Med.* **1995**, *151*, 129–135. [CrossRef]
114. Bell, L.C.K.; Noursadeghi, M. Pathogenesis of HIV-1 and Mycobacterium Tuberculosis Co-Infection. *Nat. Rev. Microbiol.* **2018**, *16*, 80–90. [CrossRef]
115. Marais, S.; Meintjes, G.; Lesosky, M.; Wilkinson, K.A.; Wilkinson, R.J. Interleukin-17 Mediated Differences in the Pathogenesis of HIV-1-Associated Tuberculous and Cryptococcal Meningitis. *AIDS* **2015**, *30*, 395–404. [CrossRef]
116. Goletti, D.; Weissman, D.; Jackson, R.W.; Graham, N.M.; Vlahov, D.; Klein, R.S.; Munsiff, S.S.; Ortona, L.; Cauda, R.; Fauci, A.S. Effect of Mycobacterium Tuberculosis on HIV Replication. Role of Immune Activation. *J. Immunol.* **1996**, *157*, 1271–1278. [CrossRef]
117. Kalsdorf, B.; Scriba, T.J.; Wood, K.; Day, C.L.; Dheda, K.; Dawson, R.; Hanekom, W.A.; Lange, C.; Wilkinson, R.J. HIV-1 Infection Impairs the Bronchoalveolar T-Cell Response to Mycobacteria. *Am. J. Respir. Crit. Care Med.* **2009**, *180*, 1262–1270. [CrossRef]
118. Sullivan, Z.A.; Wong, E.B.; Ndung'u, T.; Kasprowicz, V.O.; Bishai, W.R. Latent and Active Tuberculosis Infection Increase Immune Activation in Individuals Co-Infected with HIV. *EBioMedicine* **2015**, *2*, 334–340. [CrossRef]
119. Ratto-Kim, S.; Schuetz, A.; Sithinamsuwan, P.; Barber, J.; Hutchings, N.; Lerdlum, S.; Fletcher, J.L.K.; Phuang-Ngern, Y.; Chuenarom, W.; Tipsuk, S.; et al. Characterization of Cellular Immune Responses in Thai Individuals with and without HIV-Associated Neurocognitive Disorders. *AIDS Res. Hum. Retroviruses* **2018**, *34*, 685–689. [CrossRef]
120. Swanta, N.; Aryal, S.; Nejtek, V.; Shenoy, S.; Ghorpade, A.; Borgmann, K. Blood-Based Inflammation Biomarkers of Neurocognitive Impairment in People Living with HIV. *J. Neurovirol.* **2020**, *26*, 358–370. [CrossRef]
121. Williams, M.E.; Stein, D.J.; Joska, J.A.; Naudé, P.J.W. Cerebrospinal Fluid Immune Markers and HIV-Associated Neurocognitive Impairments: A Systematic Review. *J. Neuroimmunol.* **2021**, *358*, 577649. [CrossRef]
122. Livelli, A.; Vaida, F.; Ellis, R.J.; Ma, Q.; Ferrara, M.; Clifford, D.B.; Collier, A.C.; Gelman, B.B.; Marra, C.M.; McArthur, J.C.; et al. Correlates of HIV RNA Concentrations in Cerebrospinal Fluid during Antiretroviral Therapy: A Longitudinal Cohort Study. *Lancet HIV* **2019**, *6*, e456–e462. [CrossRef]
123. Li, H.; McLaurin, K.A.; Mactutus, C.F.; Booze, R.M. A Rat Model of EcoHIV Brain Infection. *J. Vis. Exp.* **2021**, *167*, e62137. [CrossRef]
124. Gumbs, S.B.H.; Berdenis van Berlekom, A.; Kübler, R.; Schipper, P.J.; Gharu, L.; Boks, M.P.; Ormel, P.R.; Wensing, A.M.J.; de Witte, L.D.; Nijhuis, M. Characterization of HIV-1 Infection in Microglia-Containing Human Cerebral Organoids. *Viruses* **2022**, *14*, 829. [CrossRef]
125. Sominsky, L.; De Luca, S.; Spencer, S.J. Microglia: Key Players in Neurodevelopment and Neuronal Plasticity. *Int. J. Biochem. Cell Biol.* **2018**, *94*, 56–60. [CrossRef]

126. Réu, P.; Khosravi, A.; Bernard, S.; Mold, J.E.; Salehpour, M.; Alkass, K.; Perl, S.; Tisdale, J.; Possnert, G.; Druid, H.; et al. The Lifespan and Turnover of Microglia in the Human Brain. *Cell Rep.* **2017**, *20*, 779–784. [CrossRef]
127. Illes, P.; Rubini, P.; Ulrich, H.; Zhao, Y.; Tang, Y. Regulation of Microglial Functions by Purinergic Mechanisms in the Healthy and Diseased CNS. *Cells* **2020**, *9*, 1108. [CrossRef]
128. Baxter, P.S.; Dando, O.; Emelianova, K.; He, X.; McKay, S.; Hardingham, G.E.; Qiu, J. Microglial Identity and Inflammatory Responses Are Controlled by the Combined Effects of Neurons and Astrocytes. *Cell Rep.* **2021**, *34*, 108882. [CrossRef]
129. Holtman, I.R.; Skola, D.; Glass, C.K. Transcriptional Control of Microglia Phenotypes in Health and Disease. *J. Clin. Investig.* **2017**, *127*, 3220–3229. [CrossRef]
130. Ryan, S.K.; Gonzalez, M.V.; Garifallou, J.P.; Bennett, F.C.; Williams, K.S.; Sotuyo, N.P.; Mironets, E.; Cook, K.; Hakonarson, H.; Anderson, S.A.; et al. Neuroinflammation and EIF2 Signaling Persist despite Antiretroviral Treatment in an HiPSC Tri-Culture Model of HIV Infection. *Stem Cell Rep.* **2020**, *14*, 703–716. [CrossRef]
131. Garcia-Mesa, Y.; Jay, T.R.; Checkley, M.A.; Luttge, B.; Dobrowolski, C.; Valadkhan, S.; Landreth, G.E.; Karn, J.; Alvarez-Carbonell, D. Immortalization of Primary Microglia: A New Platform to Study HIV Regulation in the Central Nervous System. *J. Neurovirol.* **2017**, *23*, 47–66. [CrossRef]
132. Desplats, P.; Dumaop, W.; Smith, D.; Adame, A.; Everall, I.; Letendre, S.; Ellis, R.; Cherner, M.; Grant, I.; Masliah, E. Molecular and Pathologic Insights from Latent HIV-1 Infection in the Human Brain. *Neurology* **2013**, *80*, 1415–1423. [CrossRef]
133. Alvarez-Carbonell, D.; Ye, F.; Ramanath, N.; Garcia-Mesa, Y.; Knapp, P.E.; Hauser, K.F.; Karn, J. Cross-Talk between Microglia and Neurons Regulates HIV Latency. *PLoS Pathog.* **2019**, *15*, e1008249. [CrossRef]
134. Zhan, L.; Krabbe, G.; Du, F.; Jones, I.; Reichert, M.C.; Telpoukhovskaia, M.; Kodama, L.; Wang, C.; Cho, S.; Sayed, F.; et al. Proximal Recolonization by Self-Renewing Microglia Re-Establishes Microglial Homeostasis in the Adult Mouse Brain. *PLoS Biol.* **2019**, *17*, e3000134. [CrossRef]
135. Chhatbar, C.; Prinz, M. The Roles of Microglia in Viral Encephalitis: From Sensome to Therapeutic Targeting. *Cell Mol. Immunol.* **2021**, *18*, 250–258. [CrossRef]
136. Alvarez-Carbonell, D.; Garcia-Mesa, Y.; Milne, S.; Das, B.; Dobrowolski, C.; Rojas, R.; Karn, J. Toll-like Receptor 3 Activation Selectively Reverses HIV Latency in Microglial Cells. *Retrovirology* **2017**, *14*, 9. [CrossRef]
137. Garland, E.F.; Hartnell, I.J.; Boche, D. Microglia and Astrocyte Function and Communication: What Do We Know in Humans? *Front. Neurosci.* **2022**, *16*, 824888. [CrossRef]
138. Pereira, C.F.; Middel, J.; Jansen, G.; Verhoef, J.; Nottet, H.S.L.M. Enhanced Expression of Fractalkine in HIV-1 Associated Dementia. *J. Neuroimmunol.* **2001**, *115*, 168–175. [CrossRef]
139. Tong, N.; Perry, S.W.; Zhang, Q.; James, H.J.; Guo, H.; Brooks, A.; Bal, H.; Kinnear, S.A.; Fine, S.; Epstein, L.G.; et al. Neuronal Fractalkine Expression in HIV-1 Encephalitis: Roles for Macrophage Recruitment and Neuroprotection in the Central Nervous System. *J. Immunol.* **2000**, *164*, 1333–1339. [CrossRef]
140. Erichsen, D.; Lopez, A.L.; Peng, H.; Niemann, D.; Williams, C.; Bauer, M.; Morgello, S.; Cotter, R.L.; Ryan, L.A.; Ghorpade, A.; et al. Neuronal Injury Regulates Fractalkine: Relevance for HIV-1 Associated Dementia. *J. Neuroimmunol.* **2003**, *138*, 144–155. [CrossRef]
141. Nicolini, A.; Ajmone-Cat, M.A.; Bernardo, A.; Levi, G.; Minghetti, L. Human Immunodeficiency Virus Type-1 Tat Protein Induces Nuclear Factor (NF)-κB Activation and Oxidative Stress in Microglial Cultures by Independent Mechanisms. *J. Neurochem.* **2001**, *79*, 713–716. [CrossRef]
142. Yeung, M.C.; Pulliam, L.; Lau, A.S. The HIV Envelope Protein Gp120 Is Toxic to Human Brain-Cell Cultures through the Induction of Interleukin-6 and Tumor Necrosis Factor-Alpha. *AIDS* **1995**, *9*, 137–143. [CrossRef]
143. Zhao, M.-L.; Kim, M.-O.; Morgello, S.; Lee, S.C. Expression of Inducible Nitric Oxide Synthase, Interleukin-1 and Caspase-1 in HIV-1 Encephalitis. *J. Neuroimmunol.* **2001**, *115*, 182–191. [CrossRef]
144. He, X.; Yang, W.; Zeng, Z.; Wei, Y.; Gao, J.; Zhang, B.; Li, L.; Liu, L.; Wan, Y.; Zeng, Q.; et al. NLRP3-Dependent Pyroptosis Is Required for HIV-1 Gp120-Induced Neuropathology. *Cell Mol. Immunol.* **2020**, *17*, 283–299. [CrossRef]
145. Chivero, E.T.; Guo, M.-L.; Periyasamy, P.; Liao, K.; Callen, S.E.; Buch, S. HIV-1 Tat Primes and Activates Microglial NLRP3 Inflammasome-Mediated Neuroinflammation. *J. Neurosci.* **2017**, *37*, 3599–3609. [CrossRef]
146. Trillo-Pazos, G.; Diamanturos, A.; Rislove, L.; Menza, T.; Chao, W.; Belem, P.; Sadiq, S.; Morgello, S.; Sharer, L.; Volsky, D.J. Detection of HIV-1 DNA in Microglia/ Macrophages, Astrocytes and Neurons Isolated from Brain Tissue with HIV-1 Encephalitis by Laser Capture Microdissection. *Brain Pathol.* **2003**, *13*, 144–154. [CrossRef]
147. Valdebenito, S.; Castellano, P.; Ajasin, D.; Eugenin, E.A. Astrocytes Are HIV Reservoirs in the Brain: A Cell Type with Poor HIV Infectivity and Replication but Efficient Cell-to-cell Viral Transfer. *J. Neurochem.* **2021**, *158*, 429–443. [CrossRef]
148. Carroll-Anzinger, D.; Al-Harthi, L. Gamma Interferon Primes Productive Human Immunodeficiency Virus Infection in Astrocytes. *J. Virol.* **2006**, *80*, 541–544. [CrossRef]
149. Narasipura, S.D.; Kim, S.; Al-Harthi, L. Epigenetic Regulation of HIV-1 Latency in Astrocytes. *J. Virol.* **2014**, *88*, 3031–3038. [CrossRef]
150. Mbonye, U.; Karn, J. The Molecular Basis for Human Immunodeficiency Virus Latency. *Annu. Rev. Virol.* **2017**, *4*, 261–285. [CrossRef]
151. Agosto, L.; Gagne, M.; Henderson, A. Impact of Chromatin on HIV Replication. *Genes* **2015**, *6*, 957–976. [CrossRef]
152. Ruelas, D.S.; Greene, W.C. An Integrated Overview of HIV-1 Latency. *Cell* **2013**, *155*, 519–529. [CrossRef]

153. Jadlowsky, J.K.; Wong, J.Y.; Graham, A.C.; Dobrowolski, C.; Devor, R.L.; Adams, M.D.; Fujinaga, K.; Karn, J. Negative Elongation Factor Is Required for the Maintenance of Proviral Latency but Does Not Induce Promoter-Proximal Pausing of RNA Polymerase II on the HIV Long Terminal Repeat. *Mol. Cell Biol.* **2014**, *34*, 1911–1928. [CrossRef]
154. Natarajan, M.; Lester, G.M.; Lee, C.; Missra, A.; Wasserman, G.A.; Steffen, M.; Gilmour, D.S.; Henderson, A.J. Negative Elongation Factor (NELF) Coordinates RNA Polymerase II Pausing, Premature Termination, and Chromatin Remodeling to Regulate HIV Transcription. *J. Biol. Chem.* **2013**, *288*, 25995–26003. [CrossRef]
155. Soriano-Sarabia, N.; Bateson, R.E.; Dahl, N.P.; Crooks, A.M.; Kuruc, J.D.; Margolis, D.M.; Archin, N.M. Quantitation of Replication-Competent HIV-1 in Populations of Resting CD4$^+$ T Cells. *J. Virol.* **2014**, *88*, 14070–14077. [CrossRef]
156. Chomont, N.; El-Far, M.; Ancuta, P.; Trautmann, L.; Procopio, F.A.; Yassine-Diab, B.; Boucher, G.; Boulassel, M.-R.; Ghattas, G.; Brenchley, J.M.; et al. HIV Reservoir Size and Persistence Are Driven by T Cell Survival and Homeostatic Proliferation. *Nat. Med.* **2009**, *15*, 893–900. [CrossRef]
157. Siliciano, R.F.; Greene, W.C. HIV Latency. *Cold Spring Harb. Perspect. Med.* **2011**, *1*, a007096. [CrossRef]
158. Shan, L.; Deng, K.; Gao, H.; Xing, S.; Capoferri, A.A.; Durand, C.M.; Rabi, S.A.; Laird, G.M.; Kim, M.; Hosmane, N.N.; et al. Transcriptional Reprogramming during Effector-to-Memory Transition Renders CD4$^+$ T Cells Permissive for Latent HIV-1 Infection. *Immunity* **2017**, *47*, 766–775. [CrossRef]
159. Kumar, N.A.; Cheong, K.; Powell, D.R.; da Fonseca Pereira, C.; Anderson, J.; Evans, V.A.; Lewin, S.R.; Cameron, P.U. The Role of Antigen Presenting Cells in the Induction of HIV-1 Latency in Resting CD4$^+$ T-Cells. *Retrovirology* **2015**, *12*, 76. [CrossRef]
160. Evans, V.A.; Kumar, N.; Filali, A.; Procopio, F.A.; Yegorov, O.; Goulet, J.-P.; Saleh, S.; Haddad, E.K.; da Fonseca Pereira, C.; Ellenberg, P.C.; et al. Myeloid Dendritic Cells Induce HIV-1 Latency in Non-Proliferating CD4$^+$ T Cells. *PLoS Pathog.* **2013**, *9*, e1003799. [CrossRef]
161. Izquierdo-Useros, N.; Lorizate, M.; Puertas, M.C.; Rodriguez-Plata, M.T.; Zangger, N.; Erikson, E.; Pino, M.; Erkizia, I.; Glass, B.; Clotet, B.; et al. Siglec-1 Is a Novel Dendritic Cell Receptor That Mediates HIV-1 Trans-Infection through Recognition of Viral Membrane Gangliosides. *PLoS Biol.* **2012**, *10*, e1001448. [CrossRef]
162. Fortin, J.-F.; Barat, C.; Beauséjour, Y.; Barbeau, B.; Tremblay, M.J. Hyper-Responsiveness to Stimulation of Human Immunodeficiency Virus-Infected CD4$^+$ T Cells Requires Nef and Tat Virus Gene Products and Results from Higher NFAT, NF-KB, and AP-1 Induction. *J. Biol. Chem.* **2004**, *279*, 39520–39531. [CrossRef]
163. Neri, F.; Giolo, G.; Potestà, M.; Petrini, S.; Doria, M. The HIV-1 Nef Protein Has a Dual Role in T Cell Receptor Signaling in Infected CD4$^+$ T Lymphocytes. *Virology* **2011**, *410*, 316–326. [CrossRef]
164. Fenard, D.; Yonemoto, W.; de Noronha, C.; Cavrois, M.; Williams, S.A.; Greene, W.C. Nef Is Physically Recruited into the Immunological Synapse and Potentiates T Cell Activation Early after TCR Engagement. *J. Immunol.* **2005**, *175*, 6050–6057. [CrossRef]
165. Thompson, K.A.; Cherry, C.L.; Bell, J.E.; McLean, C.A. Brain Cell Reservoirs of Latent Virus in Presymptomatic HIV-Infected Individuals. *Am. J. Pathol.* **2011**, *179*, 1623–1629. [CrossRef]
166. Sadowski, I.; Hashemi, F.B. Strategies to Eradicate HIV from Infected Patients: Elimination of Latent Provirus Reservoirs. *Cell. Mol. Life Sci.* **2019**, *76*, 3583–3600. [CrossRef]
167. Nühn, M.M.; Gumbs, S.B.H.; Buchholtz, N.V.E.J.; Jannink, L.M.; Gharu, L.; de Witte, L.D.; Wensing, A.M.J.; Lewin, S.R.; Nijhuis, M.; Symons, J. Shock and Kill within the CNS: A Promising HIV Eradication Approach? *J. Leukoc. Biol.* **2022**, *112*, 1297–1315. [CrossRef]
168. Sarabia, I.; Bosque, A. HIV-1 Latency and Latency Reversal: Does Subtype Matter? *Viruses* **2019**, *11*, 1104. [CrossRef]
169. Kim, Y.; Anderson, J.L.; Lewin, S.R. Getting the "Kill" into "Shock and Kill": Strategies to Eliminate Latent HIV. *Cell Host Microbe* **2018**, *23*, 14–26. [CrossRef]
170. Abner, E.; Jordan, A. HIV "Shock and Kill" Therapy: In Need of Revision. *Antivir. Res.* **2019**, *166*, 19–34. [CrossRef]
171. Vansant, G.; Bruggemans, A.; Janssens, J.; Debyser, Z. Block-And-Lock Strategies to Cure HIV Infection. *Viruses* **2020**, *12*, 84. [CrossRef]
172. Kessing, C.F.; Nixon, C.C.; Li, C.; Tsai, P.; Takata, H.; Mousseau, G.; Ho, P.T.; Honeycutt, J.B.; Fallahi, M.; Trautmann, L.; et al. In Vivo Suppression of HIV Rebound by Didehydro-Cortistatin A, a "Block-and-Lock" Strategy for HIV-1 Treatment. *Cell Rep.* **2017**, *21*, 600–611. [CrossRef]
173. Mousseau, G.; Kessing, C.F.; Fromentin, R.; Trautmann, L.; Chomont, N.; Valente, S.T. The Tat Inhibitor Didehydro-Cortistatin A Prevents HIV-1 Reactivation from Latency. *mBio* **2015**, *6*, e00465-15. [CrossRef]
174. Mousseau, G.; Clementz, M.A.; Bakeman, W.N.; Nagarsheth, N.; Cameron, M.; Shi, J.; Baran, P.; Fromentin, R.; Chomont, N.; Valente, S.T. An Analog of the Natural Steroidal Alkaloid Cortistatin A Potently Suppresses Tat-Dependent HIV Transcription. *Cell Host Microbe* **2012**, *12*, 97–108. [CrossRef]
175. Li, C.; Mousseau, G.; Valente, S.T. Tat Inhibition by Didehydro-Cortistatin A Promotes Heterochromatin Formation at the HIV-1 Long Terminal Repeat. *Epigenetics Chromatin* **2019**, *12*, 23. [CrossRef]
176. Christ, F.; Shaw, S.; Demeulemeester, J.; Desimmie, B.A.; Marchand, A.; Butler, S.; Smets, W.; Chaltin, P.; Westby, M.; Debyser, Z.; et al. Small-Molecule Inhibitors of the LEDGF/P75 Binding Site of Integrase Block HIV Replication and Modulate Integrase Multimerization. *Antimicrob. Agents Chemother.* **2012**, *56*, 4365–4374. [CrossRef]

177. Christ, F.; Voet, A.; Marchand, A.; Nicolet, S.; Desimmie, B.A.; Marchand, D.; Bardiot, D.; Van der Veken, N.J.; Van Remoortel, B.; Strelkov, S.V.; et al. Rational Design of Small-Molecule Inhibitors of the LEDGF/P75-Integrase Interaction and HIV Replication. *Nat. Chem. Biol.* **2010**, *6*, 442–448. [CrossRef]
178. Jurado, K.A.; Wang, H.; Slaughter, A.; Feng, L.; Kessl, J.J.; Koh, Y.; Wang, W.; Ballandras-Colas, A.; Patel, P.A.; Fuchs, J.R.; et al. Allosteric Integrase Inhibitor Potency Is Determined through the Inhibition of HIV-1 Particle Maturation. *Proc. Natl. Acad. Sci. USA* **2013**, *110*, 8690–8695. [CrossRef]
179. Nabel, G.; Baltimore, D. An Inducible Transcription Factor Activates Expression of Human Immunodeficiency Virus in T Cells. *Nature* **1987**, *326*, 711–713. [CrossRef]
180. Anderson, I.; Low, J.S.; Weston, S.; Weinberger, M.; Zhyvoloup, A.; Labokha, A.A.; Corazza, G.; Kitson, R.A.; Moody, C.J.; Marcello, A.; et al. Heat Shock Protein 90 Controls HIV-1 Reactivation from Latency. *Proc. Natl. Acad. Sci. USA* **2014**, *111*, E1528–E1537. [CrossRef]
181. Kim, H.; Choi, M.-S.; Inn, K.-S.; Kim, B.-J. Inhibition of HIV-1 Reactivation by a Telomerase-Derived Peptide in a HSP90-Dependent Manner. *Sci. Rep.* **2016**, *6*, 28896. [CrossRef]
182. Joshi, P.; Maidji, E.; Stoddart, C.A. Inhibition of Heat Shock Protein 90 Prevents HIV Rebound. *J. Biol. Chem.* **2016**, *291*, 10332–10346. [CrossRef]
183. Nandini, D.; Rao, R.; Deepak, B.; Reddy, P. Sulforaphane in Broccoli: The Green Chemoprevention!! Role in Cancer Prevention and Therapy. *J. Oral Maxillofac. Pathol.* **2020**, *24*, 405. [CrossRef]
184. Li, Y.; Buckhaults, P.; Li, S.; Tollefsbol, T. Temporal Efficacy of a Sulforaphane-Based Broccoli Sprout Diet in Prevention of Breast Cancer through Modulation of Epigenetic Mechanisms. *Cancer Prev. Res.* **2018**, *11*, 451–464. [CrossRef]
185. Jamal, I.; Paudel, A.; Thompson, L.; Abdelmalek, M.; Khan, I.A.; Singh, V.B. Sulforaphane Prevents the Reactivation of HIV-1 by Suppressing NFκB Signaling. *J. Virus Erad.* **2023**, *9*, 100341. [CrossRef]
186. Mollace, V.; Nottet, H.S.L.M.; Clayette, P.; Turco, M.C.; Muscoli, C.; Salvemini, D.; Perno, C.F. Oxidative Stress and NeuroAIDS: Triggers, Modulators and Novel Antioxidants. *Trends Neurosci.* **2001**, *24*, 411–416. [CrossRef]
187. Tufekci, K.U.; Ercan, I.; Isci, K.B.; Olcum, M.; Tastan, B.; Gonul, C.P.; Genc, K.; Genc, S. Sulforaphane Inhibits NLRP3 Inflammasome Activation in Microglia through Nrf2-Mediated MiRNA Alteration. *Immunol. Lett.* **2021**, *233*, 20–30. [CrossRef]
188. Tozzi, V.; Balestra, P.; Bellagamba, R.; Corpolongo, A.; Salvatori, M.F.; Visco-Comandini, U.; Vlassi, C.; Giulianelli, M.; Galgani, S.; Antinori, A.; et al. Persistence of Neuropsychologic Deficits Despite Long-Term Highly Active Antiretroviral Therapy in Patients with HIV-Related Neurocognitive Impairment. *JAIDS J. Acquir. Immune Defic. Syndr.* **2007**, *45*, 174–182. [CrossRef]
189. Robertson, K.R.; Su, Z.; Margolis, D.M.; Krambrink, A.; Havlir, D.V.; Evans, S.; Skiest, D.J. Neurocognitive Effects of Treatment Interruption in Stable HIV-Positive Patients in an Observational Cohort. *Neurology* **2010**, *74*, 1260–1266. [CrossRef]
190. Caniglia, E.C.; Cain, L.E.; Justice, A.; Tate, J.; Logan, R.; Sabin, C.; Winston, A.; van Sighem, A.; Miro, J.M.; Podzamczer, D.; et al. Antiretroviral Penetration into the CNS and Incidence of AIDS-Defining Neurologic Conditions. *Neurology* **2014**, *83*, 134–141. [CrossRef]
191. Perkins, D.O.; Stern, R.A.; Golden, R.N.; Murphy, C.; Naftolowitz, D.; Evans, D.L. Mood Disorders in HIV Infection: Prevalence and Risk Factors in a Nonepicenter of the AIDS Epidemic. *Am. J. Psychiatry* **1994**, *151*, 233–236. [CrossRef]
192. Rabkin, J.G.; Rabkin, R.; Wagner, G. Effects of Fluoxetine on Mood and Immune Status in Depressed Patients with HIV Illness. *J. Clin. Psychiatry* **1994**, *55*, 92–97.
193. D'Antoni, M.L.; Paul, R.H.; Mitchell, B.I.; Kohorn, L.; Fischer, L.; Lefebvre, E.; Seyedkazemi, S.; Nakamoto, B.K.; Walker, M.; Kallianpur, K.J.; et al. Improved Cognitive Performance and Reduced Monocyte Activation in Virally Suppressed Chronic HIV After Dual CCR2 and CCR5 Antagonism. *JAIDS J. Acquir. Immune Defic. Syndr.* **2018**, *79*, 108–116. [CrossRef]

MDPI AG
Grosspeteranlage 5
4052 Basel
Switzerland
Tel.: +41 61 683 77 34

International Journal of Molecular Sciences Editorial Office
E-mail: ijms@mdpi.com
www.mdpi.com/journal/ijms

www.ingramcontent.com/pod-product-compliance
Lightning Source LLC
Chambersburg PA
CBHW061441210926
51850CB00008B/1576
9783725855957